Springer Undergraduate Mathematics Series

Advisory Editors
M. A. J. Chaplain, *St. Andrews, UK*
Angus MacIntyre, *Edinburgh, UK*
Simon Scott, *London, UK*
Nicole Snashall, *Leicester, UK*
Endre Süli, *Oxford, UK*
Michael R. Tehranchi, *Cambridge, UK*
John F. Toland, *Bath, UK*

T0172034

More information about this series at http://www.springer.com/series/3423

Michael S. Ruderman

Fluid Dynamics and Linear Elasticity

A First Course in Continuum Mechanics

Springer

Prof. Michael S. Ruderman
School of Mathematics and Statistics
University of Sheffield
Sheffield, UK

ISSN 1615-2085 ISSN 2197-4144 (electronic)
Springer Undergraduate Mathematics Series
ISBN 978-3-030-19296-9 ISBN 978-3-030-19297-6 (eBook)
https://doi.org/10.1007/978-3-030-19297-6

Mathematics Subject Classification (2010): 74B05, 74J05, 74J15, 75Bxx, 76Dxx

© Springer Nature Switzerland AG 2019
This work is subject to copyright. All rights are reserved by the Publisher, whether the whole or part of the material is concerned, specifically the rights of translation, reprinting, reuse of illustrations, recitation, broadcasting, reproduction on microfilms or in any other physical way, and transmission or information storage and retrieval, electronic adaptation, computer software, or by similar or dissimilar methodology now known or hereafter developed.
The use of general descriptive names, registered names, trademarks, service marks, etc. in this publication does not imply, even in the absence of a specific statement, that such names are exempt from the relevant protective laws and regulations and therefore free for general use.
The publisher, the authors and the editors are safe to assume that the advice and information in this book are believed to be true and accurate at the date of publication. Neither the publisher nor the authors or the editors give a warranty, expressed or implied, with respect to the material contained herein or for any errors or omissions that may have been made. The publisher remains neutral with regard to jurisdictional claims in published maps and institutional affiliations.

This Springer imprint is published by the registered company Springer Nature Switzerland AG
The registered company address is: Gewerbestrasse 11, 6330 Cham, Switzerland

Preface

This text is based on an introductory course in Continuum Mechanics given at the University of Sheffield over a period of fifteen years. The main aim of the course, as well as this book, is to give a unified treatment of the topic. I have tried to restrict the required background to that normally familiar to second-year undergraduate mathematics students in the UK. This includes calculus, mathematical analysis, and the basics of the theory of systems of first-order differential equations. I have included a chapter on the theory of tensors since it is commonly used in continuum mechanics but not usually covered in general calculus courses. Its main aim is to introduce tensors and various operations involving them. Tensors are introduced in a non-traditional way as bilinear and polylinear functions of vector variables. An advantage of this definition is that it immediately shows that, similar to vectors, they are invariant objects independent of coordinate systems. Although operations on vectors and tensors in Cartesian, cylindrical, and spherical coordinates are considered, general curvilinear coordinates and covariant differentiation are not included because I consider these topics to be much more advanced and not suitable for an introductory course.

The book can be divided into four unequal parts. The first part consists of a very short introductory chapter (Chap. 1) followed by a chapter detailing the mathematical preliminaries (Chap. 2). The second part consists of Chaps. 3 and 4, where we consider the general theory of continuum motion. In these chapters we study continuum kinematics and dynamics, and derive the equations of mass and momentum conservation. The third fundamental law of physics is energy conservation. We postpone the derivation of this law until Chap. 8, where it is first used.

The aim of the third part is to give examples of applications of the general theory to particular models of continuum mechanics. To do this we chose the simplest models, one of which is related to fluid motion and uses the Eulerian continuum description, and the other describes the deformation of solids and uses the Lagrangian description. These are the models of an ideal incompressible fluid and linear elasticity, respectively. The first model is described in Chap. 5. We derive Bernoulli's integral and then use it to obtain d'Alembert's paradox. We also consider the dry leaf phenomenon in football and present a qualitative analysis explaining the nature of the lifting force keeping an airplane in the air. The second model is discussed in Chap. 6. There we formulate the main postulates of linear elasticity and derive the general form of Hooke's law, valid for any elastic materials. Then we simplify it in the case of isotropic materials using an analysis of invariance with respect to the coordinate transformation. We show that in this case Hooke's law contains only two arbitrary constants. While for the most part Chap. 6 concerns linear elasticity, we also briefly consider viscoelasticity and plasticity in the last two sections. The main aim is to show how linear elasticity should be modified when we go beyond the limits of its applicability. The first six chapters of the book can be used as a basis for a one semester module for third-year undergraduate students. The whole book can be used as a basis for a two-semester course, which is also suitable for students doing a Masters degree.

The fourth part of the book presents three models of fluids. In Chap. 7 we study the motion of a viscous incompressible fluid. The motion of this fluid is described by the Navier–Stokes equation. We show that, in dimensionless variables, this equation contains only one dimensionless parameter, the Reynolds number Re. We consider two limiting cases: $\mathrm{Re} \ll 1$ and $\mathrm{Re} \gg 1$. In the first case we can neglect the term describing inertia in the Navier–Stokes equation. Then we arrive at the Stokes approximation. As an example of using this approximation we consider the slow motion of a ball in a highly viscous fluid. In the second case we discuss the theory of boundary layers and study the boundary layer at an infinite plate. In Chap. 8 we consider the motion of an ideal compressible gas. We study the propagation of sound waves and nonlinear simple waves. We also study shock waves. Finally, we consider nonlinear acoustics. In the last chapter we study the motion of a compressible, viscous, and thermal conducting gas, we consider the damping of sound waves caused by viscosity and thermal conduction, and we investigate the solution describing the structure of a shock. In the last section of this chapter we derive Burgers' equation and use it to study the evolution and damping of a nonlinear sound wave. In general, the material in Chaps. 7–9 is more advanced than the material in the previous chapters.

The reason for including only models of fluid mechanics in this part is twofold. First, it is related to my personal interests. Secondly, although the two well-known models related to solid mechanics and elasticity are viscoelasticity and plasticity, there are no generally accepted models of viscoelasticity and plasticity. There are a few increasingly complicated models of viscoelasticity, and likewise for plasticity. Hence, we only briefly describe viscoelasticity and plasticity at the end of Chap. 6.

There is a second classical model of continuum mechanics with generally accepted postulates that describes the deformation of solids, namely nonlinear elasticity. The reason why this model has not been included in the book is the following. I believe that it is natural to study the large displacements which feature in nonlinear elasticity by using the concomitant coordinate system with the coordinate lines frozen in a deformed continuum. The description of continuum motion in the concomitant coordinates is directly related to covariant differentiation, which, as mentioned earlier, I have avoided in this course.

The most important formulae appear in boxes. Each chapter includes problems of various levels of difficulty. Altogether there are 79 problems. Detailed solutions to these problems are given in the Appendix.

Sheffield, UK
February 2019

Michael S. Ruderman

Acknowledgements

I am indebted very much to a former head of the department of Applied Mathematics of University of Sheffield Prof. P. C. Chatwin who suggested that I would teach the course of Continuum Mechanics. He also helped me very much when I was preparing this course. I am also indebted to the members of staff of School of Mathematics and statistics of the University of Sheffield for a very friendly atmosphere facilitating research and excellence in teaching.

Finally, I thank my wife Larisa for partient encouragement.

Contents

Chapter 1
Introduction

1.1 The Subject of Continuum Mechanics

Continuum mechanics is the part of Applied Mathematics which studies motions of gases, liquids and deformed solids. It provides a way to study things as diverse as the lift of the wing of an airplane, or the shocks created by earthquakes. Below we give a non-exhaustive list of problems studied by continuum mechanics.

Examples of problems studied by continuum mechanics

(i) Sea waves;
(ii) Air motion near airplanes;
(iii) The structure of the Sun and other stars;
(iv) Seismic waves caused by earthquakes;
(v) Propagation of shocks caused by explosions (e.g., atomic);
(vi) Global atmospheric circulation (related to the weather forecast);
(vii) The motion of mixtures of reacting fluids (applications to the chemical industry);
(viii) Deformation of metallic, concrete, etc., constructions (applications to mechanical engineering, shipbuilding, construction, etc.);
(ix) Blood flow in vessels (applications to medicine).

As an example of the application of continuum mechanics we consider an *abdominal aortic aneurysm*. The aorta is the main artery in the human body, originating from the left ventricle of the heart and extending down to the abdomen, where it splits into two smaller arteries. It distributes oxygenated blood to all parts of the body. An abdominal aortic aneurysm is a localised enlargement of the abdominal aorta such that the diameter is greater than 3 cm, or more than 50% larger than the normal diameter. A large aneurysm can rupture. Very often this causes death.

© Springer Nature Switzerland AG 2019
M. S. Ruderman, *Fluid Dynamics and Linear Elasticity*, Springer Undergraduate Mathematics Series, https://doi.org/10.1007/978-3-030-19297-6_1

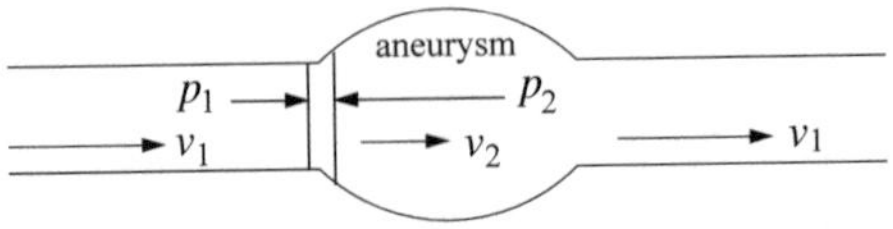

Fig. 1.1 Schematic picture of a blood vessel with an aneurysm. A volume of blood moving in the aneurysm is decelerated by the difference of the blood pressure in the aneurysm and the rest of the blood vessel

Hydrodynamics, which is a branch of continuum mechanics, can explain how an aneurysm develops. The blood can be considered as an incompressible fluid with constant density. The amount of the blood that passes through a blood vessel cross-section of area S in one second is vS, where v is the blood speed. Since the amount of blood passing through any blood vessel cross-section must be the same, we conclude that $vS =$ constant. Let us now assume that a part of the blood vessel has become wider, that is, its cross-section has increased. Then it follows that the velocity in this part is smaller than in the rest of the blood vessel. Hence, a small volume of blood moving into the wider part of the blood vessel must decelerate. The acceleration of the blood volume is determined by the difference between the pressure force acting on the rear and front surfaces of the volume (see Fig. 1.1). Since the volume decelerates it follows that the force acting on the front surface is bigger than that acting on the rear surface. We might think that this occurs solely because the front surface is bigger than the rear one. However, we will show in Sect. 5.3 that the pressure at the front surface exceeds the pressure at the rear surface. Increased pressure stretches the blood vessel wall making it wider. This further decelerates the blood flow in the aneurysm, which, in turn, further increases the pressure, and so on. Eventually the blood vessel wall becomes so thin because of the stretching, and the pressure in the aneurysm increases so much, that the aneurysm ruptures.

1.2 The Main Postulates and Equations of Continuum Mechanics

Usually material bodies consist of molecules and atoms. Some of them can also include charged particles: electrons and ions, which are molecules and atoms either with a deficiency or with an excess of electrons. However in continuum mechanics we disregard the microscopic structure and consider all media as consisting of a continuously distributed mass. To understand why and when we can do this we provide below some numbers related to the microscopic structure of various media.

Data on elementary particles, atoms and molecules

The radius of an atomic nucleus is about 10^{-15} m;
The radius of a hydrogen atom is about 10^{-10} m;

The mass of an electron is $9.1055 \cdot 10^{-31}$ kg;
The mass of a proton is $1.6724 \cdot 10^{-27}$ kg;
There are about $8.6 \cdot 10^{28}$ atoms in 1m^3 of iron;
There are about $2.7 \cdot 10^{25}$ molecules in 1m^3 of air at sea level;
There are about $8 \cdot 10^{21}$ molecules in 1m^3 of air at a height of 60 km;
There are about 10^6 ions in 1m^3 of interplanetary space;
We see that in 1m^3 of almost any matter there are a huge number of particles.

At normal temperatures (not to close to absolute zero) molecules and atoms in gases are in a state of permanent chaotic motion. For example, air molecules at a moderate temperature are moving with an average speed of about 500 m/s $=$ 1800 km/h. In 1 s they collide with each other more than $6 \cdot 10^9$ times.

In continuum mechanics we usually consider the motion of macroscopic bodies with masses at least of the order of a gram and sizes not smaller than one millimeter. We can see that there is a huge number of atoms or molecules even in very small volumes of any media. For example, the mean weight of one air molecule is $29m_p$, where m_p is the proton mass. Hence, in one gram of air there are more than 10^{21} molecules. The mean distance between two air molecules at a height of 60 km is less than 10^{-7} m. The mean distance between two charged particles in interplanetary space is about 10^{-2} m, but there we consider motions of volumes of interplanetary plasma which are many kilometers across. The chaotic motion of atoms and molecules makes their distribution in relatively small macroscopic volumes practically homogeneous. This discussion implies that, when studying macroscopic motions, we can neglect the fact that all the mass is concentrated in tiny volumes with sizes of the order of a nuclear radius and consider it to be homogeneously distributed in any small macroscopic volume of a medium.

Statistical and phenomenological approaches

It is impossible to trace the motion of each separate molecule. Indeed, to trace the motion of molecules in 1m^3 of air we need to solve about 10^{26} equations of motion, where we have taken into account that to describe the motion of one molecule we need to solve at least three second-order ordinary differential equations. It is difficult to imagine such a supercomputer that can solve 10^{26} equations in a reasonable time. Even if it were possible, what would we do with this hugely excessive information? In fact, for all practical purposes we only need to know some average or global characteristics of motion.

One possible approach is *statistical*, where we use a *probabilistic* description of motion. This approach is based on the use of a so-called distribution function which describes the probability of finding a definite number of molecules or atoms constituting the gas in a fixed small volume of space and having velocities in a small interval. The distribution function depends on at least seven variables. To study the evolution of this function in time is very difficult. This problem can only be solved for the simplest physical models and under numerous simplifying assumptions.

Another approach is the development of *phenomenological macroscopic theories*. It is based on two types on laws. The first type consists of three fundamental laws of physics, which are the laws of mass, momentum, and energy conservation.

These laws are applicable to any medium. The second kind of laws are so-called *constitutive equations*. They are obtained from observations and experiments. Each constitutive equation is only applicable to a particular model of continuum mechanics. For example, constitutive equations that are applicable to fluids do not work in the mechanics of deformed solids, and visa versa. Macroscopic theories give very efficient methods for solving important practical problems. In what follows we will use this second approach.

The continuum hypothesis

Let us introduce the concept of a *continuum medium*. All material bodies consist of separate particles. However, there are so many particles in any macroscopic volume that we can consider any material body as a medium continuously filling a spatial domain. In particular, this means that we can split this domain in half as many times as we want. This idealisation is necessary because we are going to use differentiation and integration when studying motions of material bodies.

Space and time

We assume that our space is *Euclidean*, so that we can introduce *Cartesian coordinates* in the whole space. We know from general relativity that this is not the case. Our space is curved and this curvature is related to the presence of mass. However, this curvature is only important either over huge distances comparable with the size of our galaxy, or near strongly concentrated masses like black holes. When studying the motion of various media on the Earth, or, for example, the motion of gas in the sun, we can safely neglect spacial curvature. We also assume that there is a *universal time* that is the same in any moving reference frame. Hence, the time is the same for a person staying on the ground, a person travelling in a train, and a person flying in a rocket in space. Again, this is only an approximation. We know from special relativity that the time is different in two different reference frames. However, this difference is only important when the relative speed of the two frames is comparable to the speed of light, which is 300,000 km s^{-2}. When we consider motions with speed much smaller than the speed of light we can neglect the time difference in different reference frames.

Chapter 2
Mathematical Preliminaries

In this chapter we introduce the mathematical tools that are used in continuum mechanics. We start by recalling those mathematical tools that are usually learned in undergraduate calculus and mathematical analysis courses. Then we proceed to the main aim of this chapter, which is the introduction of tensors and operations on them.

2.1 Scalar and Vector Fields and Their Main Properties

We introduce Cartesian coordinates $x,\ y,\ z$. If a function $\phi(x, y, z)$ with domain $\mathcal{D}$ is given, then we say that a *scalar field* $\phi(x, y, z)$ is defined in the domain $\mathcal{D}$.

If a vector-function $\boldsymbol{v}(x, y, z)$ with domain $\mathcal{D}$ is given, then we say that a *vector field* $\boldsymbol{v}(x, y, z)$ is defined in the domain $\mathcal{D}$. Both $\phi(x, y, z)$ and $\boldsymbol{v}(x, y, z)$ can depend on time t. Note that defining one vector field $\boldsymbol{v}(x, y, z)$ is equivalent to defining three scalar fields $v_x(x, y, z)$, $v_y(x, y, z)$, and $v_z(x, y, z)$.

Operations with Scalar and Vector Fields

$$\operatorname{grad}\phi = \nabla\phi = \frac{\partial\phi}{\partial x}\boldsymbol{i} + \frac{\partial\phi}{\partial y}\boldsymbol{j} + \frac{\partial\phi}{\partial z}\boldsymbol{k}, \tag{2.1.1}$$

$$\operatorname{div}\boldsymbol{v} = \nabla\cdot\boldsymbol{v} = \frac{\partial v_x}{\partial x} + \frac{\partial v_y}{\partial y} + \frac{\partial v_z}{\partial z}, \tag{2.1.2}$$

© Springer Nature Switzerland AG 2019

M. S. Ruderman, *Fluid Dynamics and Linear Elasticity*, Springer Undergraduate Mathematics Series, https://doi.org/10.1007/978-3-030-19297-6_2

$$\operatorname{curl}\boldsymbol{v} = \nabla \times \boldsymbol{v} = \left(\frac{\partial v_z}{\partial y} - \frac{\partial v_y}{\partial z}\right)\boldsymbol{i} + \left(\frac{\partial v_x}{\partial z} - \frac{\partial v_z}{\partial x}\right)\boldsymbol{j} + \left(\frac{\partial v_y}{\partial x} - \frac{\partial v_x}{\partial y}\right)\boldsymbol{k}$$
$$= \begin{vmatrix} \boldsymbol{i} & \boldsymbol{j} & \boldsymbol{k} \\ \partial/\partial x & \partial/\partial y & \partial/\partial z \\ v_x & v_y & v_z \end{vmatrix}, \tag{2.1.3}$$

where $\boldsymbol{i}$, $\boldsymbol{j}$, and $\boldsymbol{k}$ are the unit vectors of Cartesian coordinates.

Gauss' Divergence Theorem

$$\int_S \boldsymbol{v} \cdot d\boldsymbol{S} = \int_V \nabla \cdot \boldsymbol{v}\, dV, \tag{2.1.4}$$

where the surface S encloses the volume V, $d\boldsymbol{S} = \boldsymbol{n}\, dS$, and $\boldsymbol{n}$ is the external normal to the surface S.

Stokes' Theorem

$$\oint_C \boldsymbol{v} \cdot d\boldsymbol{l} = \int_S (\nabla \times \boldsymbol{v}) \cdot d\boldsymbol{S}, \tag{2.1.5}$$

where the contour C is the boundary of the surface S, $\boldsymbol{l}$ is tangent to C, and $\boldsymbol{n}$ is in the positive direction, i.e. from the end of $\boldsymbol{n}$ we see C traversed in the counter-clockwise direction. Here again $d\boldsymbol{S} = \boldsymbol{n}\, dS$, and $\boldsymbol{n}$ is the external normal to the surface S.

Bijective Mappings and the Jacobian

Consider two domains, $\mathcal{D}$ and $\widetilde{\mathcal{D}}$, and a bijection or one-to-one mapping $f : \mathcal{D} \to \widetilde{\mathcal{D}}$ defined by

$$\begin{cases} u = f(x, y, z), \\ v = g(x, y, z), \qquad (x, y, z) \in \mathcal{D}, \quad (u, v, w) \in \widetilde{\mathcal{D}}. \\ w = h(x, y, z), \end{cases}$$

We assume that the functions f, g and h have continuous derivatives. The Jacobian $\mathcal{J}$ of the mapping is defined by

$$\mathcal{J}(x, y, z) = \frac{D(u, v, w)}{D(x, y, z)} = \begin{vmatrix} \frac{\partial f}{\partial x} & \frac{\partial f}{\partial y} & \frac{\partial f}{\partial z} \\ \frac{\partial g}{\partial x} & \frac{\partial g}{\partial y} & \frac{\partial g}{\partial z} \\ \frac{\partial h}{\partial x} & \frac{\partial h}{\partial y} & \frac{\partial h}{\partial z} \end{vmatrix}. \tag{2.1.6}$$

It follows from the condition that the mapping is a bijection that $\mathcal{J}(x, y, z) \neq 0$.

Systems of Ordinary Differential Equations

A system of first-order differential equations has the form

$$\frac{dx_j}{dt} = f_j(t, x_1, \ldots, x_n), \tag{2.1.7}$$

where $j = 1, \ldots, n$. A solution to this system is a set of functions $x_1(t), \ldots, x_n(t)$ satisfying Eq. (2.1.7). If we add the condition

$$x_j(t_0) = \xi_j, \tag{2.1.8}$$

where $\xi_1, \ldots, \xi_n$ are constants, then we obtain the initial value problem for the system of Eq. (2.1.7). Let the functions $f_j(t, x_1, \ldots, x_n)$ be defined for $t \in [a, b]$ and $(x_1, \ldots, x_2) \in \mathcal{D}$, where $\mathcal{D}$ is a closed simply connected domain in the n-dimensional space $\mathbb{R}^n$, and assume they are continuous in $[a, b] \times \mathcal{D}$. Let the functions $f_j(t, x_1, \ldots, x_n)$ also satisfy a so-called Lipschitz condition,

$$|f_j(t, y_1, \ldots, y_n) - f_j(t, x_1, \ldots, x_n)| \leq L \sum_{j=1}^{n} |y_j - x_j|,$$

where L is a universal constant, that is, it is independent of t, $x_1, \ldots, x_n$, and $y_1, \ldots, y_n$. Then the initial value problem has a unique solution on the interval $[t_0 - \epsilon, t_0 + \epsilon]$ for some ϵ such that $t_0 - \epsilon \geq a$ and $t_0 + \epsilon \leq b$ (Coddington and Levinson, see Further Reading).

The curve in $\mathbb{R}^n$ defined by the equations $x_j = x_j(t)$ is called an integral curve. Since there is a unique solution for the initial value problem, there is always an integral curve passing through the point $(\xi_1, \ldots, \xi_n) \in \mathcal{D}$ at $t = t_0$. Moreover, any integral curve can be extended to the boundary of the domain $[a, b] \times \mathcal{D}$.

The general solution to the system of Eq. (2.1.7) contains n arbitrary constants $C_1, \ldots, C_n$, so that it has the form $x_j = x_j(t, C_1, \ldots, C_n)$. Once we have the general solution, we need to solve the system of algebraic equations

$$x_j(t_0, C_1, \ldots, C_n) = \xi_j \tag{2.1.9}$$

to obtain the solution to the initial value problem. Since there is always a unique solution to the initial value problem, it follows that the system of algebraic Eq. (2.1.9) uniquely defines the constants $C_1, \ldots, C_n$.

2.2 Euclidean Vector Spaces

In what follows we consider three-dimensional vectors only. They can be added together and multiplied by a number, meaning that they constitute a three-dimensional *vector space*. We can also calculate the scalar product of two vectors,

$\boldsymbol{u} \cdot \boldsymbol{v} = \|\boldsymbol{u}\| \|\boldsymbol{v}\| \cos \alpha$, where α is the angle between $\boldsymbol{u}$ and $\boldsymbol{v}$, and $\|\boldsymbol{u}\|$ indicates the Euclidean norm of $\boldsymbol{u}$. This implies that the vector space is Euclidean. In what follows we denote the three-dimensional Euclidean vector space by $\mathfrak{V}$. Vectors are denoted by bold small or capital Latin or Greek letters.

Recall the rules of operations with vectors:

(i) $$\boldsymbol{a} + \boldsymbol{b} = \boldsymbol{b} + \boldsymbol{a}$$

(ii) $$(\boldsymbol{a} + \boldsymbol{b}) + \boldsymbol{c} = \boldsymbol{a} + (\boldsymbol{b} + \boldsymbol{c})$$

(iii) $$\exists \boldsymbol{0} : \quad \boldsymbol{a} + \boldsymbol{0} = \boldsymbol{0} + \boldsymbol{a} = \boldsymbol{a}$$

(iv) $$\forall \boldsymbol{a} \ \exists (-\boldsymbol{a}) : \quad \boldsymbol{a} + (-\boldsymbol{a}) = \boldsymbol{a} - \boldsymbol{a} = \boldsymbol{0}$$

(v) $$1\boldsymbol{a} = \boldsymbol{a}$$

(vi) $$(\alpha + \beta)\boldsymbol{a} = \alpha\boldsymbol{a} + \beta\boldsymbol{a}$$

(vii) $$\alpha(\boldsymbol{a} + \boldsymbol{b}) = \alpha\boldsymbol{a} + \alpha\boldsymbol{b}$$

(viii) $$\alpha(\beta\boldsymbol{a}) = (\alpha\beta)\boldsymbol{a}$$

It is easy to prove that there is only one $\boldsymbol{0}$, only one $-\boldsymbol{a}$ for each $\boldsymbol{a}$, that $0\boldsymbol{a} = \boldsymbol{0}$, and all other familiar properties of operations with vectors.

2.3 Second-Order Tensors in Euclidean Space

After a brief reminder of mathematics familiar to second-year students we proceed to the main topic of this chapter, the theory of tensors. We will see below that they are widely used in continuum mechanics. We start with the simplest type of tensors, namely second-order tensors.

Any vector $\boldsymbol{u}$ can be considered as a linear function defined on $\mathfrak{V}$, $\boldsymbol{u} : \mathfrak{V} \longmapsto \mathbb{R}$, where $\mathbb{R}$ is the set of real numbers. This function is defined as follows:

$$\boldsymbol{u}(\boldsymbol{x}) = \boldsymbol{u} \cdot \boldsymbol{x}, \quad \forall \boldsymbol{x} \in \mathfrak{V}. \tag{2.3.1}$$

This function is linear because

$$\boldsymbol{u}(\alpha\boldsymbol{x} + \beta\boldsymbol{y}) = \alpha\boldsymbol{u}(\boldsymbol{x}) + \beta\boldsymbol{u}(\boldsymbol{y}).$$

Lemma 2.1 *If $f(\boldsymbol{x})$ is a linear function on $\mathfrak{V}$, then $\exists\, \boldsymbol{u} \in \mathfrak{V}$ such that*

$$f(\boldsymbol{x}) = \boldsymbol{u} \cdot \boldsymbol{x}.$$

Proof Consider the set of vectors $\boldsymbol{x}$ satisfying $f(\boldsymbol{x}) = 0$. Obviously it is a vector subspace of $\mathfrak{V}$. We denote it by $\mathfrak{V}_0$. If its dimension is 3, then $\mathfrak{V}_0 = \mathfrak{V}$, $f \equiv 0$, and we can take $u = 0$. Now assume that f is not identically equal to 0. In this case the dimension of $\mathfrak{V}_0$ is less than 3, and its orthogonal compliment is not empty. Let $v \neq \mathbf{0}$ belong to this orthogonal compliment. Then $v \cdot y = 0$ for any $y \in \mathfrak{V}_0$. Since $v \notin \mathfrak{V}_0$, $f(v) = a \neq 0$. Let $u = v/a$, so $f(u) = 1$. We take an arbitrary $x \in \mathfrak{V}$ and define $x_\perp = x - bu$, where $b = f(x)$. Then $f(x_\perp) = f(x) - bf(u) = 0$, which implies that $x_\perp \in \mathfrak{V}_0$ and we obtain

$$f(x) = f(x_\perp + bu) = b = u \cdot (x - x_\perp) = u \cdot x.$$

The lemma is proved. □

We see that there is a one-to-one correspondence between vectors and linear functions defined on $\mathfrak{V}$. It is straightforward to see that, if $f(x) = u \cdot x$ and $g(x) = v \cdot x$, then $\alpha f(x) + \beta g(x) = (\alpha u + \beta v) \cdot x$, implying that the correspondence between vectors and linear functions is an *isomorphism* of vector spaces. Hence, we can identify vectors and linear functions and, from now on, consider a vector as a linear function defined on $\mathfrak{V}$.

We now consider the direct product of two identical vector spaces, $\mathfrak{V} \times \mathfrak{V}$. It consists of all pairs (x, y) such that $x, y \in \mathfrak{V}$. Any two vectors, u and v, define a function uv on the direct product $\mathfrak{V} \times \mathfrak{V}$, $uv : \mathfrak{V} \times \mathfrak{V} \longmapsto \mathbb{R}$, i.e. a function defined on pairs (x, y), where $x, y \in \mathfrak{V}$, or, what is the same, $(x, y) \in \mathfrak{V} \times \mathfrak{V}$:

$$uv(x, y) = (u \cdot x)(v \cdot y). \tag{2.3.2}$$

The function uv is called the *dyadic product* of vectors u and v. This is a bilinear function because

$$\begin{aligned} uv(\alpha x_1 + \beta x_2, y) &= (u \cdot (\alpha x_1 + \beta x_2))(v \cdot y) = (\alpha u \cdot x_1 + \beta u \cdot x_2)(v \cdot y) \\ &= \alpha(u \cdot x_1)(v \cdot y) + \beta(u \cdot x_2)(v \cdot y) = \alpha uv(x_1, y) + \beta uv(x_2, y), \end{aligned}$$

and similarly

$$uv(x, \alpha y_1 + \beta y_2) = \alpha uv(x, y_1) + \beta uv(x, y_2).$$

Warning: $uv \neq vu$ because $uv(x, y) = (u \cdot x)(v \cdot y)$, while $vu(x, y) = (v \cdot x)(u \cdot y) = (u \cdot y)(v \cdot x)$.

Let us now consider an arbitrary **bilinear** function $\mathsf{T} : \mathfrak{V} \times \mathfrak{V} \longmapsto \mathbb{R}$. This function is called a *second-order tensor*. We can define a linear combination of two tensors, $\alpha\mathsf{T} + \beta\mathsf{U}$, by

$$(\alpha\mathsf{T} + \beta\mathsf{U})(x, y) = \alpha\mathsf{T}(x, y) + \beta\mathsf{U}(x, y).$$

It is easily shown that the set of all second-order tensors satisfies all the axioms of a vector space, (i)–(viii), so that the set of second-order tensors is a vector space. In what follows tensors are denoted by bold capital sans serif letters. We see below that the second-order tensors play a very important role in continuum mechanics. In particular, they are used to determine the deformation of a continuum and internal forces inside a continuum.

A second-order tensor T is called *symmetric* if

$$\mathsf{T}(\boldsymbol{x}, \boldsymbol{y}) = \mathsf{T}(\boldsymbol{y}, \boldsymbol{x}), \quad \forall \boldsymbol{x}, \boldsymbol{y} \in \mathfrak{V}. \tag{2.3.3}$$

A second-order tensor U is called *skew-symmetric* if

$$\mathsf{U}(\boldsymbol{x}, \boldsymbol{y}) = -\mathsf{U}(\boldsymbol{y}, \boldsymbol{x}), \quad \forall \boldsymbol{x}, \boldsymbol{y} \in \mathfrak{V}. \tag{2.3.4}$$

Let us fix a vector $\boldsymbol{u}$ and consider the function $\boldsymbol{v}(\boldsymbol{x}) = \mathsf{T}(\boldsymbol{u}, \boldsymbol{x})$. Obviously, $\boldsymbol{v}$ is a linear function of $\boldsymbol{x}$, which means that $\boldsymbol{v}$ is a vector. Since $\boldsymbol{u}$ is arbitrary, we obtain a mapping $\mathsf{T}_l : \mathfrak{V} \longmapsto \mathfrak{V}$, which is given by $\boldsymbol{v} = \mathsf{T}_l(\boldsymbol{u})$.

Let $\mathsf{T}_l(\boldsymbol{u}_1) = \boldsymbol{v}_1$, $\mathsf{T}_l(\boldsymbol{u}_2) = \boldsymbol{v}_2$, $\mathsf{T}_l(\alpha\boldsymbol{u}_1 + \beta\boldsymbol{u}_2) = \boldsymbol{v}_3$. Then, by definition

$$\boldsymbol{v}_1(\boldsymbol{x}) = \mathsf{T}(\boldsymbol{u}_1, \boldsymbol{x}), \quad \boldsymbol{v}_2(\boldsymbol{x}) = \mathsf{T}(\boldsymbol{u}_2, \boldsymbol{x}),$$

$$\begin{aligned} \boldsymbol{v}_3(\boldsymbol{x}) &= \mathsf{T}(\alpha\boldsymbol{u}_1 + \beta\boldsymbol{u}_2, \boldsymbol{x}) = \alpha\mathsf{T}(\boldsymbol{u}_1, \boldsymbol{x}) + \beta\mathsf{T}(\boldsymbol{u}_2, \boldsymbol{x}) \\ &= \alpha\boldsymbol{v}_1(\boldsymbol{x}) + \beta\boldsymbol{v}_2(\boldsymbol{x}), \qquad \forall \boldsymbol{x} \in \mathfrak{V}. \end{aligned}$$

Then it follows that

$$\mathsf{T}_l(\alpha\boldsymbol{u}_1 + \beta\boldsymbol{u}_2) = \alpha\mathsf{T}_l(\boldsymbol{u}_1) + \beta\mathsf{T}_l(\boldsymbol{u}_2),$$

meaning that T_l is a *linear operator* on $\mathfrak{V}$. Summarising, by definition

$$\boldsymbol{v} = \mathsf{T}_l(\boldsymbol{u}) \iff \boldsymbol{v}(\boldsymbol{x}) \equiv \boldsymbol{v} \cdot \boldsymbol{x} = \mathsf{T}(\boldsymbol{u}, \boldsymbol{x}), \quad \forall \boldsymbol{x} \in \mathfrak{V}. \tag{2.3.5}$$

In the same way we define a second linear operator on $\mathfrak{V}$, $\mathsf{T}_r(\boldsymbol{u})$:

$$\boldsymbol{w} = \mathsf{T}_r(\boldsymbol{u}) \iff \boldsymbol{w}(\boldsymbol{x}) \equiv \boldsymbol{w} \cdot \boldsymbol{x} = \mathsf{T}(\boldsymbol{x}, \boldsymbol{u}), \quad \forall \boldsymbol{x} \in \mathfrak{V}. \tag{2.3.6}$$

If T is symmetric, then

$$\boldsymbol{w}(\boldsymbol{x}) = \mathsf{T}(\boldsymbol{x}, \boldsymbol{u}) = \mathsf{T}(\boldsymbol{u}, \boldsymbol{x}) = \boldsymbol{v}(\boldsymbol{x}),$$

meaning that $\mathsf{T}_l = \mathsf{T}_r$. If T is skew-symmetric, then

$$\boldsymbol{w}(\boldsymbol{x}) = \mathsf{T}(\boldsymbol{x}, \boldsymbol{u}) = -\mathsf{T}(\boldsymbol{u}, \boldsymbol{x}) = -\boldsymbol{v}(\boldsymbol{x}),$$

meaning that $\mathsf{T}_l = -\mathsf{T}_r$.

Example Consider the tensor I defined by

$$\mathsf{I}(\boldsymbol{x}, \boldsymbol{y}) = \boldsymbol{x} \cdot \boldsymbol{y}. \tag{2.3.7}$$

It is symmetric, so that $\mathsf{I}_l = \mathsf{I}_r$. Let $\boldsymbol{v} = \mathsf{I}_l(\boldsymbol{u})$. Then it follows that

$$\boldsymbol{v} \cdot \boldsymbol{x} = \boldsymbol{v}(\boldsymbol{x}) = \mathsf{I}(\boldsymbol{u}, \boldsymbol{x}) = \boldsymbol{u} \cdot \boldsymbol{x},$$

which implies

$$(\boldsymbol{v} - \boldsymbol{u}) \cdot \boldsymbol{x} = 0, \quad \forall \boldsymbol{x} \in \mathfrak{V}.$$

Taking $\boldsymbol{x} = \boldsymbol{v} - \boldsymbol{u}$ we obtain $\|\boldsymbol{v} - \boldsymbol{u}\|^2 = 0$, so $\boldsymbol{v} = \boldsymbol{u}$. As a result we obtain

$$\mathsf{I}_l(\boldsymbol{u}) = \mathsf{I}_r(\boldsymbol{u}) = \boldsymbol{u},$$

i.e. I_l and I_r are equal to the identity operator. The tensor I is called the *unit* tensor.

2.4 Higher-Order Tensors in Euclidean Space

Although we will mainly use second-order tensors, sometimes we need tensors of higher orders. For example, the general relation between the deformation and internal forces in solids involves a fourth-order tensor. Any n vectors, $\boldsymbol{u}_1, \ldots, \boldsymbol{u}_n$, define a function $\boldsymbol{u}_1 \ldots \boldsymbol{u}_n$ on the direct product $\underbrace{\mathfrak{V} \times \cdots \times \mathfrak{V}}_{n \text{ multipliers}}$, $\boldsymbol{u}_1 \ldots \boldsymbol{u}_n : \mathfrak{V} \times \cdots \times \mathfrak{V} \longmapsto \mathbb{R}$, i.e. a function defined on sequences of n elements, $(\boldsymbol{x}_1, \ldots, \boldsymbol{x}_n)$ such that $\boldsymbol{x}_1, \ldots, \boldsymbol{x}_n \in \mathfrak{V}$, or, what is the same, $(\boldsymbol{x}_1, \ldots, \boldsymbol{x}_n) \in \mathfrak{V} \times \cdots \times \mathfrak{V}$:

$$\boldsymbol{u}_1 \ldots \boldsymbol{u}_n(\boldsymbol{x}_1, \ldots, \boldsymbol{x}_n) = (\boldsymbol{u}_1 \cdot \boldsymbol{x}_1) \ldots (\boldsymbol{u}_n \cdot \boldsymbol{x}_n). \tag{2.4.1}$$

The function $\boldsymbol{u}_1 \ldots \boldsymbol{u}_n$ is called a *polyadic product* of vectors $\boldsymbol{u}_1, \ldots, \boldsymbol{u}_n$. This is a polylinear function because

$$\begin{aligned}
&\boldsymbol{u}_1 \ldots \boldsymbol{u}_n(\boldsymbol{x}_1, \ldots, \boldsymbol{x}_{j-1}, \alpha \boldsymbol{x}_{j1} + \beta \boldsymbol{x}_{j2}, \boldsymbol{x}_{j+1}, \ldots, \boldsymbol{x}_n) \\
&\quad = (\boldsymbol{u}_1 \cdot \boldsymbol{x}_1) \ldots (\boldsymbol{u}_{j-1} \cdot \boldsymbol{x}_{j-1})(\boldsymbol{u}_j \cdot (\alpha \boldsymbol{x}_{j1} + \beta \boldsymbol{x}_{j2}))(\boldsymbol{u}_{j+1} \cdot \boldsymbol{x}_{j+1}) \ldots (\boldsymbol{u}_n \cdot \boldsymbol{x}_n) \\
&\quad = (\boldsymbol{u}_1 \cdot \boldsymbol{x}_1) \ldots (\boldsymbol{u}_{j-1} \cdot \boldsymbol{x}_{j-1})(\alpha \boldsymbol{u}_j \cdot \boldsymbol{x}_{j1} + \beta \boldsymbol{u}_j \cdot \boldsymbol{x}_{j2})(\boldsymbol{u}_{j+1} \cdot \boldsymbol{x}_{j+1}) \ldots (\boldsymbol{u}_n \cdot \boldsymbol{x}_n) \\
&\quad = \alpha (\boldsymbol{u}_1 \cdot \boldsymbol{x}_1) \ldots (\boldsymbol{u}_j \cdot \boldsymbol{x}_{j1}) \ldots (\boldsymbol{u}_n \cdot \boldsymbol{x}_n) + \beta (\boldsymbol{u}_1 \cdot \boldsymbol{x}_1) \ldots (\boldsymbol{u}_j \cdot \boldsymbol{x}_{j2}) \ldots (\boldsymbol{u}_n \cdot \boldsymbol{x}_n) \\
&\quad = \alpha \boldsymbol{u}_1 \ldots \boldsymbol{u}_n(\boldsymbol{x}_1, \ldots, \boldsymbol{x}_{j-1}, \boldsymbol{x}_{j1}, \boldsymbol{x}_{j+1}, \ldots, \boldsymbol{x}_n) \\
&\qquad + \beta \boldsymbol{u}_1 \ldots \boldsymbol{u}_n(\boldsymbol{x}_1, \ldots, \boldsymbol{x}_{j-1}, \boldsymbol{x}_{j2}, \boldsymbol{x}_{j+1}, \ldots, \boldsymbol{x}_n).
\end{aligned}$$

Warning: the order of multiplication is important. For example,

$$\boldsymbol{u}_1\boldsymbol{u}_2\ldots\boldsymbol{u}_n \neq \boldsymbol{u}_2\boldsymbol{u}_1\ldots\boldsymbol{u}_n.$$

Let us now consider an arbitrary **polylinear** function $\mathbf{T} : \mathfrak{V} \times \cdots \times \mathfrak{V} \longmapsto \mathbb{R}$. This function is called an *nth-order tensor*. We can define a linear combination of two tensors, $\alpha\mathbf{T} + \beta\mathbf{U}$, by

$$(\alpha\mathbf{T} + \beta\mathbf{U})(\boldsymbol{x}_1, \ldots \boldsymbol{x}_n) = \alpha\mathbf{T}(\boldsymbol{x}_1, \ldots, \boldsymbol{x}_n) + \beta\mathbf{U}(\boldsymbol{x}_1, \ldots, \boldsymbol{x}_n).$$

It is easily shown that the set of all nth-order tensors satisfies all the axioms of a vector space, (i)–(viii), so that the set of nth-order tensors is a linear space.

Example Consider a third-order tensor $\mathbf{S}$ defined by

$$\mathbf{S}(\boldsymbol{x}, \boldsymbol{y}, \boldsymbol{z}) = \boldsymbol{x} \cdot (\boldsymbol{y} \times \boldsymbol{z}). \tag{2.4.2}$$

Since $\boldsymbol{x} \cdot (\boldsymbol{y} \times \boldsymbol{z})$ changes sign when we swap any two vectors, the tensor $\mathbf{S}$ is called *completely skew-symmetric*. It changes sign when we swap any two of its arguments:

$$\mathbf{S}(\boldsymbol{x}, \boldsymbol{y}, \boldsymbol{z}) = -\mathbf{S}(\boldsymbol{y}, \boldsymbol{x}, \boldsymbol{z}), \quad \mathbf{S}(\boldsymbol{x}, \boldsymbol{y}, \boldsymbol{z}) = -\mathbf{S}(\boldsymbol{z}, \boldsymbol{y}, \boldsymbol{x}), \quad \mathbf{S}(\boldsymbol{x}, \boldsymbol{y}, \boldsymbol{z}) = -\mathbf{S}(\boldsymbol{x}, \boldsymbol{z}, \boldsymbol{y}). \tag{2.4.3}$$

It follows from these relations that a cyclic substitution of arguments does not change the sign of $\mathbf{S}(\boldsymbol{x}, \boldsymbol{y}, \boldsymbol{z})$:

$$\mathbf{S}(\boldsymbol{x}, \boldsymbol{y}, \boldsymbol{z}) = \mathbf{S}(\boldsymbol{y}, \boldsymbol{z}, \boldsymbol{x}) = \mathbf{S}(\boldsymbol{z}, \boldsymbol{x}, \boldsymbol{y}). \tag{2.4.4}$$

$\mathbf{S}$ is not a true tensor. The sign of $\boldsymbol{y} \times \boldsymbol{z}$ changes when we change the orientation of the coordinate system (from right to left or vice versa). Therefore, $\boldsymbol{y} \times \boldsymbol{z}$ is not a true vector. "Vectors" with such a property are called *pseudovectors*. It follows from (2.4.2) that $\mathbf{S}$ also changes sign when we change the orientation of coordinate system. "Tensors" with such a property are called *pseudotensors*.

2.5 The Direct Product of Tensors

We can consider a vector as a first-order tensor, and a scalar as a zero-order tensor. Then the dyadic product of two vectors is used to construct a second-order tensor from two first-order tensors. We generalize this operation and construct an $(n + m)$th-order tensor from nth-order and mth-order tensors. This operation is called the *direct product* of two tensors. Let $\mathbf{T}$ and $\mathbf{U}$ be the nth- and the mth-order tensors respectively. The direct product of these tensors, $\mathbf{W} = \mathbf{TU}$, is defined by

$$\mathbf{W}(\boldsymbol{x}_1, \ldots, \boldsymbol{x}_n, \boldsymbol{x}_{n+1}, \ldots, \boldsymbol{x}_{n+m}) = \mathbf{T}(\boldsymbol{x}_1, \ldots, \boldsymbol{x}_n)\mathbf{U}(\boldsymbol{x}_{n+1}, \ldots, \boldsymbol{x}_{n+m}). \tag{2.5.1}$$

Warning: in general, $\mathbf{TU} \neq \mathbf{UT}$.

Example Let $\boldsymbol{v}$ and $\mathbf{T}$ be a vector and a second-order tensor. Then we can construct two third-order tensors: $\boldsymbol{v}\mathbf{T}$ and $\mathbf{T}\boldsymbol{v}$.

2.6 Cartesian Coordinate Transformation

The vector and tensor notation is very convenient for the general theory because it enables us to write the equations of continuum mechanics in a very concise form. However, when solving particular problems we usually need to rewrite these equations in terms of components of a particular coordinate system. Very often we need to write the same equation in various coordinate systems. Writing equations in various coordinate systems is also very important when using the notion of invariance. In particular, we will see in Chap. 6 that this notion will help us to enormously simplify the equations describing the motion of deformable solids.

To be able to write the equations of continuum mechanics in various coordinate systems we first need to study coordinate transformations. Hence, we start by considering the transformation of Cartesian coordinates. Assume that we rotate the coordinate system x_1, x_2, x_3 to obtain the new coordinate system x'_1, x'_2, x'_3 (see Fig. 2.1). We can expand the unit vectors $\boldsymbol{e}'_1, \boldsymbol{e}'_2, \boldsymbol{e}'_3$ of the new coordinate system with respect to the unit vectors $\boldsymbol{e}_1, \boldsymbol{e}_2, \boldsymbol{e}_3$ of the old coordinate system,

$$
\begin{aligned}
\boldsymbol{e}'_1 &= a_{11}\boldsymbol{e}_1 + a_{12}\boldsymbol{e}_2 + a_{13}\boldsymbol{e}_3,\\
\boldsymbol{e}'_2 &= a_{21}\boldsymbol{e}_1 + a_{22}\boldsymbol{e}_2 + a_{23}\boldsymbol{e}_3,\\
\boldsymbol{e}'_3 &= a_{31}\boldsymbol{e}_1 + a_{32}\boldsymbol{e}_2 + a_{33}\boldsymbol{e}_3,
\end{aligned}
$$

where

$$a_{ij} = \boldsymbol{e}'_i \cdot \boldsymbol{e}_j \quad (i, j = 1, 2, 3).$$

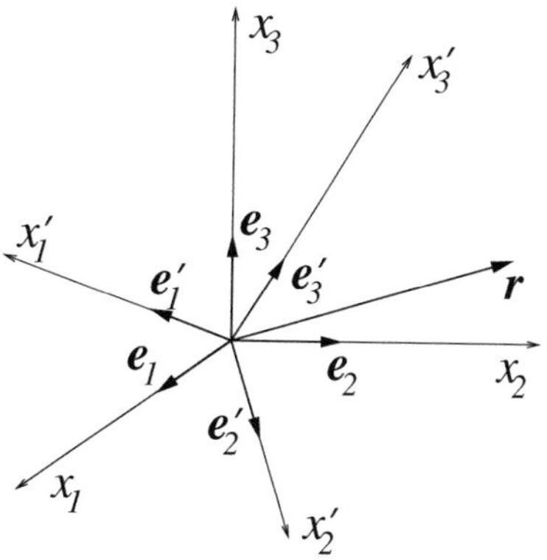

Fig. 2.1 The old coordinate system x_1, x_2, x_3 with the unit vectors $\boldsymbol{e}_1, \boldsymbol{e}_2, \boldsymbol{e}_3$ and the new coordinate system x'_1, x'_2, x'_3 with the unit vectors $\boldsymbol{e}'_1, \boldsymbol{e}'_2, \boldsymbol{e}'_3$

We also can expand the unit vectors of the old coordinate system with respect to the unit vectors of the new coordinate system,

$$\begin{aligned}
\boldsymbol{e}_1 &= a'_{11}\boldsymbol{e}'_1 + a'_{12}\boldsymbol{e}'_2 + a'_{13}\boldsymbol{e}'_3,\\
\boldsymbol{e}_2 &= a'_{21}\boldsymbol{e}'_1 + a'_{22}\boldsymbol{e}'_2 + a'_{23}\boldsymbol{e}'_3,\\
\boldsymbol{e}_3 &= a'_{31}\boldsymbol{e}'_1 + a'_{32}\boldsymbol{e}'_2 + a'_{33}\boldsymbol{e}'_3,
\end{aligned}$$

where

$$a'_{ij} = \boldsymbol{e}_i \cdot \boldsymbol{e}'_j \quad (i, j = 1, 2, 3).$$

In what follows matrices are denoted by bold capital sans serif letters with a hat. The matrices of the direct and inverse transformations are given by

$$\hat{\mathbf{A}} = \begin{pmatrix} a_{11} & a_{12} & a_{13} \\ a_{21} & a_{22} & a_{23} \\ a_{31} & a_{32} & a_{33} \end{pmatrix}, \qquad \hat{\mathbf{A}}^{-1} = \begin{pmatrix} a'_{11} & a'_{12} & a'_{13} \\ a'_{21} & a'_{22} & a'_{23} \\ a'_{31} & a'_{32} & a'_{33} \end{pmatrix}.$$

Since $a'_{ij} = \boldsymbol{e}_i \cdot \boldsymbol{e}'_j = \boldsymbol{e}'_j \cdot \boldsymbol{e}_i = a_{ji}$ it follows that

$$\hat{\mathbf{A}}^{-1} = \begin{pmatrix} a_{11} & a_{21} & a_{31} \\ a_{12} & a_{22} & a_{32} \\ a_{13} & a_{23} & a_{33} \end{pmatrix} = \hat{\mathbf{A}}^{T}. \tag{2.6.1}$$

A matrix $\hat{\mathbf{A}}$ satisfying $\hat{\mathbf{A}}^{-1} = \hat{\mathbf{A}}^{T}$ is called *orthogonal.*

Let us introduce the Kronecker delta-symbol, defined as

$$\delta_{ij} = \begin{cases} 0, \ i \neq j, \\ 1, \ i = j. \end{cases}$$

Since $\hat{\mathbf{A}}\hat{\mathbf{A}}^{-1} = \hat{\mathbf{I}}$, where $\hat{\mathbf{I}}$ is the *unit* matrix with entries δ_{ij}, it follows that $\sum_{l=1}^{3} a_{il}\, a'_{lj} = \delta_{ij}$. Below we use the *Einstein summation rule* with respect to a repeating index,

$$\sum_{l=1}^{3} a_{il}a'_{lj} = a_{il}a'_{lj}.$$

Since $a'_{lj} = a_{jl}$, we obtain

$$a_{il}a_{jl} = \delta_{ij}. \tag{2.6.2}$$

Using the relation $\hat{\mathbf{A}}^{-1}\hat{\mathbf{A}} = \hat{\mathbf{I}}$ we obtain in a similar way

$$a_{li}a_{lj} = \delta_{ij}. \tag{2.6.3}$$

Using $\hat{\mathbf{A}}^{-1} = \hat{\mathbf{A}}^T$ yields $\hat{\mathbf{A}}\hat{\mathbf{A}}^T = \hat{\mathbf{I}}$. Since the determinant of a product of two matrices is equal to the product of their determinants it follows that $\det(\hat{\mathbf{A}}\hat{\mathbf{A}}^T) = 1$. Since $\det(\hat{\mathbf{A}}) = \det(\hat{\mathbf{A}}^T)$ we eventually obtain $[\det(\hat{\mathbf{A}})]^2 = 1$. This result implies that

$$\det(\hat{\mathbf{A}}) = \pm 1. \tag{2.6.4}$$

When $\det(\hat{\mathbf{A}}) = 1$ the old and new systems have the same orientations, and the matrix $\hat{\mathbf{A}}$ is called *proper* orthogonal, while when $\det(\hat{\mathbf{A}}) = -1$ the old and new systems have opposite orientations, and the matrix $\hat{\mathbf{A}}$ is called *improper* orthogonal.

Consider a position vector $\boldsymbol{r}$. It can be written as a linear combination of the old as well as new unit coordinate vectors, $\boldsymbol{r} = x_i\boldsymbol{e}_i = x'_i\boldsymbol{e}'_i$. Then the old and new components of the vector $\boldsymbol{r}$ are given by

$$x_i = \boldsymbol{r}\cdot\boldsymbol{e}_i, \quad x'_i = \boldsymbol{r}\cdot\boldsymbol{e}'_i, \quad (i = 1, 2, 3).$$

Now, using the relation $\boldsymbol{e}'_i = a_{ij}\boldsymbol{e}_j$ we obtain

$$x'_i = \boldsymbol{r}\cdot\boldsymbol{e}'_i = \boldsymbol{r}\cdot(a_{ij}\boldsymbol{e}_j) = a_{ij}\boldsymbol{r}\cdot\boldsymbol{e}_j = a_{ij}x_j. \tag{2.6.5}$$

Hence, the transformation from (x_1, x_2, x_3) to (x'_1, x'_2, x'_3) is given by the same matrix $\hat{\mathbf{A}}$ as the transformation from $(\boldsymbol{e}_1, \boldsymbol{e}_2, \boldsymbol{e}_3)$ to $(\boldsymbol{e}'_1, \boldsymbol{e}'_2, \boldsymbol{e}'_3)$.

2.7 Cartesian Components of Vectors and Tensors

We know that any vector can be written in the form $\boldsymbol{v} = v_i\boldsymbol{e}_i$, where v_i, $i = 1, 2, 3$, are called the components of the vector $\boldsymbol{v}$ with respect to the basis $(\boldsymbol{e}_1, \boldsymbol{e}_2, \boldsymbol{e}_3)$. Consider dyadic products of the basis vectors, $\boldsymbol{e}_i\boldsymbol{e}_j$. There are 9 dyadic products of this type. By definition

$$\boldsymbol{e}_i\boldsymbol{e}_j(\boldsymbol{x}, \boldsymbol{y}) = (\boldsymbol{e}_i\cdot\boldsymbol{x})(\boldsymbol{e}_j\cdot\boldsymbol{y}) = x_iy_j.$$

Let $\mathbf{T}$ be a second-order tensor. Then

$$\mathbf{T}(\boldsymbol{x}, \boldsymbol{y}) = \mathbf{T}(x_i\boldsymbol{e}_i, y_j\boldsymbol{e}_j) = x_iy_j\mathbf{T}(\boldsymbol{e}_i, \boldsymbol{e}_j) = T_{ij}\boldsymbol{e}_i\boldsymbol{e}_j(\boldsymbol{x}, \boldsymbol{y}),$$

where $T_{ij} = \mathbf{T}(\boldsymbol{e}_i, \boldsymbol{e}_j)$ are called the *components* of the tensor $\mathbf{T}$ with respect to the basis $(\boldsymbol{e}_1, \boldsymbol{e}_2, \boldsymbol{e}_3)$. Then it follows that

$$\mathbf{T} = T_{ij}\boldsymbol{e}_i\boldsymbol{e}_j, \quad T_{ij} = \mathbf{T}(\boldsymbol{e}_i, \boldsymbol{e}_j). \tag{2.7.1}$$

Hence any second-order tensor is a linear combinations of nine tensors $\boldsymbol{e}_i\boldsymbol{e}_j$. These tensors are linearly independent, i.e. none of them is equal to a linear combination of the others. Consequently, the tensors $\boldsymbol{e}_i\boldsymbol{e}_j$ constitute a basis in the vector space of second-order tensors, and this space is 9-dimensional. The components of the tensor $\mathbf{T}$ constitute a matrix, which we denote by the same symbol, but with a hat:

$$\hat{\mathbf{T}} = \begin{pmatrix} T_{11} & T_{12} & T_{13} \\ T_{21} & T_{22} & T_{23} \\ T_{31} & T_{32} & T_{33} \end{pmatrix}. \tag{2.7.2}$$

We can repeat this analysis for an nth-order tensor. Now we consider the polyadic products $\boldsymbol{e}_{i_1} \dots \boldsymbol{e}_{i_n}$. There are 3^n of them. By definition

$$\boldsymbol{e}_{i_1} \dots \boldsymbol{e}_{i_n}(\boldsymbol{x}_1, \dots, \boldsymbol{x}_n) = (\boldsymbol{e}_{i_1} \cdot \boldsymbol{x}_1) \dots (\boldsymbol{e}_{i_n} \cdot \boldsymbol{x}_n) = x_{1i_1} \dots x_{ni_n}.$$

Let $\mathbf{T}$ be an nth-order tensor. Then

$$\begin{aligned}\mathbf{T}(\boldsymbol{x}_1, \dots, \boldsymbol{x}_n) &= \mathbf{T}(x_{1i_1}\boldsymbol{e}_{i_1}, \dots, x_{ni_n}\boldsymbol{e}_{i_n}) \\ &= x_{1i_1} \dots x_{ni_n}\mathbf{T}(\boldsymbol{e}_{i_1}, \dots, \boldsymbol{e}_{i_n}) = T_{i_1 \dots i_n}\boldsymbol{e}_{i_1} \dots \boldsymbol{e}_{i_n}(\boldsymbol{x}_1, \dots, \boldsymbol{x}_n).\end{aligned}$$

$T_{i_1 \dots i_n} = \mathbf{T}(\boldsymbol{e}_{i_1}, \dots, \boldsymbol{e}_{i_n})$ are called the *components* of the tensor $\mathbf{T}$ with respect to the basis $(\boldsymbol{e}_1, \boldsymbol{e}_2, \boldsymbol{e}_3)$. Then it follows that

$$\mathbf{T} = T_{i_1 \dots i_n}\boldsymbol{e}_{i_1} \dots \boldsymbol{e}_{i_n}, \quad T_{i_1 \dots i_n} = \mathbf{T}(\boldsymbol{e}_{i_1}, \dots, \boldsymbol{e}_{i_n}). \tag{2.7.3}$$

Now instead of "tensor $\mathbf{T}$" we can write "tensor $T_{i_1 \dots i_n}\boldsymbol{e}_{i_1} \dots \boldsymbol{e}_{i_n}$". However, to save space, we will sometimes simply write "tensor $T_{i_1 \dots i_n}$".

Examples (i) The second-order unit tensor $\mathbf{I}$:

$$I_{ij} = \mathbf{I}(\boldsymbol{e}_i, \boldsymbol{e}_j) = \boldsymbol{e}_i \cdot \boldsymbol{e}_j = \delta_{ij}. \tag{2.7.4}$$

Hence, the components of $\mathbf{I}$ constitute the unit matrix $\hat{\mathbf{I}}$.

(ii) The third-order fully skew-symmetric pseudotensor $\mathbf{S}$. For the components of this tensor we use the special notation ε_{ijk} instead of S_{ijk}:

$$\varepsilon_{ijk} = \mathbf{S}(\boldsymbol{e}_i, \boldsymbol{e}_j, \boldsymbol{e}_k) = \boldsymbol{e}_i \cdot (\boldsymbol{e}_j \times \boldsymbol{e}_k).$$

We see that $\varepsilon_{ijk} = 0$ if at least two indices are equal. If all of them are different, then $\boldsymbol{e}_j \times \boldsymbol{e}_k = \pm\boldsymbol{e}_i$, and $\varepsilon_{ijk} = \pm 1$, where the sign is '+' if the substitution i, j, k is even, and '−' if the substitution i, j, k is odd. Hence

$$\varepsilon_{123} = \varepsilon_{231} = \varepsilon_{312} = 1, \quad \varepsilon_{213} = \varepsilon_{321} = \varepsilon_{132} = -1. \tag{2.7.5}$$

The quantities ε_{ijk} are called *Levi-Civita symbols*.

If $\mathbf{T}$ and $\mathbf{U}$ are nth-order tensors, and $\mathbf{W} = \alpha\mathbf{T} + \beta\mathbf{U}$, then

$$W_{i_1\ldots i_n} = \alpha T_{i_1\ldots i_n} + \beta U_{i_1\ldots i_n}. \tag{2.7.6}$$

If $\mathbf{T}$ is an nth-order tensor, $\mathbf{U}$ is an mth-order tensor, and $\mathbf{W}$ is their direct product, $\mathbf{W} = \mathbf{TU}$, then

$$W_{i_1\ldots i_n i_{n+1}\ldots i_{n+m}} = T_{i_1\ldots i_n} U_{i_{n+1}\ldots i_{n+m}}. \tag{2.7.7}$$

2.8 Transformation of Cartesian Components of Vectors and Tensors

In Sect. 2.6 we derived the expression (2.6.5) for the components of the position vector $\boldsymbol{r}$ in the new coordinate system in terms of its components in the old coordinate system. The same expression is valid for the components of any vector $\boldsymbol{v}$. Let us now do the same for the components of a tensor. We start with second-order tensors. Let $\mathbf{T}$ be a second-order tensor. Then, using $\boldsymbol{e}'_i = a_{ij}\boldsymbol{e}_j$, we obtain

$$T'_{ij} = \mathbf{T}(\boldsymbol{e}'_i, \boldsymbol{e}'_j) = \mathbf{T}(a_{ik}\boldsymbol{e}_k, a_{jl}\boldsymbol{e}_l) = a_{ik}a_{jl}\mathbf{T}(\boldsymbol{e}_k, \boldsymbol{e}_l) = a_{ik}a_{jl}T_{kl}.$$

Hence,

$$\boxed{T'_{ij} = a_{ik}a_{jl}T_{kl}.} \tag{2.8.1}$$

Now let $\mathbf{T}$ be an nth-order tensor. Then

$$\begin{aligned} T'_{i_1\ldots i_n} &= \mathbf{T}(\boldsymbol{e}'_{i_1}, \ldots, \boldsymbol{e}'_{i_n}) = \mathbf{T}(a_{i_1 j_1}\boldsymbol{e}_{j_1}, \ldots, a_{i_n j_n}\boldsymbol{e}_{j_n}) \\ &= a_{i_1 j_1}\ldots a_{i_n j_n}\mathbf{T}(\boldsymbol{e}_{j_1}, \ldots, \boldsymbol{e}_{j_n}) = a_{i_1 j_1}\ldots a_{i_n j_n} T_{j_1\ldots j_n}. \end{aligned}$$

Hence,

$$\boxed{T'_{i_1\ldots i_n} = a_{i_1 j_1}\ldots a_{i_n j_n} T_{j_1\ldots j_n}.} \tag{2.8.2}$$

Examples (i) Let us verify that expression (2.7.4) for the components of $\mathbf{I}$ is consistent with (2.8.1). Using (2.6.2) we obtain that, in the new coordinate system, the components of $\mathbf{I}$ are:

$$a_{ik}a_{jl}\delta_{kl} = a_{ik}a_{jk} = \delta_{ij}.$$

(ii) Now we verify that the expression for the components of the third-order fully skew-symmetric pseudotensor $\mathbf{S}$ is consistent with (2.8.2). In the old coordinate system the components of $\mathbf{S}$ are $S_{ijk} = \varepsilon_{ijk}$. Then in the new coordinate system they are given by

$$S'_{ijk} = \det(\hat{\mathbf{A}}) a_{il} a_{jm} a_{kn} \varepsilon_{lmn}.$$

We introduced the multiplier $\det(\hat{\mathbf{A}})$ because $\mathbf{S}$ is a pseudotensor that changes sign when the orientation of coordinate system changes. The multiplier $\det(\hat{\mathbf{A}})$ provides this change of sign because $\det(\hat{\mathbf{A}}) = 1$ when the old and new coordinate systems have the same orientation, while $\det(\hat{\mathbf{A}}) = -1$ when the old and new coordinate systems have different orientations. Using (2.7.5) we expand the expression for S'_{ijk}:

$$\begin{aligned} S'_{ijk} &= \det(\hat{\mathbf{A}})(a_{i1}a_{j2}a_{k3} + a_{i2}a_{j3}a_{k1} + a_{i3}a_{j1}a_{k2} \\ &\quad - a_{i2}a_{j1}a_{k3} - a_{i1}a_{j3}a_{k2} - a_{i3}a_{j2}a_{k1}). \end{aligned}$$

Recall the expression for $\det(\hat{\mathbf{A}})$:

$$\begin{aligned} \det(\hat{\mathbf{A}}) &= a_{11}a_{22}a_{33} + a_{12}a_{23}a_{31} + a_{13}a_{21}a_{32} \\ &\quad - a_{12}a_{21}a_{33} - a_{11}a_{23}a_{32} - a_{13}a_{22}a_{31}. \end{aligned}$$

Using this expression we easily verify that

$$\begin{aligned} S'_{123} = S'_{231} = S'_{312} &= [\det(\hat{\mathbf{A}})]^2 = 1, \\ S'_{213} = S'_{132} = S'_{321} &= -[\det(\hat{\mathbf{A}})]^2 = -1, \end{aligned}$$

and $S'_{ijk} = 0$ if two indices are equal. Hence $S'_{ijk} = \varepsilon_{ijk}$.

2.9 Tensor Contraction

Let us introduce one more operation on tensors called *tensor contraction*. Consider an nth-order tensor $\mathbf{T}$. The $(n-2)$th-order tensor $\mathbf{U}$ with components

$$U_{i_1 \ldots i_{n-2}} = T_{i_1 \ldots i_{n-2} i_{n-1} i_{n-1}} \tag{2.9.1}$$

is called the *contraction* of the tensor $\mathbf{T}$. In accordance with Einstein's rule

$$T_{i_1 \ldots i_{n-2} i_{n-1} i_{n-1}} = T_{i_1 \ldots i_{n-2} 11} + T_{i_1 \ldots i_{n-2} 22} + T_{i_1 \ldots i_{n-2} 33}.$$

Tensor contraction is equivalent to taking the scalar product of the corresponding basis vectors in the tensor expansion with respect to multiadic products:

$$\begin{aligned} U_{i_1 \ldots i_{n-2}} \mathbf{e}_{i_1} \ldots \mathbf{e}_{i_{n-2}} &= T_{i_1 \ldots i_{n-2} i_{n-1} i_n} \delta_{i_{n-1} i_n} \mathbf{e}_{i_1} \ldots \mathbf{e}_{i_{n-2}} \\ &= T_{i_1 \ldots i_{n-2} i_{n-1} i_n} \mathbf{e}_{i_1} \ldots \mathbf{e}_{i_{n-2}} (\mathbf{e}_{i_{n-1}} \cdot \mathbf{e}_{i_n}). \end{aligned} \tag{2.9.2}$$

Comment: We may contract a tensor with respect to any two indices, not necessarily with respect to the last two. For example, the tensor $\mathbf{W}$ with components

$$W_{i_3\ldots i_n} = T_{i_1 i_1 i_3\ldots i_n}$$

is also a contraction of the tensor $\mathbf{T}$. In general, $\mathbf{W} \neq \mathbf{U}$.

Examples (i) Contraction of the dyadic product of two vectors, $\boldsymbol{u}$ and $\boldsymbol{v}$, gives their scalar product:

$$u_i v_i = \boldsymbol{u} \cdot \boldsymbol{v}.$$

(ii) Contraction of $\mathbf{I}$ is a scalar, and it is equal to

$$\delta_{ii} = \delta_{11} + \delta_{22} + \delta_{33} = 3.$$

(iii) Consider a vector $\boldsymbol{u}$ and a second-order tensor $\mathbf{T}$. Taking their direct product we obtain two third-order tensors: $\boldsymbol{u}\mathbf{T}$ and $\mathbf{T}\boldsymbol{u}$. Let us now take the contraction of the first tensor with respect to the first two indices, and the second tensor with respect to the last two indices. Then we obtain two vectors, $\boldsymbol{v}$ and $\boldsymbol{w}$. We introduce the notation

$$\boldsymbol{v} = \boldsymbol{u} \cdot \mathbf{T}, \quad \boldsymbol{w} = \mathbf{T} \cdot \boldsymbol{u}.$$

The components of these vectors are

$$v_i = T_{ji} u_j, \quad w_i = T_{ij} u_j.$$

Let us now calculate the components of the vectors $\tilde{\boldsymbol{v}} = \mathbf{T}_l(\boldsymbol{u})$ and $\tilde{\boldsymbol{w}} = \mathbf{T}_r(\boldsymbol{u})$, where the linear operators $\mathbf{T}_l(\boldsymbol{u})$ and $\mathbf{T}_r(\boldsymbol{u})$ are determined by (2.3.5) and (2.3.6) respectively. Using (2.3.5) we obtain

$$\tilde{v}_i = \tilde{\boldsymbol{v}} \cdot \boldsymbol{e}_i = \mathbf{T}(\boldsymbol{u}, \boldsymbol{e}_i) = \mathbf{T}(u_j \boldsymbol{e}_j, \boldsymbol{e}_i) = u_j \mathbf{T}(\boldsymbol{e}_j, \boldsymbol{e}_i) = T_{ji} u_j = v_i.$$

Hence $\tilde{\boldsymbol{v}} = \boldsymbol{v}$, which implies that $\mathbf{T}_l(\boldsymbol{u}) = \boldsymbol{u} \cdot \mathbf{T}$.

Similarly, using (2.3.6) we obtain

$$\tilde{w}_i = \tilde{\boldsymbol{w}} \cdot \boldsymbol{e}_i = \mathbf{T}(\boldsymbol{e}_i, \boldsymbol{u}) = \mathbf{T}(\boldsymbol{e}_i, u_j \boldsymbol{e}_j) = u_j \mathbf{T}(\boldsymbol{e}_i, \boldsymbol{e}_j) = T_{ij} u_j = w_i.$$

Hence $\tilde{\boldsymbol{w}} = \boldsymbol{w}$, which implies that $\mathbf{T}_r(\boldsymbol{u}) = \mathbf{T} \cdot \boldsymbol{u}$. Below we use $\boldsymbol{u} \cdot \mathbf{T}$ and $\mathbf{T} \cdot \boldsymbol{u}$ instead of $\mathbf{T}_l(\boldsymbol{u})$ and $\mathbf{T}_r(\boldsymbol{u})$. We can write the components of the vectors $\boldsymbol{u} \cdot \mathbf{T}$ and $\mathbf{T} \cdot \boldsymbol{u}$ in the matrix form:

$$(v_1, v_2, v_3) = (u_1, u_2, u_3) \begin{pmatrix} T_{11} & T_{12} & T_{13} \\ T_{21} & T_{22} & T_{23} \\ T_{31} & T_{32} & T_{33} \end{pmatrix}, \tag{2.9.3}$$

$$\begin{pmatrix} w_1 \\ w_2 \\ w_3 \end{pmatrix} = \begin{pmatrix} T_{11} & T_{12} & T_{13} \\ T_{21} & T_{22} & T_{23} \\ T_{31} & T_{32} & T_{33} \end{pmatrix} \begin{pmatrix} u_1 \\ u_2 \\ u_3 \end{pmatrix}. \tag{2.9.4}$$

(iv) Consider the direct product of S and two vectors, $\boldsymbol{u}$ and $\boldsymbol{v}$. We obtain the fifth-order pseudotensor

$$\mathsf{S}\boldsymbol{u}\boldsymbol{v} = \varepsilon_{ijk} u_l v_m \boldsymbol{e}_i \boldsymbol{e}_j \boldsymbol{e}_k \boldsymbol{e}_l \boldsymbol{e}_m.$$

Let us contract this pseudotensor with respect to the second and fourth indices, and with respect to the third and fifth indices. We obtain the pseudovector $\boldsymbol{w} = \varepsilon_{ijk} u_j v_k \boldsymbol{e}_i = \boldsymbol{u} \times \boldsymbol{v}$. We can write this symbolically as

$$\boldsymbol{u} \times \boldsymbol{v} = \mathsf{S} : \boldsymbol{u}\boldsymbol{v}, \tag{2.9.5}$$

where $:$ indicates double contraction.

(v) Consider the direct product of S with itself. We obtain the sixth-order tensor

$$\mathsf{SS} = \varepsilon_{ijm} \varepsilon_{kln} \boldsymbol{e}_i \boldsymbol{e}_j \boldsymbol{e}_m \boldsymbol{e}_k \boldsymbol{e}_l \boldsymbol{e}_n.$$

Let us contract this tensor with respect to the third and the sixth indices. We obtain a fourth-order tensor that we denote by

$$\mathsf{S} \cdot \mathsf{S} = \varepsilon_{ijm} \varepsilon_{klm} \boldsymbol{e}_i \boldsymbol{e}_j \boldsymbol{e}_k \boldsymbol{e}_l.$$

It can be verified with the use of the definitions of δ_{ij} and ε_{ijk} that

$$\varepsilon_{ijm} \varepsilon_{klm} = \delta_{ik} \delta_{jl} - \delta_{il} \delta_{jk}. \tag{2.9.6}$$

Hence, $\mathsf{S} \cdot \mathsf{S} = (\delta_{ik} \delta_{jl} - \delta_{il} \delta_{jk}) \boldsymbol{e}_i \boldsymbol{e}_j \boldsymbol{e}_k \boldsymbol{e}_l$.

2.10 Some Properties of Second-Order Tensors and Linear Operators

As we have seen, a second-order tensor T is associated with the 3×3 matrix of its components. In what follows we will use the relation between the old and new components of the tensor T written in matrix form. It follows from (2.8.1) that

$$\boxed{\hat{\mathsf{T}}' = \hat{\mathsf{A}} \hat{\mathsf{T}} \hat{\mathsf{A}}^T.} \tag{2.10.1}$$

The quantity

$$I_1 = \operatorname{tr} \hat{\mathsf{T}} = T_{ii} = T_{11} + T_{22} + T_{33} \tag{2.10.2}$$

is called the *trace* of the matrix $\hat{\mathbf{T}}$ and tensor $\mathbf{T}$. This quantity is an invariant of the tensor $\mathbf{T}$:

$$\operatorname{tr}\hat{\mathbf{T}}' = a_{ik}a_{il}T_{kl} = \delta_{kl}T_{kl} = T_{kk} = \operatorname{tr}\hat{\mathbf{T}}.$$

The fourth-order tensor $\mathbf{TT}$ is the direct product of $\mathbf{T}$ with itself. Consider its contraction with respect to the second and third indices. We denote this contraction by $\mathbf{T}^2$. Hence, $(T^2)_{ij} = T_{ik}T_{kj}$ and

$$\begin{aligned}\operatorname{tr}\left(\hat{\mathbf{T}}'^2\right) &= T'_{ik}T'_{ki} = a_{ij}a_{kl}T_{jl}a_{km}a_{in}T_{mn} = a_{ij}a_{kl}a_{km}a_{in}T_{jl}T_{mn}\\ &= \delta_{jn}\delta_{lm}T_{jl}T_{mn} = T_{nl}T_{ln} = \operatorname{tr}\left(\hat{\mathbf{T}}^2\right).\end{aligned}$$

We see that $\operatorname{tr}\left(\hat{\mathbf{T}}^2\right)$ is an invariant of the tensor $\mathbf{T}$. Then it follows that

$$I_2 = \tfrac{1}{2}I_1^2 - \tfrac{1}{2}\operatorname{tr}\left(\hat{\mathbf{T}}^2\right) \tag{2.10.3}$$

is also an invariant of the tensor $\mathbf{T}$. We will see later why it is convenient to use I_2 rather than simply $\operatorname{tr}\left(\hat{\mathbf{T}}^2\right)$. Finally,

$$\begin{aligned}\det(\hat{\mathbf{T}}') &= \det\left(\hat{\mathbf{A}}\hat{\mathbf{T}}\hat{\mathbf{A}}^T\right) = \det(\hat{\mathbf{A}})\cdot\det(\hat{\mathbf{T}})\cdot\det\left(\hat{\mathbf{A}}^T\right)\\ &= [\det(\hat{\mathbf{A}})]^2\det(\hat{\mathbf{T}}) = \det(\hat{\mathbf{T}}).\end{aligned}$$

This implies that

$$I_3 = \det(\hat{\mathbf{T}}) \tag{2.10.4}$$

is an invariant of $\mathbf{T}$. I_1, I_2, and I_3 are the *principal invariants* of $\mathbf{T}$.

We have also seen that the tensor $\mathbf{T}$ is associated with two linear operators in the three-dimensional vector space, $\boldsymbol{v} = \boldsymbol{u}\cdot\mathbf{T}$ and $\boldsymbol{w} = \mathbf{T}\cdot\boldsymbol{u}$. Since the relation $\boldsymbol{v} = \boldsymbol{u}\cdot\mathbf{T}$ is equivalent to $v_i = T_{ji}u_j$, it can be rewritten as $\boldsymbol{v} = \mathbf{T}^T\cdot\boldsymbol{u}$, where the *transpose* $\mathbf{T}^T$ of the tensor $\mathbf{T}$ is defined by

$$\mathbf{T}^T(\boldsymbol{x},\boldsymbol{y}) = \mathbf{T}(\boldsymbol{y},\boldsymbol{x}). \tag{2.10.5}$$

It follows from this relation that the components of $\mathbf{T}^T$ are given by

$$T_{ij}^T = \mathbf{T}^T(\boldsymbol{e}_i,\boldsymbol{e}_j) = \mathbf{T}(\boldsymbol{e}_j,\boldsymbol{e}_i) = T_{ji},$$

so that the matrix of coefficients of $\mathbf{T}^T$ is equal to the transposed matrix of coefficients of $\mathbf{T}$. The relation $\boldsymbol{u}\cdot\mathbf{T} = \mathbf{T}^T\cdot\boldsymbol{u}$ enables us to study only operators of the form $\boldsymbol{w} = \mathbf{T}\cdot\boldsymbol{u}$. A vector $\boldsymbol{u}$ is called an *eigenvector* of the operator $\mathbf{T}$ if

$$\mathbf{T}\cdot\boldsymbol{u} = \lambda\boldsymbol{u}. \tag{2.10.6}$$

The quantity λ is called the *eigenvalue* of $\mathbf{T}$ corresponding to the eigenvector $\boldsymbol{u}$.

If we write Eq. (2.10.6) in matrix form, then we obtain a system of three linear homogeneous algebraic equations for the three components of the vector $\boldsymbol{u}$. This system has a non-trivial solution only if its determinant is zero. This condition gives the equation determining λ:

$$\det(\hat{\mathbf{T}} - \lambda\hat{\mathbf{I}}) = 0. \tag{2.10.7}$$

Equation (2.10.7) is called the *characteristic equation* for the matrix $\hat{\mathbf{T}}$. Simple calculations show that it can be written in the form

$$\lambda^3 - I_1\lambda^2 + I_2\lambda - I_3 = 0. \tag{2.10.8}$$

We see that the coefficient of λ is I_2. This explains why we defined the second principal invariant by Eq. (2.10.3), rather than simply as $\mathrm{tr}\left(\mathbf{T}^2\right)$.

Cayley-Hamilton Theorem (without proof) *A square matrix satisfies its own characteristic equation. In particular, for a* 3×3 *matrix* $\hat{\mathbf{T}}$ *we obtain*

$$\hat{\mathbf{T}}^3 - I_1\hat{\mathbf{T}}^2 + I_2\hat{\mathbf{T}} - I_3\hat{\mathbf{I}} = 0. \tag{2.10.9}$$

Theorem 2.1 *Two eigenvectors of a symmetric operator corresponding to different eigenvalues are orthogonal.*

Proof Let $\mathbf{T}$ be a symmetric tensor, and $\boldsymbol{u}$ and $\boldsymbol{v}$ be eigenvectors of the linear operator associated with this tensor corresponding to different eigenvalues:

$$\mathbf{T}\cdot\boldsymbol{u} = \lambda\boldsymbol{u}, \quad \mathbf{T}\cdot\boldsymbol{v} = \mu\boldsymbol{v}, \quad \lambda \neq \mu.$$

Then

$$\lambda\boldsymbol{u}\cdot\boldsymbol{v} = \boldsymbol{v}\cdot\mathbf{T}\cdot\boldsymbol{u},$$

$$\mu\boldsymbol{u}\cdot\boldsymbol{v} = \boldsymbol{u}\cdot\mathbf{T}\cdot\boldsymbol{v} = \boldsymbol{v}\cdot\mathbf{T}^T\cdot\boldsymbol{u} = \boldsymbol{v}\cdot\mathbf{T}\cdot\boldsymbol{u}.$$

Subtracting these two equations we obtain

$$(\lambda - \mu)\boldsymbol{u}\cdot\boldsymbol{v} = 0 \quad\Longrightarrow\quad \boldsymbol{u}\cdot\boldsymbol{v} = 0 \quad\Longrightarrow\quad \boldsymbol{u}\perp\boldsymbol{v}.$$

The theorem is proved. □

Theorem 2.2 (without proof) *Any symmetric operator* $\mathbf{T}$ *has three mutually orthogonal eigenvectors.*

Since eigenvectors are defined up to multiplication by an arbitrary constant, we can always assume that they are unit vectors. Let us denote the unit eigenvectors of $\mathbf{T}$ by $\boldsymbol{e}'_1$, $\boldsymbol{e}'_2$, and $\boldsymbol{e}'_3$. We introduce a new orthogonal coordinate system (x'_1, x'_2, x'_3) with the basis $\boldsymbol{e}'_1, \boldsymbol{e}'_2, \boldsymbol{e}'_3$. Let $\hat{\mathbf{A}}$ with entries a_{ij} be the transformation matrix from the old coordinates (x_1, x_2, x_3) to the new coordinates (x_1', x_2', x_3'). Then $\boldsymbol{e}_i' = a_{ij}\boldsymbol{e}_j$ and, in accordance with (2.10.6) with $\boldsymbol{e}_i'$ substituted for $\boldsymbol{u}$,

$$T_{jk}a_{ik} = \lambda_{[i]}a_{[i]j}.$$

Here λ_i is the eigenvalue corresponding to the eigenvector $\boldsymbol{e}_i'$, and the square brackets at index i indicate that there is no summation with respect to i. Then the new components of $\mathbf{T}$ are given by

$$\begin{aligned} T_{ij}' &= a_{ik}a_{jl}T_{kl} = a_{ik}(T_{kl}a_{jl}) = a_{ik}(\lambda_{[j]}a_{[j]k}) \\ &= \lambda_{[j]}a_{ik}a_{[j]k} = \lambda_{[j]}\delta_{i[j]} = \lambda_{[i]}\delta_{[i]j}, \end{aligned}$$

where we have used the condition of orthogonality of $\hat{\mathbf{A}}$, (2.6.2). This relation can be written in matrix form as

$$\hat{\mathbf{T}}' = \hat{\mathbf{A}}\hat{\mathbf{T}}\hat{\mathbf{A}}^T = \begin{pmatrix} \lambda_1 & 0 & 0 \\ 0 & \lambda_2 & 0 \\ 0 & 0 & \lambda_3 \end{pmatrix}.$$

The axes x_1', x_2', and x_3' are called the *principal axes* of the tensor and operator $\mathbf{T}$. It follows from (2.10.2)–(2.10.4) that

$$I_1 = \lambda_1 + \lambda_2 + \lambda_3, \quad I_2 = \lambda_1\lambda_2 + \lambda_1\lambda_3 + \lambda_2\lambda_3, \quad I_3 = \lambda_1\lambda_2\lambda_3. \qquad (2.10.10)$$

Definition An operator $\mathbf{U}$ is called *positive-definite* if $\boldsymbol{u} \cdot \mathbf{U} \cdot \boldsymbol{u} > 0$ for any $\boldsymbol{u} \neq 0$.

Definition An operator $\mathbf{B}$ is called *unitary* if $\|\mathbf{B} \cdot \boldsymbol{u}\| = \|\boldsymbol{u}\|$ for $\forall\, \boldsymbol{u} \in \mathfrak{V}$.

Let $\hat{\mathbf{B}}$ be the matrix of an operator $\mathbf{B}$ with respect to the basis $\boldsymbol{e}_1, \boldsymbol{e}_2, \boldsymbol{e}_3$. If $\boldsymbol{u} = u_i\boldsymbol{e}_i$, $\boldsymbol{v} = v_i\boldsymbol{e}_i$, and $\boldsymbol{v} = \mathbf{B} \cdot \boldsymbol{u}$, then $v_i = B_{ij}u_j$, where B_{ij} are the entries of the matrix $\hat{\mathbf{B}}$. We have

$$\begin{aligned} \|\boldsymbol{v}\|^2 &= v_iv_i = B_{ij}u_jB_{ik}u_k = B_{ij}B_{ik}u_ju_k \\ &= \|\boldsymbol{u}\|^2 = u_ju_j = u_ju_k\delta_{jk} \implies B_{ij}B_{ik} = \delta_{jk} \\ &\implies \hat{\mathbf{B}}^T\hat{\mathbf{B}} = \hat{\mathbf{I}} \implies \hat{\mathbf{B}}^{-1} = \hat{\mathbf{B}}^T. \end{aligned}$$

This implies that $\hat{\mathbf{B}}$ is an orthogonal matrix.

In the new coordinates the operator matrix is $\hat{\mathbf{B}}' = \hat{\mathbf{A}}\hat{\mathbf{B}}\hat{\mathbf{A}}^T$, where $\hat{\mathbf{A}}$ is the transformation matrix from old to new coordinates (recall that $\hat{\mathbf{A}}$ is orthogonal: $\hat{\mathbf{A}}^{-1} = \hat{\mathbf{A}}^T$). We have

$$(\hat{\mathbf{B}}')^{-1} = (\hat{\mathbf{A}}^T)^{-1}\hat{\mathbf{B}}^{-1}\hat{\mathbf{A}}^{-1} = \hat{\mathbf{A}}\hat{\mathbf{B}}^T\hat{\mathbf{A}}^T = (\hat{\mathbf{B}}')^T,$$

which implies that $\hat{\mathbf{B}}'$ is also orthogonal.

Since $\hat{\mathbf{B}}$ is orthogonal, it follows that $\det(\hat{\mathbf{B}}) = \pm 1$. When $\det(\hat{\mathbf{B}}) = 1$ the operator $\mathbf{B}$ does not change the vector space orientation and it is called *proper unitary*.

When $\det(\hat{\mathbf{B}}) = -1$ the operator $\mathbf{B}$ changes the vector space orientation and it is called *improper unitary*.

Theorem 2.3 ***The polar decomposition theorem*** *(without proof). Let the operator* $\mathbf{T}$ *be invertible, i.e. the matrix of its coefficients, which is denoted by* $\hat{\mathbf{T}}$*, is non-singular,* $\det(\hat{\mathbf{T}}) \neq 0$*. Then* $\mathbf{T}$ *can be expressed in the form*

$$\mathbf{T} = \mathbf{BU} = \mathbf{VB}, \tag{2.10.11}$$

where $\mathbf{B}$ *is a proper unitary operator, and the operators* $\mathbf{U}$ *and* $\mathbf{V}$ *are positive-definite and symmetric.*

We will use this theorem later to describe continuum displacement.

Example Let the matrix of coefficients of $\mathbf{T}$ be given in a coordinate system by

$$\hat{\mathbf{T}} = \begin{pmatrix} 4 & 4 & 1 \\ \sqrt{6} & -\sqrt{6} & 0 \\ 2\sqrt{2} & 2\sqrt{2} & -\sqrt{2} \end{pmatrix}.$$

We obtain $\det(\hat{\mathbf{T}}) = 24\sqrt{3}$. This implies that the operator $\mathbf{T}$ is invertible. Consider

$$\hat{\mathbf{B}} = \begin{pmatrix} \frac{1}{3}\sqrt{3} & \frac{1}{3}\sqrt{3} & \frac{1}{3}\sqrt{3} \\ \frac{1}{2}\sqrt{2} & -\frac{1}{2}\sqrt{2} & 0 \\ \frac{1}{6}\sqrt{6} & \frac{1}{6}\sqrt{6} & -\frac{1}{3}\sqrt{6} \end{pmatrix}.$$

It is easy to verify that $\hat{\mathbf{B}}$ is orthogonal. We also consider

$$\hat{\mathbf{U}} = \begin{pmatrix} 3\sqrt{3} & \sqrt{3} & 0 \\ \sqrt{3} & 3\sqrt{3} & 0 \\ 0 & 0 & \sqrt{3} \end{pmatrix}, \quad \hat{\mathbf{V}} = \begin{pmatrix} 3\sqrt{3} & 0 & \sqrt{6} \\ 0 & 2\sqrt{3} & 0 \\ \sqrt{6} & 0 & 2\sqrt{3} \end{pmatrix}.$$

We obtain

$$\boldsymbol{u} \cdot \mathbf{U} \cdot \boldsymbol{u} = \sqrt{3}\left[2u_1^2 + 2u_2^2 + u_3^2 + (u_1 + u_2)^2\right] > 0,$$

$$\boldsymbol{u} \cdot \mathbf{V} \cdot \boldsymbol{u} = \sqrt{3}\left[u_1^2 + 2u_2^2 + u_3^2 + (u_1\sqrt{2} + u_3)^2\right] > 0,$$

meaning that the operators $\mathbf{U}$ and $\mathbf{V}$ are positive definite. Finally, it is straightforward to verify that $\mathbf{T} = \mathbf{BU} = \mathbf{VB}$.

2.11 Differentiation of Vectors and Tensors in Cartesian Coordinates

The description of continuum motion is based on the use of partial differential equations. This implies that we need to study how to differentiate vectors and tensors.

The linear part of the variation of a function f is called the *differential* of f and is denoted by df. Hence, if $\Delta f = f(\boldsymbol{r} + d\boldsymbol{r}) - f(\boldsymbol{r})$, where $\boldsymbol{r}$ is an arbitrary vector, then

$$\Delta f = f(\boldsymbol{r} + d\boldsymbol{r}) - f(\boldsymbol{r}) = df + o(\|d\boldsymbol{r}\|).$$

Here $o(\|d\boldsymbol{r}\|)$ is a quantity that tends to zero faster than $\|d\boldsymbol{r}\|$ when $\|d\boldsymbol{r}\| \to 0$. In Cartesian coordinates $df = d\boldsymbol{r} \cdot \nabla f$. Since $d\boldsymbol{r}$ is an arbitrary vector, df is a linear function defined in three-dimensional vector space. Hence, ∇f is a vector. In Cartesian coordinates $\nabla f = \dfrac{\partial f}{\partial x_i} \boldsymbol{e}_i$. We can consider ∇ as a symbolic vector

$$\nabla = \boldsymbol{e}_i \frac{\partial}{\partial x_i} \tag{2.11.1}$$

with components $(\partial_1, \partial_2, \partial_3)$, where $\partial_i = \dfrac{\partial}{\partial x_i}$.

We can calculate the components of ∇ in a new coordinate system either using the rule for transformation of components of vectors and tensors (2.8.2) with $n = 1$ for vectors, or using the chain rule. Let us show that the two methods give the same result. The transformation from new to old coordinates is given by $x_j = a'_{ji} x'_i$, where a'_{ji} are the elements of the inverse matrix of the transformation, $\hat{\mathbf{A}}^{-1}$. But $\hat{\mathbf{A}}$ is orthogonal, $\hat{\mathbf{A}}^{-1} = \hat{\mathbf{A}}^{T}$, meaning that $a'_{ji} = a_{ij}$. Then $x_j = a_{ij} x'_i$. Using this relation and the chain rule we obtain

$$\frac{\partial}{\partial x'_i} = \frac{\partial x_j}{\partial x'_i} \frac{\partial}{\partial x_j} = a_{ij} \frac{\partial}{\partial x_j}.$$

This is exactly what follows from (2.8.2) with $n = 1$.

We can apply ∇ not only to scalars, but also to vectors and tensors. The dyadic product of two vectors, ∇ and $\boldsymbol{v}$, gives the second-order tensor

$$\nabla \boldsymbol{v} = \frac{\partial v_j}{\partial x_i} \boldsymbol{e}_i \boldsymbol{e}_j. \tag{2.11.2}$$

If we take the contraction of this tensor, then we obtain the *divergence* of $\boldsymbol{v}$, $\nabla \cdot \boldsymbol{v} = \dfrac{\partial v_i}{\partial x_i}$. Applying ∇ to tensors in Cartesian coordinates is simple because $\boldsymbol{e}_1, \boldsymbol{e}_2, \boldsymbol{e}_3$ are constant. Hence we have to differentiate only the components of tensors. The direct product of the vector ∇ and tensor $\mathbf{T}$ is given by

$$\nabla \mathbf{T} = \frac{\partial T_{j_1 \ldots j_n}}{\partial x_i} \boldsymbol{e}_i \boldsymbol{e}_{j_1} \ldots \boldsymbol{e}_{j_n} = \partial_i T_{j_1 \ldots j_n} \boldsymbol{e}_i \boldsymbol{e}_{j_1} \ldots \boldsymbol{e}_{j_n}. \tag{2.11.3}$$

In particular, when $\mathbf{T}$ is a second-order tensor, we obtain the third-order tensor

$$\nabla \mathbf{T} = \frac{\partial T_{jk}}{\partial x_i} \boldsymbol{e}_i \boldsymbol{e}_j \boldsymbol{e}_k = \partial_i T_{jk} \boldsymbol{e}_i \boldsymbol{e}_j \boldsymbol{e}_k. \tag{2.11.4}$$

This tensor can be contracted in three different ways. If we contract with respect to j and k, then T_{jj} is a scalar function, and ∇T_{jj} is its gradient. We can also contract with respect to i and j. We denote this contraction by

$$\nabla \cdot \mathbf{T} = \frac{\partial T_{ik}}{\partial x_i} \boldsymbol{e}_k, \tag{2.11.5}$$

and call it the *divergence* of the second-order tensor $\mathbf{T}$. We see that $\nabla \cdot \mathbf{T}$ is a vector.

Finally we can contract with respect to i and k to obtain $\partial_i T_{ji} \boldsymbol{e}_j$. If $\mathbf{T}$ is symmetric, then the latter two contractions are the same,

$$\frac{\partial T_{ji}}{\partial x_i} \boldsymbol{e}_j = \frac{\partial T_{ij}}{\partial x_i} \boldsymbol{e}_j = \nabla \cdot \mathbf{T}.$$

2.12 Cylindrical and Spherical Coordinates

Cartesian coordinates are very useful when developing the general theory of continuum motion and also when solving many particular problems. However, it is often much more convenient to use other coordinates to solve particular problems. Among them cylindrical coordinates are appropriate for solving problems with axial symmetry, and spherical coordinates for solving problems with spherical symmetry. Below we consider these coordinate systems and learn how to differentiate vectors and tensors in these coordinates.

A. Cylindrical Coordinates

The coordinates of a point A are (r_0, ϕ_0, z_0). The coordinate lines are

$$\boldsymbol{r}_r = \boldsymbol{r}(r, \phi_0, z_0), \quad \boldsymbol{r}_\phi = \boldsymbol{r}(r_0, \phi, z_0), \quad \boldsymbol{r}_z = \boldsymbol{r}(r_0, \phi_0, z).$$

The unit vectors $\boldsymbol{e}_r$, $\boldsymbol{e}_\phi$, $\boldsymbol{e}_z$ are tangent to the coordinate lines at A and point in the direction of increase of r, ϕ, and z, respectively. They constitute an *orthogonal local basis*. Hence, cylindrical coordinates are *orthogonal*. Note that the basis is right-oriented. We can consider components of vectors and tensors with respect to this local basis. If we introduce the notation $\boldsymbol{e}_1 = \boldsymbol{e}_r$, $\boldsymbol{e}_2 = \boldsymbol{e}_\phi$, $\boldsymbol{e}_3 = \boldsymbol{e}_z$, then they are given by (2.7.3) (see Fig. 2.2).

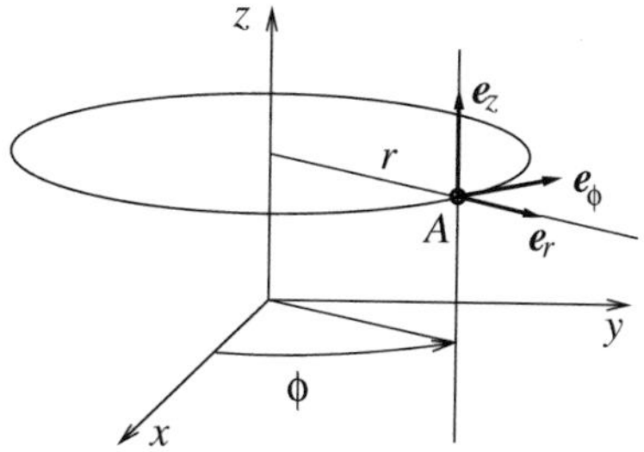

Fig. 2.2 Cylindrical coordinates. The two straight lines passing through the point A show the r and z coordinate lines, while arrows show the unit vectors tangent to these lines

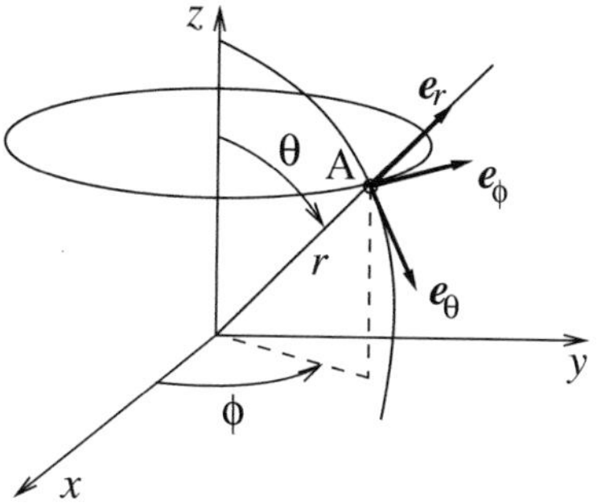

Fig. 2.3 Spherical coordinates

Cartesian coordinates are related to cylindrical coordinates by

$$x = r\cos\phi, \quad y = r\sin\phi, \quad z = z. \tag{2.12.1}$$

B. Spherical Coordinates

Spherical coordinates are r, θ, ϕ, where $0 \le \theta \le \pi$, $0 \le \phi < 2\pi$. The coordinates of a point A are (r_0, θ_0, ϕ_0). The coordinate lines are

$$\boldsymbol{r}_r = \boldsymbol{r}(r, \theta_0, \phi_0), \quad \boldsymbol{r}_\theta = \boldsymbol{r}(r_0, \theta, \phi_0), \quad \boldsymbol{r}_\phi = \boldsymbol{r}(r_0, \theta_0, \phi).$$

The unit vectors $\boldsymbol{e}_r$, $\boldsymbol{e}_\theta$, $\boldsymbol{e}_\phi$ are tangent to the coordinates lines at A and point in the direction of increase of r, θ, and ϕ, respectively. They constitute a right-oriented *orthogonal local basis*. Hence, spherical coordinates are *orthogonal*. Again we can consider components of vectors and tensors with respect to this local basis. If we introduce the notation $\boldsymbol{e}_1 = \boldsymbol{e}_r$, $\boldsymbol{e}_2 = \boldsymbol{e}_\theta$, $\boldsymbol{e}_3 = \boldsymbol{e}_\phi$, then they are given by (2.7.3) (see Fig. 2.3).

Cartesian coordinates are related to spherical coordinates by

$$x = r\sin\theta\cos\phi, \quad y = r\sin\theta\sin\phi, \quad z = r\cos\theta. \tag{2.12.2}$$

2.13 Differentiation of Vectors and Tensors in Cylindrical and Spherical Coordinates

As we have already stated before, the vectors $\boldsymbol{e}_r$, $\boldsymbol{e}_\phi$, $\boldsymbol{e}_z$ in cylindrical, and $\boldsymbol{e}_r$, $\boldsymbol{e}_\theta$, $\boldsymbol{e}_\phi$ in spherical coordinates constitute right-oriented orthogonal bases, and we can calculate the components of a tensor with respect to these bases. However, now the basis vectors depend on coordinates and we have to take this into account when differentiating tensors.

A. Cylindrical Coordinates

We start by calculating the components of the vector ∇ (see Fig. 2.4):

$$d\boldsymbol{r} = \boldsymbol{e}_r\, dr + \boldsymbol{e}_\phi\, r\, d\phi + \boldsymbol{e}_z\, dz,$$

$$\begin{aligned} df &= f(r+dr, \phi+d\phi, z+dz) - f(r,\phi,z) = \frac{\partial f}{\partial r}\, dr + \frac{\partial f}{\partial \phi}\, d\phi + \frac{\partial f}{\partial z}\, dz \\ &= d\boldsymbol{r} \cdot \nabla f = (\nabla f)_r\, dr + (\nabla f)_\phi\, r\, d\phi + (\nabla f)_z\, dz. \end{aligned}$$

Comparing the two expressions for df we obtain

$$(\nabla f)_r = \frac{\partial f}{\partial r}, \quad (\nabla f)_\phi = \frac{1}{r}\frac{\partial f}{\partial \phi}, \quad (\nabla f)_z = \frac{\partial f}{\partial z}.$$

Using these results yields the expression for the gradient of a function f,

$$\nabla f = \frac{\partial f}{\partial r}\, \boldsymbol{e}_r + \frac{1}{r}\frac{\partial f}{\partial \phi}\, \boldsymbol{e}_\phi + \frac{\partial f}{\partial z}\, \boldsymbol{e}_z, \tag{2.13.1}$$

and also the expression for the vector operator ∇,

$$\nabla = \boldsymbol{e}_r\, \frac{\partial}{\partial r} + \frac{\boldsymbol{e}_\phi}{r}\, \frac{\partial}{\partial \phi} + \boldsymbol{e}_z\, \frac{\partial}{\partial z}. \tag{2.13.2}$$

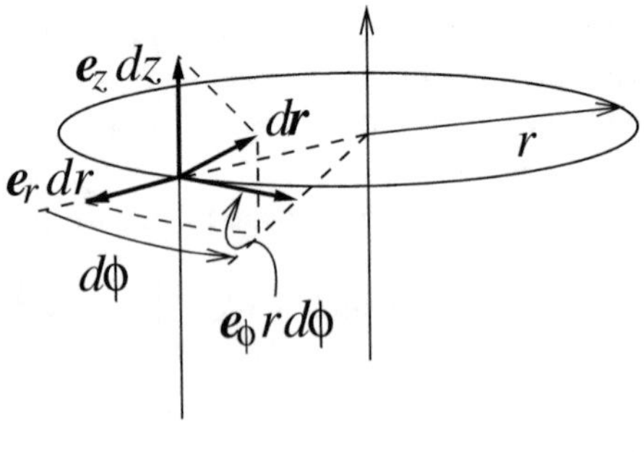

Fig. 2.4 Calculation of components of the vector $d\boldsymbol{r}$ in cylindrical coordinates

To differentiate vectors and tensors we need to calculate the components of the second-order tensors $\nabla \boldsymbol{e}_r$, $\nabla \boldsymbol{e}_\phi$, and $\nabla \boldsymbol{e}_z$. Firstly we note that $\boldsymbol{e}_z$ is constant, implying that $\nabla \boldsymbol{e}_z = 0$. We also note that both $\boldsymbol{e}_r$ and $\boldsymbol{e}_\phi$ are independent of r and z. Then we obtain

$$\nabla \boldsymbol{e}_r = \boldsymbol{e}_r \frac{\partial \boldsymbol{e}_r}{\partial r} + \frac{\boldsymbol{e}_\phi}{r} \frac{\partial \boldsymbol{e}_r}{\partial \phi} + \boldsymbol{e}_z \frac{\partial \boldsymbol{e}_r}{\partial z} = \frac{\boldsymbol{e}_\phi}{r} \frac{\partial \boldsymbol{e}_r}{\partial \phi}, \tag{2.13.3}$$

$$\nabla \boldsymbol{e}_\phi = \boldsymbol{e}_r \frac{\partial \boldsymbol{e}_\phi}{\partial r} + \frac{\boldsymbol{e}_\phi}{r} \frac{\partial \boldsymbol{e}_\phi}{\partial \phi} + \boldsymbol{e}_z \frac{\partial \boldsymbol{e}_\phi}{\partial z} = \frac{\boldsymbol{e}_\phi}{r} \frac{\partial \boldsymbol{e}_\phi}{\partial \phi}. \tag{2.13.4}$$

In the auxiliary Cartesian coordinates $x,\ y,\ z$ (see Fig. 2.2) the components of $\boldsymbol{e}_r$ and $\boldsymbol{e}_\phi$ are given by

$$\boldsymbol{e}_r = (\cos\phi, \sin\phi, 0), \quad \boldsymbol{e}_\phi = (-\sin\phi, \cos\phi, 0).$$

Differentiating these expressions we obtain

$$\frac{\partial \boldsymbol{e}_r}{\partial \phi} = (-\sin\phi, \cos\phi, 0) = \boldsymbol{e}_\phi, \quad \frac{\partial \boldsymbol{e}_\phi}{\partial \phi} = (-\cos\phi, -\sin\phi, 0) = -\boldsymbol{e}_r.$$

Using these results and Eqs. (2.13.3) and (2.13.4) yields

$$\nabla \boldsymbol{e}_r = \frac{1}{r} \boldsymbol{e}_\phi \boldsymbol{e}_\phi, \quad \nabla \boldsymbol{e}_\phi = -\frac{1}{r} \boldsymbol{e}_\phi \boldsymbol{e}_r. \tag{2.13.5}$$

Using Eq. (2.13.5) we obtain

$$\begin{aligned}
\nabla \boldsymbol{v} &= \nabla(v_r \boldsymbol{e}_r + v_\phi \boldsymbol{e}_\phi + v_z \boldsymbol{e}_z) \\
&= (\nabla v_r)\boldsymbol{e}_r + (\nabla v_\phi)\boldsymbol{e}_\phi + (\nabla v_z)\boldsymbol{e}_z + v_r \nabla \boldsymbol{e}_r + v_\phi \nabla \boldsymbol{e}_\phi \\
&= (\nabla v_r)\boldsymbol{e}_r + (\nabla v_\phi)\boldsymbol{e}_\phi + (\nabla v_z)\boldsymbol{e}_z + \frac{v_r}{r} \boldsymbol{e}_\phi \boldsymbol{e}_\phi - \frac{v_\phi}{r} \boldsymbol{e}_\phi \boldsymbol{e}_r.
\end{aligned} \tag{2.13.6}$$

Introducing the notation $\mathbf{V} = \nabla \boldsymbol{v}$ we obtain the following expression for the matrix of the second-order tensor $\mathbf{V}$:

$$\hat{\mathbf{V}} = \begin{pmatrix}
\dfrac{\partial v_r}{\partial r} & \dfrac{\partial v_\phi}{\partial r} & \dfrac{\partial v_z}{\partial r} \\
\dfrac{1}{r}\dfrac{\partial v_r}{\partial \phi} - \dfrac{v_\phi}{r} & \dfrac{1}{r}\dfrac{\partial v_\phi}{\partial \phi} + \dfrac{v_r}{r} & \dfrac{1}{r}\dfrac{\partial v_z}{\partial \phi} \\
\dfrac{\partial v_r}{\partial z} & \dfrac{\partial v_\phi}{\partial z} & \dfrac{\partial v_z}{\partial z}
\end{pmatrix}. \tag{2.13.7}$$

The contraction of $\mathbf{V}$ gives $\nabla \cdot \boldsymbol{v}$. This contraction is equal to the trace of $\mathbf{V}$, which is the sum of the diagonal elements of the matrix $\hat{\mathbf{V}}$:

$$\nabla \cdot \boldsymbol{v} = \frac{\partial v_r}{\partial r} + \frac{1}{r}\frac{\partial v_\phi}{\partial \phi} + \frac{v_r}{r} + \frac{\partial v_z}{\partial z} = \frac{1}{r}\frac{\partial (r v_r)}{\partial r} + \frac{1}{r}\frac{\partial v_\phi}{\partial \phi} + \frac{\partial v_z}{\partial z}. \tag{2.13.8}$$

Let us now calculate the components of $\nabla \times \boldsymbol{v}$ in cylindrical coordinates:

$$(\nabla \times \boldsymbol{v})_r = \varepsilon_{rij}(\nabla \boldsymbol{v})_{ij} = (\nabla \boldsymbol{v})_{\phi z} - (\nabla \boldsymbol{v})_{z\phi} = \frac{1}{r}\frac{\partial v_z}{\partial \phi} - \frac{\partial v_\phi}{\partial z},$$

$$(\nabla \times \boldsymbol{v})_\phi = \varepsilon_{\phi ij}(\nabla \boldsymbol{v})_{ij} = (\nabla \boldsymbol{v})_{zr} - (\nabla \boldsymbol{v})_{rz} = \frac{\partial v_r}{\partial z} - \frac{\partial v_z}{\partial r},$$

$$\begin{aligned}(\nabla \times \boldsymbol{v})_z &= \varepsilon_{zij}(\nabla \boldsymbol{v})_{ij} = (\nabla \boldsymbol{v})_{r\phi} - (\nabla \boldsymbol{v})_{\phi r} \\ &= \frac{\partial v_\phi}{\partial r} - \left(\frac{1}{r}\frac{\partial v_r}{\partial \phi} - \frac{v_\phi}{r}\right) = \frac{1}{r}\frac{\partial (r v_\phi)}{\partial r} - \frac{1}{r}\frac{\partial v_r}{\partial \phi}.\end{aligned}$$

Hence,

$$\nabla \times \boldsymbol{v} = \left(\frac{1}{r}\frac{\partial v_z}{\partial \phi} - \frac{\partial v_\phi}{\partial z}\right)\boldsymbol{e}_r + \left(\frac{\partial v_r}{\partial z} - \frac{\partial v_z}{\partial r}\right)\boldsymbol{e}_\phi + \frac{1}{r}\left(\frac{\partial (r v_\phi)}{\partial r} - \frac{\partial v_r}{\partial \phi}\right)\boldsymbol{e}_z. \tag{2.13.9}$$

Consider the third-order tensor $\boldsymbol{v}\nabla\boldsymbol{v}$, which is a triadic product of vectors $\boldsymbol{v}$, ∇ and $\boldsymbol{v}$. Its contraction with respect to the first and second indices is $(\boldsymbol{v} \cdot \nabla)\boldsymbol{v}$. In Cartesian coordinates

$$[(\boldsymbol{v} \cdot \nabla)\boldsymbol{v}]_i = v_j \frac{\partial v_i}{\partial x_j}.$$

Let us calculate the components of $(\boldsymbol{v} \cdot \nabla)\boldsymbol{v}$ in cylindrical coordinates. Using Eq. (2.13.7) we obtain

$$\begin{aligned}[(\boldsymbol{v} \cdot \nabla)\boldsymbol{v}]_r &= v_r \frac{\partial v_r}{\partial r} + \frac{v_\phi}{r}\frac{\partial v_r}{\partial \phi} + v_z \frac{\partial v_r}{\partial z} - \frac{v_\phi^2}{r}, \\ [(\boldsymbol{v} \cdot \nabla)\boldsymbol{v}]_\phi &= v_r \frac{\partial v_\phi}{\partial r} + \frac{v_\phi}{r}\frac{\partial v_\phi}{\partial \phi} + v_z \frac{\partial v_\phi}{\partial z} + \frac{v_r v_\phi}{r}, \\ [(\boldsymbol{v} \cdot \nabla)\boldsymbol{v}]_z &= v_r \frac{\partial v_z}{\partial r} + \frac{v_\phi}{r}\frac{\partial v_z}{\partial \phi} + v_z \frac{\partial v_z}{\partial z}.\end{aligned} \tag{2.13.10}$$

B. Spherical Coordinates

Again we start by calculating the components of the vector ∇ (see Fig. 2.5).

$$d\boldsymbol{r} = \boldsymbol{e}_r\, dr + \boldsymbol{e}_\theta r\, d\theta + \boldsymbol{e}_\phi r \sin\theta\, d\phi.$$

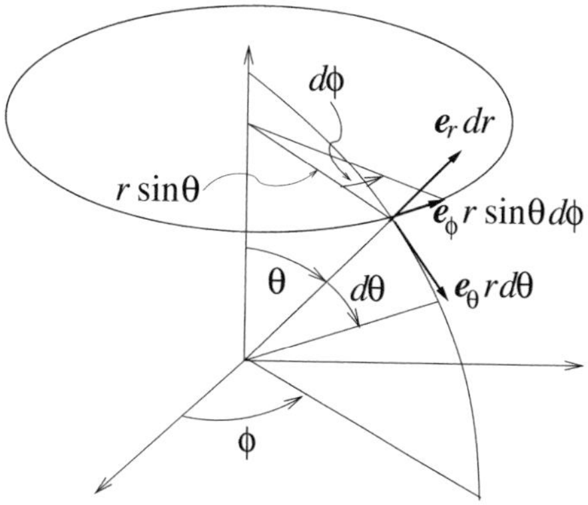

Fig. 2.5 Calculation of $d\boldsymbol{r}$ in spherical coordinates

$$
\begin{aligned}
df &= f(r+dr, \theta+d\theta, \phi+d\phi) - f(r,\theta,\phi) = \frac{\partial f}{\partial r}\,dr + \frac{\partial f}{\partial \theta}\,d\theta + \frac{\partial f}{\partial \phi}\,d\phi \\
&= d\boldsymbol{r}\cdot\nabla f = (\nabla f)_r\,dr + (\nabla f)_\theta\,r\,d\theta + (\nabla f)_\phi\,r\sin\theta\,d\phi.
\end{aligned}
$$

Comparing the two expressions for df we obtain

$$
(\nabla f)_r = \frac{\partial f}{\partial r}, \quad (\nabla f)_\theta = \frac{1}{r}\frac{\partial f}{\partial \theta}, \quad (\nabla f)_\phi = \frac{1}{r\sin\theta}\frac{\partial f}{\partial \phi}.
$$

Hence

$$
\nabla f = \frac{\partial f}{\partial r}\,\boldsymbol{e}_r + \frac{1}{r}\frac{\partial f}{\partial \theta}\,\boldsymbol{e}_\theta + \frac{1}{r\sin\theta}\frac{\partial f}{\partial \phi}\,\boldsymbol{e}_\phi. \tag{2.13.11}
$$

This equation determines the expression for the differential operator ∇,

$$
\nabla = \boldsymbol{e}_r\,\frac{\partial}{\partial r} + \frac{\boldsymbol{e}_\theta}{r}\,\frac{\partial}{\partial \theta} + \frac{\boldsymbol{e}_\phi}{r\sin\theta}\,\frac{\partial}{\partial \phi}. \tag{2.13.12}
$$

The expressions for $\nabla\boldsymbol{e}_r$, $\nabla\boldsymbol{e}_\theta$, and $\nabla\boldsymbol{e}_\phi$ are given by (*without derivation*):

$$
\begin{aligned}
\nabla\boldsymbol{e}_r &= \frac{1}{r}\,\boldsymbol{e}_\theta\boldsymbol{e}_\theta + \frac{1}{r}\,\boldsymbol{e}_\phi\boldsymbol{e}_\phi, \\
\nabla\boldsymbol{e}_\theta &= -\frac{1}{r}\,\boldsymbol{e}_\theta\boldsymbol{e}_r + \frac{\cot\theta}{r}\,\boldsymbol{e}_\phi\boldsymbol{e}_\phi, \\
\nabla\boldsymbol{e}_\phi &= -\frac{1}{r}\,\boldsymbol{e}_\phi\boldsymbol{e}_r - \frac{\cot\theta}{r}\,\boldsymbol{e}_\phi\boldsymbol{e}_\theta.
\end{aligned} \tag{2.13.13}
$$

Using Eq. (2.13.13) we obtain

$$\begin{aligned}\nabla \boldsymbol{v} &= \nabla(v_r \boldsymbol{e}_r + v_\theta \boldsymbol{e}_\theta + v_\phi \boldsymbol{e}_\phi)\\ &= (\nabla v_r)\boldsymbol{e}_r + (\nabla v_\theta)\boldsymbol{e}_\theta + (\nabla v_\phi)\boldsymbol{e}_\phi + v_r \nabla \boldsymbol{e}_r + v_\theta \nabla \boldsymbol{e}_\theta + v_\phi \nabla \boldsymbol{e}_\phi\\ &= (\nabla v_r)\boldsymbol{e}_r + (\nabla v_\theta)\boldsymbol{e}_\theta + (\nabla v_\phi)\boldsymbol{e}_\phi + \frac{v_r}{r}(\boldsymbol{e}_\theta \boldsymbol{e}_\theta + \boldsymbol{e}_\phi \boldsymbol{e}_\phi)\\ &- \frac{v_\theta}{r}(\boldsymbol{e}_\theta \boldsymbol{e}_r - \cot\theta\, \boldsymbol{e}_\phi \boldsymbol{e}_\phi) - \frac{v_\phi}{r}(\boldsymbol{e}_\phi \boldsymbol{e}_r + \cot\theta\, \boldsymbol{e}_\phi \boldsymbol{e}_\theta). \qquad (2.13.14)\end{aligned}$$

Then the matrix of the tensor $\mathbf{V} = \nabla \boldsymbol{v}$ is

$$\hat{\mathbf{V}} = \begin{pmatrix} \dfrac{\partial v_r}{\partial r} & \dfrac{\partial v_\theta}{\partial r} & \dfrac{\partial v_\phi}{\partial r} \\ \dfrac{1}{r}\dfrac{\partial v_r}{\partial \theta} - \dfrac{v_\theta}{r} & \dfrac{1}{r}\dfrac{\partial v_\theta}{\partial \theta} + \dfrac{v_r}{r} & \dfrac{1}{r}\dfrac{\partial v_\phi}{\partial \theta} \\ V_{31} & V_{32} & V_{33} \end{pmatrix}, \qquad (2.13.15)$$

where

$$\begin{aligned} V_{31} &= \frac{1}{r\sin\theta}\frac{\partial v_r}{\partial \phi} - \frac{v_\phi}{r},\\ V_{32} &= \frac{1}{r\sin\theta}\frac{\partial v_\theta}{\partial \phi} - \frac{v_\phi \cot\theta}{r}, \qquad (2.13.16)\\ V_{33} &= \frac{1}{r\sin\theta}\frac{\partial v_\phi}{\partial \phi} + \frac{v_r + v_\theta \cot\theta}{r}. \end{aligned}$$

The contraction of $\mathbf{V}$, equal to the trace of $\mathbf{V}$, gives:

$$\begin{aligned}\nabla \cdot \boldsymbol{v} &= \frac{\partial v_r}{\partial r} + \frac{1}{r}\frac{\partial v_\theta}{\partial \theta} + \frac{v_r}{r} + \frac{1}{r\sin\theta}\frac{\partial v_\phi}{\partial \phi} + \frac{v_r + v_\theta\cot\theta}{r}\\ &= \frac{1}{r^2}\frac{\partial (r^2 v_r)}{\partial r} + \frac{1}{r\sin\theta}\frac{\partial (v_\theta \sin\theta)}{\partial \theta} + \frac{1}{r\sin\theta}\frac{\partial v_\phi}{\partial \phi}. \qquad (2.13.17)\end{aligned}$$

For $\nabla \times \boldsymbol{v}$ we obtain

$$\begin{aligned}(\nabla \times \boldsymbol{v})_r &= \varepsilon_{rij}(\nabla \boldsymbol{v})_{ij} = (\nabla \boldsymbol{v})_{\theta\phi} - (\nabla \boldsymbol{v})_{\phi\theta} = \frac{1}{r}\frac{\partial v_\phi}{\partial \theta}\\ &- \frac{1}{r\sin\theta}\frac{\partial v_\theta}{\partial \phi} + \frac{v_\phi \cot\theta}{r} = \frac{1}{r\sin\theta}\left(\frac{\partial (v_\phi \sin\theta)}{\partial \theta} - \frac{\partial v_\theta}{\partial \phi}\right),\end{aligned}$$

$$\begin{aligned}(\nabla \times \boldsymbol{v})_\theta &= \varepsilon_{\theta ij}(\nabla \boldsymbol{v})_{ij} = (\nabla \boldsymbol{v})_{\phi r} - (\nabla \boldsymbol{v})_{r\phi}\\ &= \frac{1}{r\sin\theta}\frac{\partial v_r}{\partial \phi} - \frac{\partial v_\phi}{\partial r} - \frac{v_\phi}{r} = \frac{1}{r\sin\theta}\frac{\partial v_r}{\partial \phi} - \frac{1}{r}\frac{\partial (r v_\phi)}{\partial r},\end{aligned}$$

$$(\nabla \times \boldsymbol{v})_\phi = \varepsilon_{\phi ij}(\nabla \boldsymbol{v})_{ij} = (\nabla \boldsymbol{v})_{r\theta} - (\nabla \boldsymbol{v})_{\theta r}$$
$$= \frac{\partial v_\theta}{\partial r} - \frac{1}{r}\frac{\partial v_r}{\partial \theta} + \frac{v_\theta}{r} = \frac{1}{r}\frac{\partial (r v_\theta)}{\partial r} - \frac{1}{r}\frac{\partial v_r}{\partial \theta}.$$

Hence,

$$\nabla \times \boldsymbol{v} = \frac{1}{r\sin\theta}\left(\frac{\partial(v_\phi \sin\theta)}{\partial\theta} - \frac{\partial v_\theta}{\partial\phi}\right)\boldsymbol{e}_r + \frac{1}{r}\left(\frac{1}{\sin\theta}\frac{\partial v_r}{\partial\phi} - \frac{\partial(r v_\phi)}{\partial r}\right)\boldsymbol{e}_\theta + \frac{1}{r}\left(\frac{\partial(r v_\theta)}{\partial r} - \frac{\partial v_r}{\partial\theta}\right)\boldsymbol{e}_\phi. \quad (2.13.18)$$

Using (2.13.15) we obtain

$$[(\boldsymbol{v}\cdot\nabla)\boldsymbol{v}]_r = v_r\frac{\partial v_r}{\partial r} + \frac{v_\theta}{r}\frac{\partial v_r}{\partial\theta} + \frac{v_\phi}{r\sin\theta}\frac{\partial v_r}{\partial\phi} - \frac{1}{r}(v_\theta^2 + v_\phi^2),$$
$$[(\boldsymbol{v}\cdot\nabla)\boldsymbol{v}]_\theta = v_r\frac{\partial v_\theta}{\partial r} + \frac{v_\theta}{r}\frac{\partial v_\theta}{\partial\theta} + \frac{v_\phi}{r\sin\theta}\frac{\partial v_\theta}{\partial\phi} + \frac{1}{r}(v_r v_\theta - v_\phi^2\cot\theta),$$
$$[(\boldsymbol{v}\cdot\nabla)\boldsymbol{v}]_\phi = v_r\frac{\partial v_\phi}{\partial r} + \frac{v_\theta}{r}\frac{\partial v_\phi}{\partial\theta} + \frac{v_\phi}{r\sin\theta}\frac{\partial v_\phi}{\partial\phi} + \frac{1}{r}(v_r v_\phi + v_\theta v_\phi\cot\theta). \quad (2.13.19)$$

Problems

2.1: Show that the matrix

$$\hat{\mathbf{A}} = \begin{pmatrix} \frac{1}{2} & -\frac{\sqrt{3}}{2} & 0 \\ \frac{\sqrt{3}}{2} & \frac{1}{2} & 0 \\ 0 & 0 & 1 \end{pmatrix}$$

is orthogonal. Is it proper or improper?

2.2: Let $\hat{\mathbf{A}}$ be the transformation matrix from old to new Cartesian coordinates. Find the components of the vector $\boldsymbol{a}$ in the new coordinates if in the old coordinates they are given by $\boldsymbol{a} = (1, -1, 2)^T$.

2.3: Write down in full the following expressions:

(a) $x_i y_i$, (b) $t_i = n_j \sigma_{ji}$, (c) T_{ii}, (d) $B_{ii} C_{jj}$, (e) $\delta_{ij} B_i C_j$, (f) $\delta_{ij} T_{ij}$.

2.4: (i) Write down the explicit expression for the trace of the square of the matrix $\hat{\mathbf{T}}$.

(ii) Show that the second invariant of the matrix $\hat{\mathbf{T}}$ is given by

$$I_2 = T_{11}T_{22} + T_{11}T_{33} + T_{22}T_{33} - T_{12}T_{21} - T_{13}T_{31} - T_{23}T_{32}.$$

(iii) Use the results obtained in (i) and (ii) to show by direct calculation that the characteristic equation of the matrix $\hat{\mathbf{T}}$ has the form

$$\lambda^3 - I_1\lambda^2 + I_2\lambda - I_3 = 0.$$

2.5: Show that the equations $x_i' = a_{ij}x_j$ are equivalent to $x_i = a_{ji}x_j'$.

2.6: Show that, if $\hat{\mathbf{T}}$ is a 3×3 matrix with the elements T_{ij}, then $\det(\hat{\mathbf{T}}) = \varepsilon_{ijk}T_{1i}T_{2j}T_{3k}$.

2.7: If $\boldsymbol{v} = \boldsymbol{v}(x_1, x_2, x_3)$ is a twice differentiable vector field, show that

$$\varepsilon_{ijk}\frac{\partial^2 v_k}{\partial x_i \partial x_j} = 0.$$

[Hint: Use the result $\varepsilon_{ijk} = -\varepsilon_{jik}$ and change dummy suffixes.]

2.8: (i) Suppose the new Cartesian coordinates x_1', x_2', x_3' are obtained from the old ones, x_1, x_2, x_3, by rotating the coordinate axes about the x_3-axis by the angle θ. Show that the matrix of transformation from the old to the new coordinates is given by

$$\hat{\mathbf{A}} = \begin{pmatrix} \cos\theta & \sin\theta & 0 \\ -\sin\theta & \cos\theta & 0 \\ 0 & 0 & 1 \end{pmatrix}.$$

(ii) You are given that the tensor $\mathbf{T}$ is invariant with respect to rotation about the x_3-axis. Show that, in Cartesian coordinates x_1, x_2, x_3, its matrix, $\hat{\mathbf{T}}$, has the form

$$\hat{\mathbf{T}} = \begin{pmatrix} T_1 & T_2 & 0 \\ -T_2 & T_1 & 0 \\ 0 & 0 & T_3 \end{pmatrix}.$$

(You need to use the relation between the matrices of the tensor $\mathbf{T}$ in the old and new coordinates, $\hat{\mathbf{T}}' = \hat{\mathbf{A}}\hat{\mathbf{T}}\hat{\mathbf{A}}^T$.)

(iii) Calculate $\mathrm{tr}(\hat{\mathbf{T}})$ and $\mathrm{tr}(\hat{\mathbf{T}}^2)$.

(iv) The principal invariants of the matrix $\hat{\mathbf{T}}$ are given by $I_1 = 0$, $I_2 = 0$, $I_3 = 8$. You are also given that $T_2 > 0$. Find T_1, T_2 and T_3.

2.9: You are given that the matrix

$$\hat{\mathbf{A}} = \begin{pmatrix} \frac{1}{2} & \frac{\sqrt{3}}{2} & 0 \\ -\frac{\sqrt{3}}{2} & \mu & 0 \\ 0 & 0 & \lambda \end{pmatrix}$$

is proper orthogonal. Determine λ and μ.

2.10: Show that the matrix

$$\hat{\mathbf{A}} = \begin{pmatrix} \frac{1}{\sqrt{2}} & \frac{-1}{\sqrt{2}} & 0 \\ \frac{1}{\sqrt{2}} & \frac{1}{\sqrt{2}} & 0 \\ 0 & 0 & 1 \end{pmatrix}$$

is proper orthogonal.

Let $\hat{\mathbf{A}}$ be the transformation matrix from old to new Cartesian coordinates. Find the components of the tensor $\mathbf{T}$ in the new coordinates if in the old coordinates its components are given by the matrix

$$\hat{\mathbf{T}} = \begin{pmatrix} 0 & 1 & 0 \\ 1 & 0 & 0 \\ 0 & 0 & 0 \end{pmatrix}.$$

2.11: Consider the two vectors, $\boldsymbol{a}$ and $\boldsymbol{b}$, with components $\boldsymbol{a} = (2, 1, -1)$ and $\boldsymbol{b} = (1, -1, 1)$.

(i) Find the components of two unit vectors, $\boldsymbol{e}_1'$ and $\boldsymbol{e}_2'$, such that they are parallel to $\boldsymbol{a}$ and $\boldsymbol{b}$ respectively.

(ii) Find the components of the third unit vector, $\boldsymbol{e}_3'$, such that it is orthogonal to $\boldsymbol{e}_1'$ and $\boldsymbol{e}_2'$, and $\boldsymbol{e}_1', \boldsymbol{e}_2', \boldsymbol{e}_3'$ form a right-hand set.

(iii) Write down the transformation matrix $\hat{\mathbf{A}}$ from the old coordinate system to the new one with the basis $\boldsymbol{e}_1', \boldsymbol{e}_2', \boldsymbol{e}_3'$.

(iv) Verify that $\det(\hat{\mathbf{A}}) = 1$.

2.12: Suppose the new Cartesian coordinates, x_1', x_2', x_3' are obtained from the old ones, x_1, x_2, x_3, by a rotation about the x_3-axis through an angle θ. It is shown in Problem **2.8** that the transformation matrix is given by

$$\hat{\mathbf{A}} = \begin{pmatrix} \cos\theta & \sin\theta & 0 \\ -\sin\theta & \cos\theta & 0 \\ 0 & 0 & 1 \end{pmatrix}.$$

(i) The components of the tensor $\mathbf{T}$ in the old coordinates are given by

$$\hat{\mathbf{T}} = \begin{pmatrix} a & b & 0 \\ b & c & 0 \\ 0 & 0 & 0 \end{pmatrix}.$$

Find the matrix $\hat{\mathbf{T}}'$ of components of this tensor in the new coordinates.

(ii) Given that $\tan 2\theta = 2b/(a-c)$ and $0 < \theta < \pi/4$, show that

$$\text{(a)}T'_{ij} = 0\ (i \neq j), \quad \text{(b)}T'_{11} = \frac{1}{2}\left\{a + c + [(a-c)^2 + 4b^2]^{1/2}\right\}.$$

2.13: Suppose the transformation matrix from old to new coordinates is given by

$$\hat{\mathbf{A}} = \begin{pmatrix} \frac{1}{\sqrt{3}} & \frac{-1}{\sqrt{3}} & \frac{1}{\sqrt{3}} \\ \frac{1}{\sqrt{2}} & \frac{1}{\sqrt{2}} & 0 \\ \frac{1}{\sqrt{6}} & \frac{-1}{\sqrt{6}} & \frac{-2}{\sqrt{6}} \end{pmatrix}.$$

(i) Verify that $\hat{\mathbf{A}}\hat{\mathbf{A}}^T = \hat{\mathbf{I}}$, where $\hat{\mathbf{I}}$ is the unit matrix, so that $\hat{\mathbf{A}}$ is orthogonal.

(ii) The matrix of components of the second-order tensor $\hat{\mathbf{T}}$ is given in the old coordinate system by

$$\hat{\mathbf{T}} = \begin{pmatrix} 2 & 0 & c \\ 0 & a & 0 \\ 0 & 0 & b \end{pmatrix}.$$

Calculate the matrix $\hat{\mathbf{T}}'$ of components of this tensor in the new coordinates using the relation between the old and new components of the tensor $\mathbf{T}$ written in the matrix form $\hat{\mathbf{T}}' = \hat{\mathbf{A}}\hat{\mathbf{T}}\hat{\mathbf{A}}^T$.

(iii) You are given that the matrix of components of the tensor $\mathbf{T}$ is the same in the old and new coordinate systems. Find a, b, and c.

2.14: Suppose the polar decomposition of the operator $\mathbf{T}$ is given by $\mathbf{T} = \mathbf{BU}$, where the matrices of the operators $\mathbf{B}$ and $\mathbf{T}$ are defined by

$$\hat{\mathbf{B}} = \frac{1}{\sqrt{30}}\begin{pmatrix} 2\sqrt{5} & -\sqrt{5} & \sqrt{5} \\ 2 & 5 & 1 \\ \sqrt{6} & 0 & -2\sqrt{6} \end{pmatrix}, \quad \hat{\mathbf{T}} = \begin{pmatrix} 6\sqrt{5} & -\sqrt{5} & 17\sqrt{5} \\ 12 & 11 & 11 \\ -4\sqrt{6} & 3\sqrt{6} & -17\sqrt{6} \end{pmatrix}.$$

Calculate the matrix $\hat{\mathbf{U}}$ of the operator $\mathbf{U}$, and show that the operator $\mathbf{U}$ is positive-definite.

2.15: Suppose the matrix of the tensor $\mathbf{U}$ is given by

$$\hat{\mathbf{U}} = \begin{pmatrix} f & 2x_1x_2 & \cos x_1 \\ 2x_1x_2 & g & \sin x_3 \\ \cos x_1 & \sin x_3 & h \end{pmatrix},$$

where f, g, and h are functions of x_1, x_2, and x_3. You are given that this tensor is divergent-free, $\nabla \cdot \mathbf{U} = 0$. You are also given that $f = 0$ at $x_1 = 0$, $g = 0$ at $x_2 = 0$, and $h = 0$ at $x_3 = 0$. Determine f, g, and h.

2.16: Using the $\varepsilon - \delta$ relation

$$\varepsilon_{ijk}\varepsilon_{lmk} = \delta_{il}\delta_{jm} - \delta_{im}\delta_{jl}$$

prove that

$$\boldsymbol{v} \times (\nabla \times \boldsymbol{v}) = \frac{1}{2}\nabla\left(\|\boldsymbol{v}\|^2\right) - (\boldsymbol{v} \cdot \nabla)\boldsymbol{v}. \qquad (*)$$

2.17: Derive Eq. (2.13.13).
[Hint: Use the expressions of the components of the basis vectors of the spherical coordinates in auxiliary Cartesian coordinates

$$\begin{aligned} \boldsymbol{e}_r &= (\sin\theta\cos\phi, \sin\theta\sin\phi, \cos\theta), \\ \boldsymbol{e}_\theta &= (\cos\theta\cos\phi, \cos\theta\sin\phi, -\sin\theta), \\ \boldsymbol{e}_\phi &= (-\sin\phi, \cos\phi, 0).] \end{aligned}$$

2.18: Suppose the components of a vector field $\boldsymbol{v}$ are given in spherical coordinates by

$$v_r = \frac{1}{r^2\sin\theta}\frac{\partial\psi}{\partial\theta}, \quad v_\theta = -\frac{1}{r\sin\theta}\frac{\partial\psi}{\partial r}, \quad v_\phi = 0. \qquad (*)$$

(i) Show that this vector field is divergence-free, i.e. $\nabla \cdot \boldsymbol{v} = 0$.

(ii) In addition, you are given that ψ is independent of ϕ. Use $(*)$ and the expressions for the gradient of a scalar field and the curl of a vector field in spherical coordinates, Eqs. (2.13.11) and (2.13.18), to express $(\boldsymbol{v} \cdot \nabla)\boldsymbol{v}$ in terms of ψ.

2.19: (i) Let $\boldsymbol{e}_1$, $\boldsymbol{e}_2$, and $\boldsymbol{e}_3$ be the unit vectors along the axes of Cartesian coordinates x_1, x_2, x_3. We can form 27 triadic products using these vectors: $\boldsymbol{e}_i\boldsymbol{e}_j\boldsymbol{e}_k$, $i = 1, 2, 3$, $j = 1, 2, 3$, $k = 1, 2, 3$. Show that these triadic products are linearly independent, i.e. if $C_{ijk}\boldsymbol{e}_i\boldsymbol{e}_j\boldsymbol{e}_k = 0$ then $C_{ijk} = 0$, $i = 1, 2, 3$, $j = 1, 2, 3$, $k = 1, 2, 3$.

(ii) Show that the 27 tensors, $\boldsymbol{e}_i\boldsymbol{e}_j\boldsymbol{e}_k$, $i = 1, 2, 3$, $j = 1, 2, 3$, $k = 1, 2, 3$, constitute a basis in the vector space of third-order tensors, i.e. any third-order tensor $\mathbf{T}$ can be written as $\mathbf{T} = T_{ijk}\boldsymbol{e}_i\boldsymbol{e}_j\boldsymbol{e}_k$.

Chapter 3
Kinematics of Continuum

Let us recall the mechanics of a material point. This system has three degrees of freedom because its spatial position is determined by three coordinates. We first consider the kinematics of the system, that is, we develop a mathematical description of its motion. When doing so we disregard the course of motion, that is, we do not consider forces applied to the material point. We do exactly the same when studying the motion of a more complex mechanical system with a finite number of degrees of freedom. We completely describe the motion of any such system if we determine the dependence of its coordinates on time. In particular, we completely describe the motion of a material point if we find its Cartesian coordinates as functions of time.

We now use the same approach when studying the motion of a continuum. Again we start by studying its kinematics, disregarding the course of motion. There is only one way to describe the kinematics of a system with a finite number of degrees of freedom: we must determine the dependence of its coordinates on time. In contrast, as we will see below, there are two ways to describe the kinematics of a continuum.

3.1 Lagrangian and Eulerian Descriptions of Continuum Motion

In this section we describe continuum motion in terms of Cartesian coordinates. Let us fix the coordinates of each point of a continuum at the initial instance $t = 0$: $x_i(t = 0) = \xi_i$. Then we trace the motion of each point as time progresses. Hence,

$$\left.\begin{aligned} x_1 &= x_1(\xi_1, \xi_2, \xi_3, t), \\ x_2 &= x_2(\xi_1, \xi_2, \xi_3, t), \\ x_3 &= x_3(\xi_1, \xi_2, \xi_3, t), \end{aligned}\right\} \quad \text{or} \quad x_i = x_i(\xi_1, \xi_2, \xi_3, t). \tag{3.1.1}$$

© Springer Nature Switzerland AG 2019
M. S. Ruderman, *Fluid Dynamics and Linear Elasticity*, Springer Undergraduate Mathematics Series, https://doi.org/10.1007/978-3-030-19297-6_3

The quantities ξ_1, ξ_2, ξ_3 and t are called *Lagrangian variables*. If we can find functions $x_i = x_i(\xi_1, \xi_2, \xi_3, t)$, then we will have completely described the motion of the continuum. When ξ_1, ξ_2, ξ_3 are fixed and t varies, these three equations ($i = 1, 2, 3$) define a curve called a *particle trajectory*.

We now make the assumption that the mapping $(\xi_1, \xi_2, \xi_3) \longrightarrow (x_1, x_2, x_3)$ is always *one-to-one*, so that we can solve the system of equations (3.1.1) with respect to ξ_i and obtain

$$\xi_i = \xi_i(x_1, x_2, x_3, t). \tag{3.1.2}$$

In particular, this implies that

$$\mathcal{J}(\boldsymbol{\xi}, t) = \frac{D(x_1, x_2, x_3)}{D(\xi_1, \xi_2, \xi_3)} = \begin{vmatrix} \dfrac{\partial x_1}{\partial \xi_1} & \dfrac{\partial x_1}{\partial \xi_2} & \dfrac{\partial x_1}{\partial \xi_3} \\ \dfrac{\partial x_2}{\partial \xi_1} & \dfrac{\partial x_2}{\partial \xi_2} & \dfrac{\partial x_2}{\partial \xi_3} \\ \dfrac{\partial x_3}{\partial \xi_1} & \dfrac{\partial x_3}{\partial \xi_2} & \dfrac{\partial x_3}{\partial \xi_3} \end{vmatrix} \neq 0, \tag{3.1.3}$$

where $\mathcal{J}(\boldsymbol{\xi}, t) = \dfrac{D(x_1, x_2, x_3)}{D(\xi_1, \xi_2, \xi_3)}$ is the *Jacobian*.

At the initial instance the continuum occupies a domain D_0. At instance t it occupies a domain D. Equation (3.1.1) can be considered as a one-to-one mapping of D_0 onto D.

Topological Theorem *If a mapping is one-to-one and continuous, then a* volume *is mapped onto a* volume, *a* surface *is mapped onto a* surface, *and a* line *is mapped onto a* line *(see Fig. 3.1).*

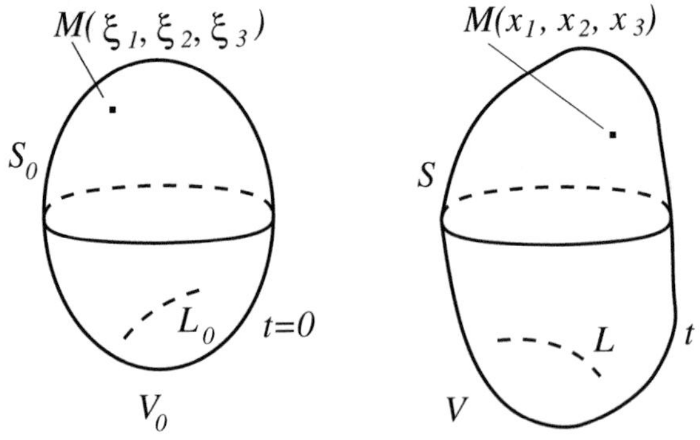

Fig. 3.1 The mapping defined by the continuum motion from its initial position to its current position. The point $M(x_1, x_2, x_3)$ is the image of the point $M(\xi_1, \xi_2, \xi_3)$, the line L is the image of the line L_0, the surface S is the image of the surface S_0, and the volume V is the image of the volume V_0

In the *Eulerian description* we study the temporal evolution of continuum parameters (e.g., velocity, temperature, pressure) at a fixed spatial position (x_1, x_2, x_3). Hence, if f is any quantity characterising the continuum, then in the Eulerian description we find it as $f = f(x_1, x_2, x_3, t)$. The variables x_1, x_2, x_3, t are called the *Eulerian variables*.

Transition from the Lagrangian to the Eulerian Description

Let us assume that we have managed to describe the continuum motion in Lagrangian variables, that is, we have found all particle trajectories $x_i = x_i(\xi_1, \xi_2, \xi_3, t)$ and all quantities characterising the continuum (velocity, pressure, temperature, etc.) in the form $f = f(\xi_1, \xi_2, \xi_3, t)$. Since, by assumption, the mapping $\boldsymbol{\xi} \to \boldsymbol{x}$ is one-to-one, we can invert it and obtain $\xi_i = \xi_i(x_1, x_2, x_3, t)$. Then any function f of the variables ξ_1, ξ_2, ξ_3 becomes a function of the variables x_1, x_2, x_3 by the substitution

$$f = f(\xi_1(x_1, x_2, x_3, t), \xi_2(x_1, x_2, x_3, t), \xi_3(x_1, x_2, x_3, t), t).$$

This gives the Eulerian description.

Example Suppose the continuum motion is described in Lagrangian variables by

$$x_1 = e^{t/t_0}\xi_1, \quad x_2 = \xi_2 \cosh(t/t_0), \quad x_3 = e^{-t/t_0}\xi_3,$$

where t_0 is a positive constant and the function $f(\xi_1, \xi_2, \xi_3)$ is given by

$$f(\xi_1, \xi_2, \xi_3) = \xi_1^2 + \xi_2^2 + \xi_3^2.$$

The Lagrangian variables are expressed in terms of the Eulerian variables by

$$\xi_1 = e^{-t/t_0}x_1, \quad \xi_2 = \frac{x_2}{\cosh(t/t_0)}, \quad \xi_3 = e^{t/t_0}x_3.$$

Then the expression of the function f in Eulerian variables is

$$f(x_1, x_2, x_3, t) = e^{-2t/t_0}x_1^2 + \frac{x_2^2}{\cosh^2(t/t_0)} + e^{2t/t_0}x_3^2.$$

Transition from the Eulerian to the Lagrangian Description

Assume now that we have managed to describe the continuum motion in Eulerian variables, that is, we have found all quantities characterising the continuum in the form $f = f(x_1, x_2, x_3, t)$. In particular, we have found the velocity $\boldsymbol{v} = \boldsymbol{v}(x_1, x_2, x_3, t)$, where

$$\boldsymbol{v} = \frac{\partial \boldsymbol{x}}{\partial t}, \quad \boldsymbol{x} = (x_1, x_2, x_3), \tag{3.1.4}$$

and the partial derivative with respect to time is taken at a fixed $\boldsymbol{\xi}$. We now use this equation to find the functions in Eq. (3.1.1). We can write $\frac{d}{dt}$ instead of $\frac{\partial}{\partial t}$ if we disregard the dependence of $\boldsymbol{x}$ on $\boldsymbol{\xi}$. Then we have the system of ordinary differential equations

$$\begin{aligned} \frac{dx_1}{dt} &= v_1(x_1, x_2, x_3, t), \\ \frac{dx_2}{dt} &= v_2(x_1, x_2, x_3, t), \\ \frac{dx_3}{dt} &= v_3(x_1, x_2, x_3, t). \end{aligned} \tag{3.1.5}$$

We solve this system to obtain the x_i as functions of time and three arbitrary constants $C_1,\ C_2,\ C_3$. Using the initial condition $\boldsymbol{x}(t=0) = \boldsymbol{\xi}$ we express $C_1,\ C_2$, and C_3 in terms of ξ_1, ξ_2, ξ_3, which gives $x_i = x_i(\xi_1, \xi_2, \xi_3, t)$. Hence, we know the trajectories of all particles. Given any function of the variables x_1, x_2, x_3, we can make it a function of ξ_1, ξ_2, ξ_3 by the substitution

$$f = f(x_1(\xi_1, \xi_2, \xi_3, t), x_2(\xi_1, \xi_2, \xi_3, t), x_3(\xi_1, \xi_2, \xi_3, t), t) = f(\xi_1, \xi_2, \xi_3, t).$$

Hence, we obtain the Lagrangian description of the continuum motion.

Example Suppose the continuum velocity is given in the Eulerian variables by

$$v_1 = \frac{x_1}{t + t_0}, \quad v_2 = \frac{x_2}{t_0}, \quad v_3 = \frac{x_3}{t_0} \tanh(t/t_0),$$

where t_0 is a positive constant and the function $f(x_1, x_2, x_3, t)$ is given by

$$f(x_1, x_2, x_3, t) = e^{-t/t_0} x_1 x_2 x_3.$$

Let us find the expression of f in Lagrangian variables. The trajectory equations are

$$\frac{dx_1}{dt} = \frac{x_1}{t + t_0}, \quad \frac{dx_2}{dt} = \frac{x_2}{t_0}, \quad \frac{dx_3}{dt} = \frac{x_3}{t_0} \tanh(t/t_0).$$

The solution to this system of equations satisfying the initial conditions $x_1 = \xi_1$, $x_2 = \xi_2$, $x_3 = \xi_3$ at $t = 0$ is

$$x_1 = \xi_1(1 + t/t_0), \quad x_2 = \xi_2 e^{t/t_0}, \quad x_3 = \xi_3 \cosh(t/t_0).$$

Substituting these expressions into the expression for f we obtain

$$f(\xi_1, \xi_2, \xi_3, t) = \xi_1 \xi_2 \xi_3 (1 + t/t_0) \cosh(t/t_0).$$

The Material Time Derivative

If we are interested in the rate of variation of the temperature T at a fixed spatial position, then it is given by

$$\left(\frac{\partial T}{\partial t}\right)_{\boldsymbol{x}},$$

where the subscript $\boldsymbol{x}$ indicates that the derivative is taken at a fixed $\boldsymbol{x}$. The rate of temperature variation in a particle that was at position $\boldsymbol{\xi}$ at $t = 0$ and is passing position $\boldsymbol{x}$ at time t is given by

$$\left(\frac{\partial T}{\partial t}\right)_{\xi}.$$

This derivative is called the *material derivative*. In the Eulerian description $T = T(x_1, x_2, x_3, t)$. To calculate the material derivative we consider T as a composite function with $\boldsymbol{x} = \boldsymbol{x}(\boldsymbol{\xi}, t)$. Then we use the chain rule:

$$\left(\frac{\partial T}{\partial t}\right)_{\xi} = \left(\frac{\partial T}{\partial t}\right)_{x} + \frac{\partial T}{\partial x_1}\left(\frac{\partial x_1}{\partial t}\right)_{\xi} + \frac{\partial T}{\partial x_2}\left(\frac{\partial x_2}{\partial t}\right)_{\xi} + \frac{\partial T}{\partial x_3}\left(\frac{\partial x_3}{\partial t}\right)_{\xi}.$$

Using Eq. (3.1.4) we finally obtain

$$\left(\frac{\partial T}{\partial t}\right)_{\xi} = \left(\frac{\partial T}{\partial t}\right)_{x} + v_i\frac{\partial T}{\partial x_i} = \frac{\partial T}{\partial t} + \boldsymbol{v}\cdot\nabla T. \qquad (3.1.6)$$

In what follows we write $\frac{DT}{Dt}$ instead of $\left(\frac{\partial T}{\partial t}\right)_{\xi}$. Of course, this formula can be applied not only to the temperature, but to any function related to the continuum.

Velocity and Acceleration

We have already introduced velocity (see Eq. (3.1.4)). In the Lagrangian description the acceleration is given by

$$\boldsymbol{a} = \frac{\partial \boldsymbol{v}}{\partial t} = \frac{\partial^2 \boldsymbol{x}}{\partial t^2}. \qquad (3.1.7)$$

In the Eulerian description it is given by

$$\boldsymbol{a} = \left(\frac{\partial \boldsymbol{v}}{\partial t}\right)_{\xi} = \frac{D\boldsymbol{v}}{Dt} = \frac{\partial \boldsymbol{v}}{\partial t} + v_i\frac{\partial \boldsymbol{v}}{\partial x_i} = \frac{\partial \boldsymbol{v}}{\partial t} + (\boldsymbol{v}\cdot\nabla)\boldsymbol{v}. \qquad (3.1.8)$$

Example Suppose the function T is given by

$$T = t\omega x_1^2 + 2x_2 + 3x_3^2,$$

and the continuum velocity is

$$\boldsymbol{v} = \omega(x_1, x_2, x_3),$$

where ω is a constant. Let us calculate DT/Dt. We obtain

$$\frac{DT}{Dt} = \omega x_1^2 + 2t\omega^2 x_1^2 + 2\omega x_2 + 6\omega x_3^2 = \omega\big[(2t\omega + 1)x_1^2 + 2x_2 + 6x_3^2\big].$$

3.2 Streamlines

A line everywhere tangent to $\boldsymbol{v}$ is called a *streamline*. Assume that a streamline is given by the parametric equations

$$x_1 = x_1(s), \quad x_2 = x_2(s), \quad x_3 = x_3(s).$$

The vector $\boldsymbol{\tau} = \left(\frac{dx_1}{ds}, \frac{dx_2}{ds}, \frac{dx_3}{ds}\right)$ is tangent to the streamline, meaning that $\boldsymbol{\tau} \parallel \boldsymbol{v}$. This implies that there is a function $h(\boldsymbol{x}, t)$ such that

$$\frac{dx_1}{ds} = h(\boldsymbol{x}, t)v_1(\boldsymbol{x}, t), \quad \frac{dx_2}{ds} = h(\boldsymbol{x}, t)v_2(\boldsymbol{x}, t), \quad \frac{dx_3}{ds} = h(\boldsymbol{x}, t)v_3(\boldsymbol{x}, t).$$

At a fixed time h is a function of s on a particular streamline. We introduce $\lambda = \int h\, ds$. Then

$$\frac{dx_1}{d\lambda} = v_1(\boldsymbol{x}, t), \quad \frac{dx_2}{d\lambda} = v_2(\boldsymbol{x}, t), \quad \frac{dx_3}{d\lambda} = v_3(\boldsymbol{x}, t). \tag{3.2.1}$$

This is a system of ordinary differential equations. Any solution to this system, $\boldsymbol{x} = \boldsymbol{x}(\lambda)$, gives a streamline. In this system t is a parameter. For different values of t, in general, we will have different pictures of streamlines. Since the general solution to Eq. (3.2.1) contains three arbitrary constants, we can satisfy the condition $\boldsymbol{x}(\lambda_0) = \boldsymbol{x}_0$ for any $\boldsymbol{x}_0$. Hence, there is a streamline passing through any point of the domain occupied by the continuum at a fixed moment of time.

Recall system (3.1.5) determining the particle trajectories:

$$\frac{d\boldsymbol{x}}{dt} = \boldsymbol{v}(\boldsymbol{x}, t).$$

While in Eq. (3.2.1) t is a parameter, in Eq. (3.1.5) t is an independent variable. As a result, in general, trajectories and streamlines do not coincide. When the motion is *stationary*, that is, $\boldsymbol{v}$ is independent of t, Eq. (3.1.5) and Eq. (3.2.1) are the same (we can write λ instead of t in Eq. (3.1.5)) meaning that in this case the trajectories and streamlines coincide.

Consider a closed contour $\mathcal{C}$. The streamlines passing through the points on $\mathcal{C}$ constitute a stream tube (see Fig. 3.2).

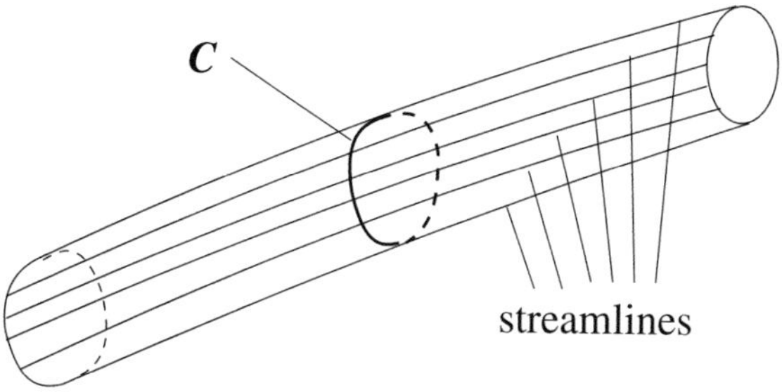

Fig. 3.2 Streamlines passing through a closed contour C constitute a stream tube

Example Let us consider a planar motion where $x_3 = \xi_3$ and $\boldsymbol{v} = (-x_2 + tx_1)\boldsymbol{e}_1 + (x_1 + tx_2)\boldsymbol{e}_2$.

Streamlines: The system of equations determining the streamlines is

$$\frac{dx_1}{d\lambda} = -x_2 + tx_1, \quad \frac{dx_2}{d\lambda} = x_1 + tx_2.$$

Introducing polar coordinates (r, ϕ) related to Cartesian coordinates by $x_1 = r\cos\phi$, $x_2 = r\sin\phi$ we obtain

$$\frac{dr}{d\lambda}\cos\phi - \frac{d\phi}{d\lambda} r\sin\phi = -r\sin\phi + tr\cos\phi,$$

$$\frac{dr}{d\lambda}\sin\phi + \frac{d\phi}{d\lambda} r\cos\phi = r\cos\phi + tr\sin\phi.$$

This system can be easily transformed to

$$\frac{dr}{d\lambda} = tr, \quad \frac{d\phi}{d\lambda} = 1,$$

and we obtain

$$r = C_1 e^{t\lambda}, \quad \phi = \lambda + C_2,$$

where C_1 and C_2 are arbitrary constants. Eliminating λ we obtain the equation for streamlines in the form

$$r = Ce^{t\phi},$$

where C is an arbitrary constant. The streamline containing the point with polar coordinates r_0, ϕ_0 is defined by the equation

$$r = r_0 e^{t(\phi-\phi_0)}. \tag{3.2.2}$$

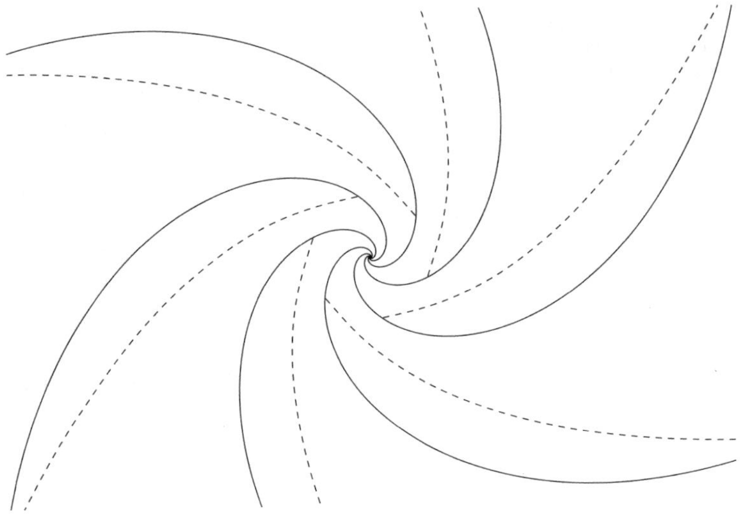

Fig. 3.3 Streamlines (solid lines) and trajectories (dashed lines)

The streamlines are depicted in Fig. 3.3 by solid lines. As $\phi \to -\infty$, all the streamlines approach the coordinate origin, which is a singular point because there $\boldsymbol{v} = 0$.

Trajectories: The equations for the trajectories are

$$\frac{dx_1}{dt} = -x_2 + tx_1, \quad \frac{dx_2}{dt} = x_1 + tx_2.$$

Once again using r and ϕ we obtain

$$\frac{dr}{dt} = tr, \quad \frac{d\phi}{dt} = 1.$$

The solution to this system of equations is

$$r = r_0 e^{t^2/2}, \quad \phi = t + \phi_0,$$

where r_0 and ϕ_0 are the polar coordinates of a point at $t = 0$. Eliminating t we can write the equation of a trajectory as

$$r = r_0 e^{(\phi - \phi_0)^2/2}. \tag{3.2.3}$$

It is clear from Eqs. (3.2.2) to (3.2.3) that the trajectories do not coincide with the streamlines. The trajectories are indicated by the dashed lines in Fig. 3.3.

3.3 The General Theory of Deformations

The Deformation-Gradient Tensor

Consider two points of a continuum that have positions $\boldsymbol{\xi}$ and $\boldsymbol{\xi} + d\boldsymbol{\xi}$ at the initial moment of time. At time t they have positions

$$\boldsymbol{x} = \boldsymbol{x}(\boldsymbol{\xi}, t), \quad \boldsymbol{x} + d\boldsymbol{x} = \boldsymbol{x}(\boldsymbol{\xi} + d\boldsymbol{\xi}, t), \quad dx_i = \frac{\partial x_i}{\partial \xi_j}\, d\xi_j. \tag{3.3.1}$$

$\dfrac{\partial x_i}{\partial \xi_j}$ are the components of transposed dyadic product of vectors ∇_{ξ} and $\boldsymbol{x}$,

$$\frac{\partial x_i}{\partial \xi_j}\, \boldsymbol{e}_i \boldsymbol{e}_j = (\nabla_{\xi}\, \boldsymbol{x})^T, \tag{3.3.2}$$

where the subscript $\boldsymbol{\xi}$ indicates that ∇ differentiates with respect to $\boldsymbol{\xi}$:

$$\nabla_{\xi} = \boldsymbol{e}_j \frac{\partial}{\partial \xi_j}.$$

The tensor $\mathbf{F} = \dfrac{\partial x_i}{\partial \xi_j}\, \boldsymbol{e}_i \boldsymbol{e}_j$ is called the *deformation-gradient tensor*.

Strain Tensors

We have

$$\det(\hat{\mathbf{F}}) = \left| \frac{\partial x_i}{\partial \xi_j} \right| = \frac{D(x_1, x_2, x_3)}{D(\xi_1, \xi_2, \xi_3)}.$$

In accordance with (3.1.3) this Jacobian is non-zero. This implies that the tensor $\mathbf{F}$ is non-singular. Then, in accordance with the polar decomposition theorem (Theorem 2.3), the linear operator associated with the tensor $\mathbf{F}$ yields the unique decompositions

$$\mathbf{F} = \mathbf{R}\mathbf{U} = \mathbf{V}\mathbf{R}, \tag{3.3.3}$$

where $\mathbf{R}$ is a proper unitary operator (implying that its matrix is proper orthogonal), and $\mathbf{U}$ and $\mathbf{V}$ are positive-definite symmetric. Now we can rewrite Eq. (3.3.1) as

$$dx_i = F_{ij}\, d\xi_j = R_{ik} U_{kj}\, d\xi_j,$$

or, equivalently, as

$$dx_i = R_{ik}\, dy_k, \quad dy_k = U_{kj}\, d\xi_j. \tag{3.3.4}$$

We see that the deformation of line element $d\boldsymbol{\xi}$ into $d\boldsymbol{x}$ caused by the motion can be split into two parts. Since $\mathbf{U}$ is a positive-definite symmetric operator, there are principal axes, with respect to which the matrix of **U** is diagonal with positive diagonal elements $U_1,\ U_2,\ U_3$. In these axes the second part of Eq. (3.3.4) takes the form

$$dy_1 = U_1\, d\xi_1, \quad dy_2 = U_2\, d\xi_2, \quad dy_3 = U_3\, d\xi_3. \tag{3.3.5}$$

This deformation represents the increase (if $U_i > 1$) or decrease (if $U_i < 1$) of the ith component of each line element. This part of the deformation results in a simple extension or contraction in three mutually perpendicular directions. The quantities U_i are called the *principal stretches*. Since **R** is unitary, $\|d\boldsymbol{x}\| = \|d\boldsymbol{y}\|$, meaning that the first equation in Eq. (3.3.4) describes a rigid-body rotation of the line element $d\boldsymbol{y}$ to the line element $d\boldsymbol{x}$.

Hence, the motion of the element $d\boldsymbol{\xi}$ can be described as a translation from a position $\boldsymbol{\xi}$ to a position $\boldsymbol{x}$, then extensions (or contractions) determined by the operator **U** in three mutually orthogonal directions, and then a rigid rotation determined by the operator **R**. To visualise this decomposition we consider the motion and deformation of an infinitesimal cube with edge equal to dl. We assume that, in the initial position, the cube edges are parallel to the principal directions of the operator **U**. The evolution of this cube is shown in Fig. 3.4.

The decomposition given by the second equation in Eq. (3.3.3) can be interpreted in the same way. The only difference is that now we first do the rotation and then the expansions. The expansions (or contractions) will be along the principal axes of the operator **V** which, in general, do not coincide with the principal axes of the operator **U**.

In practice, it is quite difficult to calculate the matrices of the operators **R**, **U**, and **V**. For this reason we introduce tensors (operators)

$$\mathbf{C} = \mathbf{F}^T\mathbf{F}, \quad \mathbf{B} = \mathbf{F}\mathbf{F}^T. \tag{3.3.6}$$

C and **B** are called the *left* and *right Cauchy–Green strain* tensors respectively. Obviously **C** and **B** are symmetric second-order tensors. Using Eq. (3.3.3) and the fact that **R** is a unitary operator, $\mathbf{R}\mathbf{R}^T = \mathbf{I}$, we obtain

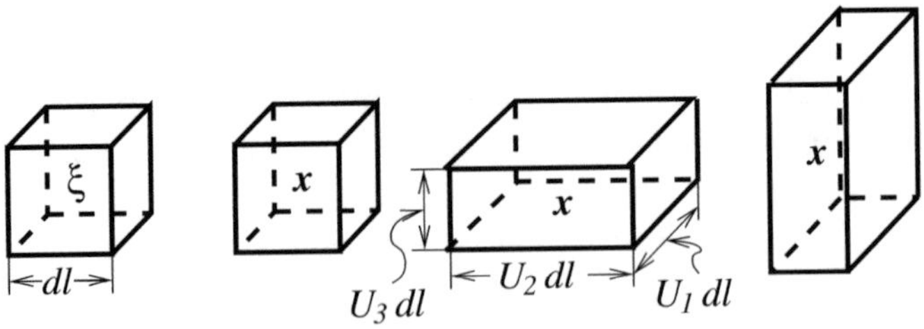

Fig. 3.4 The evolution of a deformed infinitesimal cube. It is a composition of a translation, expansion (or contraction) in three mutually orthogonal directions, and a rigid rotation

$$\begin{aligned}\mathbf{C} &= \mathbf{U}^T\mathbf{R}^T\mathbf{R}\mathbf{U} = \mathbf{U}^T\mathbf{U} = \mathbf{U}^2,\\ \mathbf{B} &= \mathbf{V}\mathbf{R}\mathbf{R}^T\mathbf{V}^T = \mathbf{V}\mathbf{V}^T = \mathbf{V}^2.\end{aligned} \tag{3.3.7}$$

(Recall that $\mathbf{U}$ and $\mathbf{V}$ are symmetric.)

Example Find $\mathbf{F}$, $\mathbf{C}$, $\mathbf{B}$, $\mathbf{U}$, $\mathbf{V}$, and $\mathbf{R}$ for the deformation

$$x_1 = \xi_1, \quad x_2 = \xi_2 - \alpha\xi_3, \quad x_3 = \xi_3 + \alpha\xi_2,$$

where $\alpha > 0$ is a constant, and interpret this deformation as a sequence of extensions and a rotation.

Solution:

$$\hat{\mathbf{F}} = \begin{pmatrix} 1 & 0 & 0 \\ 0 & 1 & -\alpha \\ 0 & \alpha & 1 \end{pmatrix}, \quad \det(\hat{\mathbf{F}}) = 1 + \alpha^2 > 0,$$

$$\hat{\mathbf{C}} = \hat{\mathbf{F}}^T\hat{\mathbf{F}} = \begin{pmatrix} 1 & 0 & 0 \\ 0 & 1 & \alpha \\ 0 & -\alpha & 1 \end{pmatrix}\begin{pmatrix} 1 & 0 & 0 \\ 0 & 1 & -\alpha \\ 0 & \alpha & 1 \end{pmatrix} = \begin{pmatrix} 1 & 0 & 0 \\ 0 & 1+\alpha^2 & 0 \\ 0 & 0 & 1+\alpha^2 \end{pmatrix},$$

$$\hat{\mathbf{U}} = \begin{pmatrix} 1 & 0 & 0 \\ 0 & (1+\alpha^2)^{1/2} & 0 \\ 0 & 0 & (1+\alpha^2)^{1/2} \end{pmatrix}.$$

It is easy to show that $\hat{\mathbf{B}} = \hat{\mathbf{C}}$ and $\hat{\mathbf{V}} = \hat{\mathbf{U}}$.

$$\begin{aligned}\hat{\mathbf{R}} = \hat{\mathbf{F}}\hat{\mathbf{U}}^{-1} &= \begin{pmatrix} 1 & 0 & 0 \\ 0 & 1 & -\alpha \\ 0 & \alpha & 1 \end{pmatrix}\begin{pmatrix} 1 & 0 & 0 \\ 0 & (1+\alpha^2)^{-1/2} & 0 \\ 0 & 0 & (1+\alpha^2)^{-1/2} \end{pmatrix}\\ &= \begin{pmatrix} 1 & 0 & 0 \\ 0 & (1+\alpha^2)^{-1/2} & -\alpha(1+\alpha^2)^{-1/2} \\ 0 & \alpha(1+\alpha^2)^{-1/2} & (1+\alpha^2)^{-1/2} \end{pmatrix}.\end{aligned}$$

Introducing $\alpha = \tan\theta$ $(0 < \theta < \pi/2)$, we obtain

$$\hat{\mathbf{R}} = \begin{pmatrix} 1 & 0 & 0 \\ 0 & \cos\theta & -\sin\theta \\ 0 & \sin\theta & \cos\theta \end{pmatrix},$$

which represents the rotation through the angle θ about the x_1-axis in the clockwise direction. We see that the deformation can be accomplished as extensions by $(1+\alpha^2)^{1/2}$ in the x_2 and x_3-directions, and then a rotation about the x_1-axis. Since $\mathbf{B} = \mathbf{C}$ and $\mathbf{V} = \mathbf{U}$, these operations can be reversed in order.

We can give another interpretation to the tensors (operators) $\mathbf{C}$ and $\mathbf{B}$. Let $dl_0 = \|d\boldsymbol{\xi}\|$ and $dl = \|d\boldsymbol{x}\|$. Then

$$(dl)^2 - (dl_0)^2 = dx_i\,dx_i - d\xi_j\,d\xi_j = \frac{\partial x_i}{\partial \xi_j}\frac{\partial x_i}{\partial \xi_k}\,d\xi_j\,d\xi_k - d\xi_j\,d\xi_j$$
$$= (C_{jk} - \delta_{jk})d\xi_j\,d\xi_k = d\boldsymbol{\xi}\cdot(\mathbf{C} - \mathbf{I})\cdot d\boldsymbol{\xi}. \tag{3.3.8}$$

We can also consider $\boldsymbol{\xi}$ as a function of $\boldsymbol{x}$. Then

$$\frac{\partial \xi_i}{\partial x_j}\frac{\partial x_j}{\partial \xi_k} = \frac{\partial \xi_i}{\partial \xi_k} = \delta_{ik}.$$

Hence $\dfrac{\partial \xi_i}{\partial x_j}$ are the entries of the matrix $\hat{\mathbf{F}}^{-1}$. We have

$$(dl)^2 - (dl_0)^2 = dx_i\,dx_i - d\xi_j\,d\xi_j = dx_i\,dx_i - \frac{\partial \xi_j}{\partial x_i}\frac{\partial \xi_j}{\partial x_k}\,dx_i\,dx_k$$
$$= dx_i\,dx_i - \left((\hat{\mathbf{F}}^T)^{-1}\right)_{ij}\left(\hat{\mathbf{F}}^{-1}\right)_{jk}dx_i\,dx_k$$
$$= dx_i\,dx_i - \left((\hat{\mathbf{F}}\hat{\mathbf{F}}^T)^{-1}\right)_{ik}dx_i\,dx_k$$
$$= \left(\delta_{ik} - \left(\hat{\mathbf{B}}^{-1}\right)_{ik}\right)dx_i\,dx_k = d\boldsymbol{x}\cdot(\mathbf{I} - \mathbf{B}^{-1})\cdot d\boldsymbol{x}. \tag{3.3.9}$$

Let us calculate the principal invariants of $\mathbf{C}$ and $\mathbf{B}$ (see Eqs. (2.10.2)–(2.10.4)). For $\mathbf{C}$ we obtain

$$I_1 = \operatorname{tr}\mathbf{C} = C_{ii} = F_{ji}F_{ji},$$

$$I_2 = \frac{1}{2}I_1^2 - \frac{1}{2}\operatorname{tr}(\mathbf{C}^2) = \frac{1}{2}I_1^2 - \frac{1}{2}C_{ij}C_{ji} = \frac{1}{2}I_1^2 - \frac{1}{2}F_{ki}F_{kj}F_{lj}F_{li},$$

$$I_3 = \det(\hat{\mathbf{C}}) = \det(\hat{\mathbf{F}}^T\hat{\mathbf{F}}) = [\det(\hat{\mathbf{F}})]^2.$$

For $\mathbf{B}$ we obtain

$$I_1 = \operatorname{tr}\mathbf{B} = B_{ii} = F_{ij}F_{ij} = F_{ji}F_{ji}$$

$$I_2 = \frac{1}{2}I_1^2 - \frac{1}{2}\operatorname{tr}(\mathbf{B}^2) = \frac{1}{2}I_1^2 - \frac{1}{2}B_{ij}B_{ji}$$
$$= \frac{1}{2}I_1^2 - \frac{1}{2}F_{ik}F_{jk}F_{jl}F_{il} = \frac{1}{2}I_1^2 - \frac{1}{2}F_{ki}F_{kj}F_{lj}F_{li}$$

$$I_3 = \det(\hat{\mathbf{B}}) = \det(\hat{\mathbf{F}}\hat{\mathbf{F}}^T) = [\det(\hat{\mathbf{F}})]^2.$$

To obtain an expression for I_2 we swap i and k, and j and l.

We see that the principal invariants of $\mathbf{C}$ and $\mathbf{B}$ are the same. This implies that the operators $\mathbf{C}$ and $\mathbf{B}$ have the same eigenvalues. In accordance with Eq. (3.3.7) the eigenvalues of $\mathbf{C}$ are equal to U_1^2, U_2^2, U_3^2, where U_1, U_2, U_3 are the principal stretches. Hence, the eigenvalues of $\mathbf{B}$ are also equal to U_1^2, U_2^2, U_3^2.

3.4 The Displacement Vector and Infinitesimal Strain Tensor

Let us introduce the displacement vector $\boldsymbol{u} = \boldsymbol{x} - \boldsymbol{\xi}$. Then $F_{ij} = \dfrac{\partial x_i}{\partial \xi_j} = \delta_{ij} + \dfrac{\partial u_i}{\partial \xi_j}$, and

$$\mathbf{F} = \mathbf{I} + \mathbf{H}, \tag{3.4.1}$$

where

$$\mathbf{H} = (\nabla_{\xi}\boldsymbol{u})^T, \quad H_{ij} = \frac{\partial u_i}{\partial \xi_j}. \tag{3.4.2}$$

The Cauchy–Green tensor can now be written as

$$\mathbf{C} = \mathbf{F}^T\mathbf{F} = (\mathbf{I} + \mathbf{H}^T)(\mathbf{I} + \mathbf{H}) = \mathbf{I} + \mathbf{H} + \mathbf{H}^T + \mathbf{H}^T\mathbf{H}, \tag{3.4.3}$$

or, in components,

$$C_{ij} = \delta_{ij} + H_{ij} + H_{ji} + H_{ki}H_{kj}. \tag{3.4.4}$$

$\mathbf{C}$ is the measure of the stretching part of the deformation. If $\mathbf{C} = \mathbf{I}$, then the continuum is moving as a rigid body. Hence, $\mathbf{C} - \mathbf{I}$ may be thought of as a measure of the change in shape.

We see that the second and third terms in Eq. (3.4.4) are proportional to $\|\boldsymbol{u}\|$, while the last term is proportional to $\|\boldsymbol{u}\|^2$. This implies that we can neglect this last term if $\|\boldsymbol{u}\|$ is small. Then we arrive at the *linear theory of elasticity*. The tensor

$$\mathbf{E} = \frac{1}{2}(\mathbf{H} + \mathbf{H}^T) \tag{3.4.5}$$

is called the *infinitesimal strain tensor*.

Example Suppose the continuum deformation is defined by

$$x_1 = a\xi_1 + b\xi_2^2, \quad x_2 = \xi_3, \quad x_3 = b\xi_1^2 + \xi_3.$$

Let us calculate the infinitesimal strain tensor. We obtain

$$\hat{\mathbf{F}} = \begin{pmatrix} a & 2b\xi_2 & 0 \\ 0 & 0 & 1 \\ 2b\xi_1 & 0 & 1 \end{pmatrix}, \qquad \hat{\mathbf{H}} = \begin{pmatrix} a-1 & 2b\xi_2 & 0 \\ 0 & -1 & 1 \\ 2b\xi_1 & 0 & 0 \end{pmatrix}.$$

Then we obtain

$$\hat{\mathbf{E}} = \begin{pmatrix} a-1 & b\xi_2 & b\xi_1 \\ b\xi_2 & -1 & 1/2 \\ b\xi_1 & 1/2 & 0 \end{pmatrix}.$$

3.5 The Rate of Deformation

In many problems of continuum mechanics it is not the change of the body shape that is most important, but the rate at which this change occurs. In particular, this is the case in fluid mechanics. Let us again consider two points that have positions $\boldsymbol{\xi}$ and $\boldsymbol{\xi} + d\boldsymbol{\xi}$ at the initial moment of time. At time t they have positions $\boldsymbol{x}$ and $\boldsymbol{x} + d\boldsymbol{x}$, where $d\boldsymbol{x}$ is given by Eq. (3.3.1). Then the rate of change of $d\boldsymbol{x}$ is given by

$$\frac{D(dx_i)}{Dt} = \frac{D}{Dt}\left(\frac{\partial x_i}{\partial \xi_j}\, d\xi_j\right).$$

Recall that D/Dt indicates the time derivative at constant $\boldsymbol{\xi}$. This equation can be rewritten as

$$\frac{D(dx_i)}{Dt} = \frac{\partial^2 x_i}{\partial \xi_j \partial t}\, d\xi_j.$$

Recalling that $\dfrac{\partial \boldsymbol{x}}{\partial t} = \boldsymbol{v}$, where $\boldsymbol{v}$ is the particle velocity, we rewrite it as

$$\frac{D(dx_i)}{Dt} = \frac{\partial v_i}{\partial \xi_j}\, d\xi_j. \tag{3.5.1}$$

Now we use the relations

$$\frac{\partial v_i}{\partial \xi_j} = \frac{\partial v_i}{\partial x_k}\frac{\partial x_k}{\partial \xi_j}, \qquad d\xi_j = \frac{\partial \xi_j}{\partial x_l}\, dx_l$$

to rewrite Eq. (3.5.1) as

$$\frac{D(dx_i)}{Dt} = \frac{\partial v_i}{\partial x_k}\frac{\partial x_k}{\partial \xi_j}\frac{\partial \xi_j}{\partial x_l}\, dx_l = \frac{\partial v_i}{\partial x_k}\delta_{kl}\, dx_l = \frac{\partial v_i}{\partial x_k}\, dx_k. \tag{3.5.2}$$

The vector form of this equation is

$$\frac{D(d\boldsymbol{x})}{Dt} = (d\boldsymbol{x} \cdot \nabla)\boldsymbol{v}. \tag{3.5.3}$$

The tensor

$$\mathbf{L} = (\nabla \boldsymbol{v})^T = \frac{\partial v_i}{\partial x_j} \boldsymbol{e}_i \boldsymbol{e}_j \tag{3.5.4}$$

is called the *velocity gradient tensor*.

Let us now calculate the relative rate of extension of a material line element:

$$\begin{aligned}
\frac{1}{\|d\boldsymbol{x}\|} \frac{D(\|d\boldsymbol{x}\|)}{Dt} &= \frac{1}{2\|d\boldsymbol{x}\|^2} \frac{D(\|d\boldsymbol{x}\|^2)}{Dt} = \frac{1}{2\|d\boldsymbol{x}\|^2} \frac{D(dx_i\, dx_i)}{Dt} = \frac{dx_i}{\|d\boldsymbol{x}\|^2} \frac{D(dx_i)}{Dt} \\
&= \frac{1}{\|d\boldsymbol{x}\|^2} \frac{\partial v_i}{\partial x_k} dx_i\, dx_k = \frac{1}{2\|d\boldsymbol{x}\|^2} \left(\frac{\partial v_i}{\partial x_k} + \frac{\partial v_k}{\partial x_i} \right) dx_i\, dx_k \\
&= \frac{1}{2} \left(\frac{\partial v_i}{\partial x_k} + \frac{\partial v_k}{\partial x_i} \right) \alpha_i\, \alpha_k,
\end{aligned}$$

where $\boldsymbol{\alpha} = d\boldsymbol{x}/\|d\boldsymbol{x}\|$ is the unit vector parallel to $d\boldsymbol{x}$. This expression can be rewritten as

$$\frac{1}{\|d\boldsymbol{x}\|} \frac{D(\|d\boldsymbol{x}\|)}{Dt} = \boldsymbol{\alpha} \cdot \mathbf{D} \cdot \boldsymbol{\alpha}, \tag{3.5.5}$$

where the tensor

$$\mathbf{D} = \frac{1}{2}(\mathbf{L} + \mathbf{L}^T) = \frac{1}{2} \left(\frac{\partial v_i}{\partial x_j} + \frac{\partial v_j}{\partial x_i} \right) \boldsymbol{e}_i\, \boldsymbol{e}_j \tag{3.5.6}$$

is the symmetric part of tensor $\mathbf{L}$. It is called the *strain-rate tensor* (other common names are the *strain-of-rate tensor* and the *rate-of-deformation tensor*).

Example Suppose the velocity is given by $\boldsymbol{v} = V \tanh x_2\, \boldsymbol{e}_1$, where V is a positive constant. Then the matrix of $\mathbf{D}$ is given by

$$\hat{\mathbf{D}} = \frac{V}{2} \begin{pmatrix} 0 & \operatorname{sech}^2 x_2 & 0 \\ \operatorname{sech}^2 x_2 & 0 & 0 \\ 0 & 0 & 0 \end{pmatrix}.$$

3.6 Velocity Distribution in a Small Volume

Consider a continuum occupying a small volume. Fix an arbitrary point O inside this volume with position vector $\boldsymbol{x}$. The velocity of this point is $\boldsymbol{v}_0$. Take another point with position vector $\boldsymbol{x} + d\boldsymbol{x}$. The velocity of this point is given by the approximate expression

$$\boldsymbol{v}_1 = \boldsymbol{v}_0 + \frac{\partial \boldsymbol{v}}{\partial x_i} dx_i = \boldsymbol{v}_0 + (d\boldsymbol{x} \cdot \nabla)\boldsymbol{v}.$$

Now we introduce the antisymmetric part of the tensor L,

$$\mathsf{W} = \frac{1}{2}\left(\mathsf{L} - \mathsf{L}^T\right) = \frac{1}{2}\left(\frac{\partial v_i}{\partial x_j} - \frac{\partial v_j}{\partial x_i}\right) \boldsymbol{e}_i \boldsymbol{e}_j, \tag{3.6.1}$$

called the *spin* or *vorticity tensor*. Then $\mathsf{L} = \mathsf{D} + \mathsf{W}$, and the expression for $\boldsymbol{v}_1$ can be rewritten as

$$\boldsymbol{v}_1 = \boldsymbol{v}_0 + \mathsf{D} \cdot d\boldsymbol{x} + \mathsf{W} \cdot d\boldsymbol{x}.$$

Let us introduce the vector $\boldsymbol{\omega}$ given by

$$\boldsymbol{\omega} = \frac{1}{2}\varepsilon_{ijk}\frac{\partial v_k}{\partial x_j}\,\boldsymbol{e}_i = \frac{1}{2}\nabla \times \boldsymbol{v}. \tag{3.6.2}$$

Then, using Eq. (2.9.6), we obtain

$$\begin{aligned}
\varepsilon_{ijk}\omega_j &= \frac{1}{2}\varepsilon_{ijk}\varepsilon_{jlm}\frac{\partial v_m}{\partial x_l} = -\frac{1}{2}\varepsilon_{ikj}\varepsilon_{lmj}\frac{\partial v_m}{\partial x_l} \\
&= -\frac{1}{2}(\delta_{il}\delta_{km} - \delta_{im}\delta_{kl})\frac{\partial v_m}{\partial x_l} = \frac{1}{2}\left(\frac{\partial v_i}{\partial x_k} - \frac{\partial v_k}{\partial x_i}\right),
\end{aligned}$$

so that

$$W_{ik} = \varepsilon_{ijk}\omega_j. \tag{3.6.3}$$

We use this relation to obtain

$$\mathsf{W} \cdot d\boldsymbol{x} = W_{ik}\,dx_k\boldsymbol{e}_i = \varepsilon_{ijk}\omega_j\,dx_k\boldsymbol{e}_i = \boldsymbol{\omega} \times d\boldsymbol{x},$$

so eventually

$$\boldsymbol{v}_1 = \boldsymbol{v}_0 + \mathsf{D} \cdot d\boldsymbol{x} + \boldsymbol{\omega} \times d\boldsymbol{x}. \tag{3.6.4}$$

The last term describes the velocity in a small volume rotating as a solid with angular velocity $\boldsymbol{\omega}$. Assume that the volume has been instantaneously frozen. Then the length of any linear element $d\boldsymbol{x}$ in the volume does not change, i.e. the right-hand side of Eq. (3.5.5) is zero for any $\boldsymbol{\alpha}$. This implies that $\mathsf{D} = 0$ and the volume moves with velocity $\boldsymbol{v}_0$ as a whole and rotates with angular velocity $\boldsymbol{\omega}$.

Now we can describe the motion of a small volume. Consider two points inside it, O_1 and O_2 (see Fig. 3.5). We change the definition of $d\boldsymbol{x}$ and take $d\boldsymbol{x} = \overrightarrow{O_1O_2}$. We also introduce $d\boldsymbol{x}_1 = \overrightarrow{OO_1}$ and $d\boldsymbol{x}_2 = \overrightarrow{OO_2}$, so that $d\boldsymbol{x} = d\boldsymbol{x}_2 - d\boldsymbol{x}_1$. Point O moves with velocity $\boldsymbol{v}_0$. The velocities of O_1 and O_2, equal to $\boldsymbol{v}_1$ and $\boldsymbol{v}_2$, are given by Eq. (3.6.4) with $d\boldsymbol{x}$ substituted by $d\boldsymbol{x}_1$ and $d\boldsymbol{x}_2$.

After a small time interval Δt the changes in positions of O, O_1 and O_2 will be $\boldsymbol{v}_0\,\Delta t$, $\boldsymbol{v}_1\,\Delta t$ and $\boldsymbol{v}_2\,\Delta t$, respectively (see Fig. 3.5). The variation of $d\boldsymbol{x}$ is

Fig. 3.5 Sketch of motion of the line element

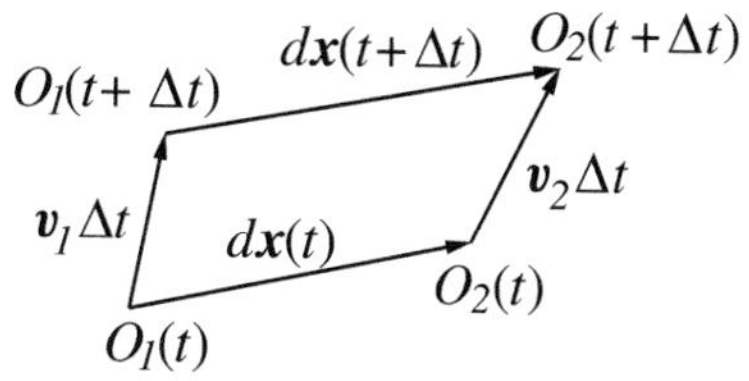

$$\Delta(dx) = dx(t+\Delta t) - dx(t) = v_2\,\Delta t - v_1\,\Delta t.$$

Using the expressions for v_1 and v_2 we obtain

$$\begin{aligned}\Delta(dx) &= (v_0 + \mathbf{D}\cdot dx_2 + \omega\times dx_2)\,\Delta t - (v_0 + \mathbf{D}\cdot dx_1 + \omega\times dx_1)\,\Delta t \\ &= (\mathbf{D}\cdot dx + \omega\times dx)\,\Delta t.\end{aligned}$$

Since $\mathbf{D}$ is a symmetric operator, it follows from Theorem 2.2 that it has three mutually orthogonal eigenvectors, d_1, d_2 and d_3, that can be taken to be unit vectors. Then we can expand dx with respect to the basis formed by these vectors: $dx = l_i d_i$. Let D_i be the eigenvalue corresponding to the eigenvector d_i. Then we can rewrite the expression for $\Delta(dx)$ as

$$\Delta(dx) = D_1 l_1 d_1\,\Delta t + D_2 l_2 d_2\,\Delta t + D_3 l_3 d_3\,\Delta t + \omega\times dx\,\Delta t.$$

Let us introduce the operators $\mathbf{D}_1$, $\mathbf{D}_2$, $\mathbf{D}_3$, and $\mathbf{\Omega}$ defined by

$$\mathbf{D}_i(x) = x + D_i\,\Delta t(x\cdot d_i)d_i, \quad i = 1, 2, 3,$$

$$\mathbf{\Omega}(x) = x + \Delta t\,\omega\times x.$$

The operator $\mathbf{D}_i$ stretches by $(1 + D_i\,\Delta t)$ the component of x parallel to d_i (contracts if $D_i < 0$) (see Fig. 3.6). The operator $\mathbf{\Omega}$ rotates x by the angle $\omega\,\Delta t$ about the vector ω. Let us successively apply operators $\mathbf{D}_1$, $\mathbf{D}_2$, $\mathbf{D}_3$ and $\mathbf{\Omega}$ to the vector dx. We introduce the notation

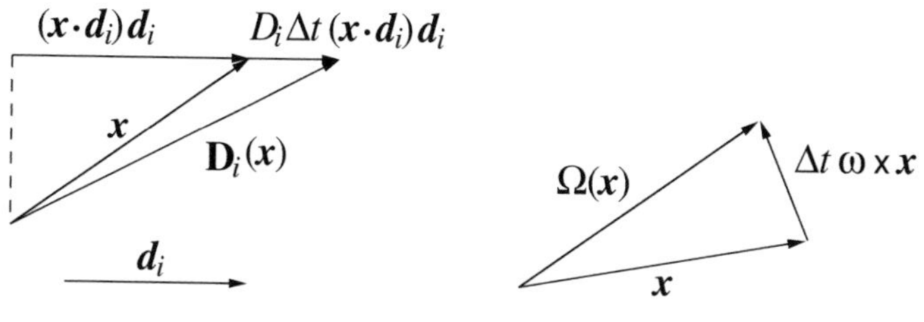

Fig. 3.6 Illustration of the operators $\mathbf{D}_i$ and $\mathbf{\Omega}$

$$d\boldsymbol{y} = \mathbf{D}_1(d\boldsymbol{x}), \quad dz = \mathbf{D}_2(d\boldsymbol{y}), \quad d\boldsymbol{u} = \mathbf{D}_3(dz), \quad d\boldsymbol{w} = \boldsymbol{\Omega}(d\boldsymbol{u}).$$

Then, using the identity $d\boldsymbol{x} \cdot \boldsymbol{d}_i = l_i$, we obtain

$$d\boldsymbol{y} = d\boldsymbol{x} + D_1\,\Delta t\, l_1 \boldsymbol{d}_1,$$

$$dz = d\boldsymbol{y} + D_2\,\Delta t (d\boldsymbol{y} \cdot \boldsymbol{d}_2)\boldsymbol{d}_2 = d\boldsymbol{x} + D_1\,\Delta t\, l_1 \boldsymbol{d}_1 + D_2\,\Delta t\, l_2 \boldsymbol{d}_2, \tag{3.6.5}$$

$$\begin{aligned} d\boldsymbol{u} &= dz + D_3\,\Delta t (dz \cdot \boldsymbol{d}_3)\boldsymbol{d}_3 \\ &= d\boldsymbol{x} + D_1\,\Delta t\, l_1 \boldsymbol{d}_1 + D_2\,\Delta t\, l_2 \boldsymbol{d}_2 + D_3\,\Delta t\, l_3 \boldsymbol{d}_3, \end{aligned} \tag{3.6.6}$$

$$\begin{aligned} d\boldsymbol{w} = d\boldsymbol{u} + \Delta t\, \boldsymbol{\omega} \times d\boldsymbol{u} = d\boldsymbol{x} + D_1\,\Delta t\, l_1 \boldsymbol{d}_1 \\ + D_2\,\Delta t\, l_2 \boldsymbol{d}_2 + D_3\,\Delta t\, l_3 \boldsymbol{d}_3 + \Delta t\, \boldsymbol{\omega} \times d\boldsymbol{x} + \mathcal{O}\big((\Delta t)^2\big). \end{aligned} \tag{3.6.7}$$

Comparing the expressions for $\Delta(d\boldsymbol{x})$ with the expression for $d\boldsymbol{w}$ we find that

$$\begin{aligned} d\boldsymbol{w} &\equiv \boldsymbol{\Omega}\,\mathbf{D}_3\mathbf{D}_2\mathbf{D}_1(d\boldsymbol{x}) = d\boldsymbol{x}(t) + \Delta(d\boldsymbol{x}) + \mathcal{O}\big((\Delta t)^2\big) \\ &= d\boldsymbol{x}(t + \Delta t) + \mathcal{O}\big((\Delta t)^2\big). \end{aligned} \tag{3.6.8}$$

Note that, up to linear terms with respect to Δt, the result is independent of the order of operators, i.e. $\mathbf{D}_3\mathbf{D}_2\mathbf{D}_1\,\boldsymbol{\Omega}(d\boldsymbol{x}) = \boldsymbol{\Omega}\,\mathbf{D}_3\mathbf{D}_2\mathbf{D}_1(d\boldsymbol{x})$, and so on. We say that the operators $\mathbf{D}_1$, $\mathbf{D}_2$, $\mathbf{D}_3$, and $\boldsymbol{\Omega}$ approximately commute. In fact, only $\boldsymbol{\Omega}$ and $\mathbf{D}_i$ $(i = 1, 2, 3)$ commute approximately. The operators $\mathbf{D}_1$, $\mathbf{D}_2$, and $\mathbf{D}_3$ commute *exactly*.

Let us assume that $d\boldsymbol{x}$ is parallel to $\boldsymbol{d}_i$. Then it follows from Eqs. (3.6.5), (3.6.6), and (3.6.7) that $\mathbf{D}_3\mathbf{D}_2\mathbf{D}_1(d\boldsymbol{x})$ is also parallel to $\boldsymbol{d}_i$, so that the operator $\mathbf{D}_3\mathbf{D}_2\mathbf{D}_1$ does not change the direction of $d\boldsymbol{x}$ and only stretches (or contracts) it. Hence, an infinitesimal motion of a small volume during the time interval Δt can be thought of as a sequence of three extensions (or contractions) in the principal directions of the tensor $\mathbf{D}$ and rotation about the vector $\boldsymbol{\omega}$ taken in an arbitrary order.

Example Suppose the velocity is given by $\boldsymbol{v} = V \tanh x_2\, \boldsymbol{e}_1$, where V is a positive constant. Describe the motion of a continuum occupying a small volume.

Solution: The expression for the matrix of the tensor $\mathbf{D}$ was obtained in the previous section. The eigenvalues of the operator $\mathbf{D}$ are defined by

$$\det(\hat{\mathbf{D}}) = \begin{vmatrix} -D & \frac{V}{2}\mathrm{sech}^2 x_2 & 0 \\ \frac{V}{2}\mathrm{sech}^2 x_2 & -D & 0 \\ 0 & 0 & -D \end{vmatrix} = D\left(\frac{V^2}{4}\,\mathrm{sech}^4 x_2 - D^2\right) = 0.$$

Hence, the eigenvalues are given by $D_1 = \frac{V}{2}\mathrm{sech}^2 x_2$, $D_2 = -\frac{V}{2}\mathrm{sech}^2 x_2$, and $D_3 = 0$. It is straightforward to show that the corresponding eigenvectors are $\boldsymbol{d}_1 = \frac{1}{\sqrt{2}}(\boldsymbol{e}_1 + \boldsymbol{e}_2)$, $\boldsymbol{d}_2 = \frac{1}{\sqrt{2}}(\boldsymbol{e}_1 - \boldsymbol{e}_2)$, and $\boldsymbol{d}_3 = \boldsymbol{e}_3$. $D_3 = 0$ corresponds to the fact that there is no motion in the x_3-direction. We also obtain $\omega = -\frac{V}{2}\boldsymbol{e}_3\,\mathrm{sech}^2 x_2$. Hence, we conclude that after time Δt an infinitesimal volume is stretched by $\big(1 + \frac{V}{2}\mathrm{sech}^2 x_2\,\Delta t\big)$

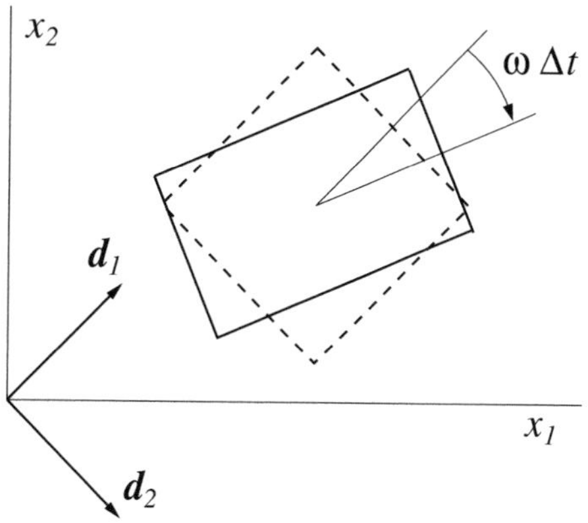

Fig. 3.7 Illustration of the motion of an infinitesimal volume. The cross-section of this volume by a plane $x_3 = \text{const}$ is shown. The dashed line corresponds to the initial time and the solid line to the cross-section after time Δt

in the direction of vector $\boldsymbol{d}_1$, compressed by $\left(1 - \frac{V}{2}\text{sech}^2 x_2\,\Delta t\right)$ in the direction of vector $\boldsymbol{d}_2$, and rotates clockwise about the e_3-direction by the angle $\frac{V}{2}\Delta t\,\text{sech}^2 x_2$. This motion of an infinitesimal volume is illustrated in Fig. 3.7.

3.7 Mass Conservation Equation in the Lagrangian Description

Let us consider a volume V frozen in a continuum. This means that V consists of the same particles at any moment of time. We have $V = V_0$ at $t = 0$ and $V = V(t)$ at an arbitrary moment of time t. Then

$$V(t) = \int_{V_0} \mathcal{J}(\boldsymbol{\xi}, t)\, dV_0.$$

Recall that

$$\mathcal{J}(\boldsymbol{\xi}, t) = \frac{D(x_1, x_2, x_3)}{D(\xi_1, \xi_2, \xi_2)}.$$

Now we assume that V is so small that we can neglect the variation of any quantity related to the continuum inside V. In particular, we can take $\mathcal{J}(\boldsymbol{\xi}, t) = \mathcal{J}(\bar{\boldsymbol{\xi}}, t)$, where $\bar{\boldsymbol{\xi}}$ is an arbitrary fixed point inside V_0. Then

$$V(t) = \mathcal{J}(\bar{\boldsymbol{\xi}}, t) \int_{V_0} dV_0 = \mathcal{J}(\bar{\boldsymbol{\xi}}, t) V_0. \tag{3.7.1}$$

In what follows we drop the bar and write $\boldsymbol{\xi}$ instead of $\bar{\boldsymbol{\xi}}$. Since V consists of the same particles at any moment of time, its mass M is independent of t. The condition that V is small, in particular, implies that it is homogeneous, i.e. if we split it into two parts with volumes $\frac{1}{2}V$ then the masses of each volume are $\frac{1}{2}M$, if we split it into three parts with volumes $\frac{1}{3}V$ then the masses of each volume are $\frac{1}{3}M$, and so on. This means that M is proportional to V:

$$M = \rho(\boldsymbol{\xi}, t)V. \tag{3.7.2}$$

The function $\rho(\boldsymbol{\xi}, t)$ is called the *density*.

Since we neglect the variation of any quantity inside V, $\boldsymbol{\xi}$ is an arbitrary point inside V. In fact, to determine the density at a point $\boldsymbol{\xi}$ we fix this point and then take the limit as the volume ΔV tends to zero, while $\boldsymbol{\xi}$ remains inside V:

$$\rho(\boldsymbol{\xi}, t) = \lim_{V\to 0} \frac{M}{V}. \tag{3.7.3}$$

When $t = 0$ Eq. (3.7.2) gives

$$M = \rho_0(\boldsymbol{\xi})V_0,$$

where $\rho_0(\boldsymbol{\xi}) = \rho(\boldsymbol{\xi}, 0)$. Using this equation, Eqs. (3.7.1), and (3.7.2) we obtain the equation of *mass conservation*

$$\boxed{\rho(\boldsymbol{\xi}, t)\mathcal{J}(\boldsymbol{\xi}, t) = \rho_0(\boldsymbol{\xi}).} \tag{3.7.4}$$

Example Suppose the continuum motion is defined by

$$x_1 = \xi_1 \cosh(\omega t) + \frac{\xi_2}{2}\sinh(\omega t), \quad x_2 = \frac{\xi_1}{2}\sinh(\omega t) + \xi_2\cosh(\omega t), \quad x_3 = \xi_3.$$

At the initial time the continuum density is $\rho = \rho_0 = \text{const}$. Find the density at time t.

Solution: First we calculate the Jacobian,

$$J = \begin{vmatrix} \cosh(\omega t) & \frac{1}{2}\sinh(\omega t) & 0 \\ \frac{1}{2}\sinh(\omega t) & \cosh(\omega t) & 0 \\ 0 & 0 & 1 \end{vmatrix} = \frac{1}{4}\left[1 + 3\cosh^2(\omega t)\right].$$

Then we obtain

$$\rho = \frac{\rho_0}{J} = \frac{4\rho_0}{1 + 3\cosh^2(\omega t)}.$$

3.8 Mass Conservation Equation in the Eulerian Description

In the Eulerian description $\rho = \rho(\boldsymbol{x}, t)$. Consider a volume V frozen in a continuum. The mass inside V is

$$M = \int_V \rho(\boldsymbol{x}, t)\, dV,$$

where the integration is with respect to x_1, x_2, x_3. Since V is frozen in the continuum, M is independent of t. Let us consider the moment of time $t + \Delta t$, where Δt is small. At that moment the part of the continuum that occupied volume V at time t occupies volume $V + \Delta V$, so that

$$M = \int_{V+\Delta V} \rho(\boldsymbol{x}, t + \Delta t)\, dV = \int_V \rho(\boldsymbol{x}, t + \Delta t)\, dV + \int_{\Delta V} \rho(\boldsymbol{x}, t + \Delta t)\, dV.$$

In what follows we neglect terms of order $(\Delta t)^2$ and of higher orders with respect to Δt. Since $\Delta V \sim \Delta t$ it follows that if we substitute $\rho(\boldsymbol{x}, t)$ for $\rho(\boldsymbol{x}, t + \Delta t)$ in the second integral, the error is of order $(\Delta t)^2$, which can be neglected. Hence, we rewrite the expression for M as

$$M = \int_V \rho(\boldsymbol{x}, t + \Delta t)\, dV + \int_{\Delta V} \rho(\boldsymbol{x}, t)\, dV.$$

The skew prism height in Fig. 3.8 is $h_i = \boldsymbol{v} \cdot \boldsymbol{n}\, \Delta t$, where $\boldsymbol{n}$ is the unit normal vector to the surface S. The prism volume is $\Delta V_i = h_i \Delta S_i = \boldsymbol{v} \cdot \boldsymbol{n}\, \Delta t\, \Delta S_i = \boldsymbol{v} \cdot \Delta \boldsymbol{S}_i \Delta t$, where $\Delta \boldsymbol{S}_i = \boldsymbol{n}\, \Delta S_i$. Then

$$\begin{aligned}
\int_{\Delta V} \rho(\boldsymbol{x}, t)\, dV &= \lim_{\max(\Delta V_i)\to 0} \sum_i \rho(\boldsymbol{x}_i, t) \Delta V_i \\
&= \lim_{\max(\Delta S_i)\to 0} \sum_i \rho(\boldsymbol{x}_i, t) \boldsymbol{v}(\boldsymbol{x}_i, t) \cdot \Delta \boldsymbol{S}_i \Delta t + \mathcal{O}\big((\Delta t)^2\big) \\
&= \Delta t \int_S \rho(\boldsymbol{x}, t) \boldsymbol{v} \cdot d\boldsymbol{S} + \mathcal{O}\big((\Delta t)^2\big).
\end{aligned}$$

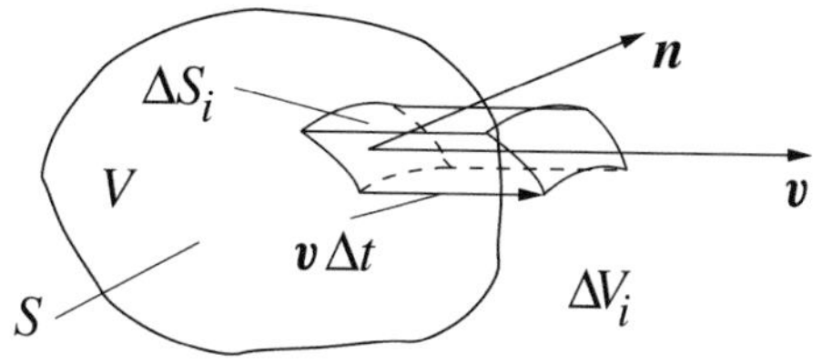

Fig. 3.8 Illustration of the derivation of the mass conservation equation in Eulerian variables

Now we use the fact that M is independent of time:

$$
\begin{aligned}
0 = \frac{dM}{dt} &= \lim_{\Delta t \to 0} \left\{ \frac{1}{\Delta t} \left(\int_{V+\Delta V} \rho(\boldsymbol{x}, t + \Delta t)\, dV - \int_V \rho(\boldsymbol{x}, t)\, dV \right) \right\} \\
&= \lim_{\Delta t \to 0} \left\{ \frac{1}{\Delta t} \left(\int_V \rho(\boldsymbol{x}, t + \Delta t)\, dV + \Delta t \int_S \rho(\boldsymbol{x}, t)\boldsymbol{v} \cdot d\boldsymbol{S} + \mathcal{O}\left((\Delta t)^2\right) - \int_V \rho(\boldsymbol{x}, t)\, dV \right) \right\} \\
&= \int_V \lim_{\Delta t \to 0} \frac{\rho(\boldsymbol{x}, t + \Delta t) - \rho(\boldsymbol{x}, t)}{\Delta t}\, dV + \int_S \rho(\boldsymbol{x}, t)\boldsymbol{v} \cdot d\boldsymbol{S} = \int_V \frac{\partial \rho}{\partial t}\, dV + \int_S \rho(\boldsymbol{x}, t)\boldsymbol{v} \cdot d\boldsymbol{S}.
\end{aligned}
$$

Using Gauss' divergence theorem we obtain

$$
\int_V \left(\frac{\partial \rho}{\partial t} + \nabla \cdot (\rho \boldsymbol{v}) \right) dV = 0. \tag{3.8.1}
$$

Lemma 3.1 *If $f(\boldsymbol{r})$ is a continuous function and $\int_V f(\boldsymbol{r})\, dV = 0$ for any V, then $f \equiv 0$.*

Proof We take an arbitrary $\boldsymbol{r}_0$. Since $f(\boldsymbol{r})$ is continuous, $\forall\ \epsilon > 0\ \exists\ \delta > 0$ such that if $|\boldsymbol{r} - \boldsymbol{r}_0| < \delta$ then $|f(\boldsymbol{r}) - f(\boldsymbol{r}_0)| < \epsilon$. We take V in the form of a ball with radius δ centred at $\boldsymbol{r}_0$. Then $\forall\ \boldsymbol{r} \in V$ we obtain $|\boldsymbol{r} - \boldsymbol{r}_0| < \delta$. Using this result we get

$$
0 = \int_V f(\boldsymbol{r})\, dV = \int_V [f(\boldsymbol{r}) - f(\boldsymbol{r}_0) + f(\boldsymbol{r}_0)]\, dV = \int_V [f(\boldsymbol{r}) - f(\boldsymbol{r}_0)]\, dV + f(\boldsymbol{r}_0)V,
$$

which can be rewritten as

$$
f(\boldsymbol{r}_0)V = -\int_V [f(\boldsymbol{r}) - f(\boldsymbol{r}_0)]\, dV.
$$

It follows from this equation that

$$
|f(\boldsymbol{r}_0)|V = \left| \int_V [f(\boldsymbol{r}) - f(\boldsymbol{r}_0)]\, dV \right| \leq \int_V |f(\boldsymbol{r}) - f(\boldsymbol{r}_0)|\, dV < \epsilon \int_V dV = \epsilon V.
$$

Hence, $|f(\boldsymbol{r}_0)| < \epsilon$ for any $\epsilon > 0$. This implies that $f(\boldsymbol{r}_0) = 0$. Since $\boldsymbol{r}_0$ is arbitrary, we obtain $f(\boldsymbol{r}) \equiv 0$. □

It follows from Lemma 3.1 and Eq. (3.8.1) that, if ρ and $\boldsymbol{v}$ are continuous together with their first derivatives, then

$$
\boxed{\frac{\partial \rho}{\partial t} + \nabla \cdot (\rho \boldsymbol{v}) = 0.} \tag{3.8.2}
$$

This is the equation of *mass conservation* in Eulerian variables. Since $\nabla \cdot (\rho \boldsymbol{v}) = \rho \nabla \cdot \boldsymbol{v} + \boldsymbol{v} \cdot \nabla \rho$, we can rewrite this equation in the equivalent form

$$\boxed{\frac{D\rho}{Dt} + \rho \nabla \cdot \boldsymbol{v} = 0.} \tag{3.8.3}$$

Example The continuum velocity is given by $\boldsymbol{v} = V(x_1/a)\boldsymbol{e}_1$, where V and a are positive constants. The continuum density at $t = 0$ is equal to $\rho_0 = \text{const}$. Calculate the dependence of the density on time.

Solution: We assume that ρ is independent of the spatial variables. Since $\nabla \cdot \boldsymbol{v} = V/a$, it follows from Eq. (3.8.2) that

$$\frac{d\rho}{dt} = -\frac{V\rho}{a}.$$

Hence, we obtain $\rho = \rho_0 e^{-Vt/a}$.

3.9 The Time Derivative of an Integral over a Moving Volume

Let us consider the integral $F = \int_V \rho f \, dV$ over a volume V frozen in a continuum, f being an arbitrary function. This means that V consists of the same particles at any moment of time. When calculating dF/dt we must take into account that $V = V(t)$. It is convenient to make the variable substitution $\boldsymbol{x} = \boldsymbol{x}(\boldsymbol{\xi}, t)$:

$$F = \int_{V_0} \rho(\boldsymbol{x}(\boldsymbol{\xi}, t), t) f(\boldsymbol{x}(\boldsymbol{\xi}, t), t)\, \mathcal{J}(\boldsymbol{\xi}, t)\, dV_0,$$

where $V_0 = V(0)$, and integration is with respect to $\boldsymbol{\xi}$. Then

$$\frac{dF}{dt} = \int_{V_0} \left(\frac{D(\rho f)}{Dt} \mathcal{J}(\boldsymbol{\xi}, t) + \rho f \frac{D\mathcal{J}}{Dt} \right) dV_0.$$

Now we use Eq. (3.7.4) to obtain

$$\frac{D\mathcal{J}}{Dt} = \rho_0(\boldsymbol{\xi}) \frac{D(1/\rho)}{Dt} = \rho \mathcal{J} \frac{D(1/\rho)}{Dt},$$

so that

$$\frac{dF}{dt} = \int_{V_0} \left(\frac{1}{\rho} \frac{D(\rho f)}{Dt} + \rho f \frac{D(1/\rho)}{Dt} \right) \rho \mathcal{J}\, dV_0 = \int_{V_0} \frac{Df}{Dt} \rho \mathcal{J}\, dV_0.$$

Then, using Eq. (3.1.6) yields

$$\boxed{\frac{dF}{dt} = \int_V \rho \left(\frac{\partial f}{\partial t} + \boldsymbol{v} \cdot \nabla f \right) dV,} \tag{3.9.1}$$

where now the time derivative is taken at a fixed $\boldsymbol{x}$. We can also rewrite this equation as

$$\boxed{\frac{dF}{dt} = \int_V \rho \frac{Df}{Dt}\, dV.} \tag{3.9.2}$$

Example The linear momentum of a continuum occupying a volume $V(t)$ is

$$M(t) = \int_{V(t)} \rho \boldsymbol{v}\, dV.$$

Then the time derivative of the momentum is

$$\dot{M}(t) = \int_{V(t)} \rho \frac{D\boldsymbol{v}}{Dt}\, dV.$$

Problems

3.1: Suppose the motion of a body is described by

$$x_1 = (1 + \alpha t)\xi_1, \quad x_2 = \xi_2, \quad x_3 = \xi_3 \quad (t \geq 0),$$

where α is a positive constant. Interpret the motion geometrically. Does the distance between neighbouring particles change? Explain.

3.2: Suppose the motion of a continuum is described in Lagrangian variables by $x_1 = \xi_1 + \xi_2(\omega t)$, $x_2 = \xi_2 - \xi_1(\omega t)$, $x_3 = \xi_3$, where $\omega > 0$ is a constant. Find the expression for the velocity $\boldsymbol{v}$ in the Lagrangian and Eulerian variables. Calculate $D\boldsymbol{v}/Dt$, $\partial \boldsymbol{v}/\partial t$, and $(\boldsymbol{v} \cdot \nabla)\boldsymbol{v}$. Verify that

$$\frac{D\boldsymbol{v}}{Dt} = \frac{\partial \boldsymbol{v}}{\partial t} + (\boldsymbol{v} \cdot \nabla)\boldsymbol{v}.$$

3.3: Suppose a body is subjected to the deformation

$$x_1 = (1 + \beta)\xi_1 + \kappa\xi_2, \quad x_2 = (1 + \beta)\xi_2, \quad x_3 = \xi_3,$$

where β and κ are non-negative constants.

(i) Interpret the deformation geometrically when $\kappa = 0$ and $\beta > 0$.

(ii) Show that if $\beta > 0$ then there is an increase in volume.

(iii) In the initial configuration, a line element of length dl_0 lies in the plane $\xi_3 = 0$ and is inclined at an acute angle θ to the ξ_1 axis so that

$$d\xi_1 = dl_0 \cos\theta, \quad d\xi_2 = dl_0 \sin\theta, \quad d\xi_3 = 0.$$

Show that the length dl of this element after deformation is given by

$$dl = \left\{(1+\beta)^2 + \kappa(1+\beta)\sin 2\theta + \kappa^2 \sin^2\theta\right\}^{1/2} dl_0.$$

Hence, show that all such elements are stretched during the deformation.

3.4: Suppose a body is subject to the deformation

$$x_1 = f(\xi_1)\cos(k\xi_2), \quad x_2 = f(\xi_1)\sin(k\xi_2), \quad x_3 = \xi_3,$$

where $f(\xi_1)$ is a positive differentiable monotonically increasing function, k is a positive constant, and $-c \leq \xi_2 \leq c$, where $c < \pi$.

(i) Show that, after the deformation, the density ρ is given by

$$\rho = \frac{\rho_0}{kf(\xi_1)f'(\xi_1)},$$

where ρ_0 is the density in the initial state.

(ii) Show that each plane that is initially normal to the ξ_1-direction is deformed into a portion of the curved surface of a circular cylinder whose axis is in the x_3-direction.

(iii) Find the equations of the planes into which planes initially normal to the ξ_2-direction are deformed.

(iv) If the body initially occupied the region

$$a \leq \xi_1 \leq b, \quad -c \leq \xi_2 \leq c, \quad 0 \leq \xi_3 \leq d,$$

where a, b, c, and d are positive constants, sketch the shape of the deformed configuration which lies in the plane $x_3 = 0$.

(v) Determine the form of the function $f(\xi_1)$ if there is no change of volume during the deformation.

3.5: Suppose the displacement of a body is given by

$$x_1 = \xi_1 + \omega^2 t^2 \xi_2, \quad x_2 = (1 + \omega^2 t^2)\xi_2, \quad x_3 = \xi_3,$$

where ω is a constant. Show that the plane Π_0 which was initially given by $\xi_1 = a$ deforms into another plane Π_t at time t defined by

$$x_1 = a + \frac{\omega^2 t^2 x_2}{(1+\omega^2 t^2)}.$$

Find the angle between Π_0 and Π_t and show that this angle tends to $\pi/4$ as $t \to \infty$.

3.6: Suppose the velocity of a planar motion is given by

$$\boldsymbol{v} = \left[-\frac{x_2(x_1^2 + x_2^2)}{T(a^2 + x_1^2 + x_2^2)} + \frac{tx_1(a^2 + x_1^2 + x_2^2)}{a^2(t^2 + T^2)}\right]\boldsymbol{e}_1 + \left[\frac{x_1(x_1^2 + x_2^2)}{T(a^2 + x_1^2 + x_2^2)} + \frac{tx_2(a^2 + x_1^2 + x_2^2)}{a^2(t^2 + T^2)}\right]\boldsymbol{e}_2,$$

where a and T are constants.

(i) Write down the equations of the streamlines. Using the variable substitution $x_1 = r\cos\phi$, $x_2 = r\sin\phi$ rewrite these equations in polar coordinates r and ϕ.

(ii) Find the equation of the streamline that contains the point of the x_1x_2-plane with polar coordinates (r_0, ϕ_0) in the form $r = r(\phi)$.

(iii) Write down the equations of the trajectories. Using the variable substitution $x_1 = r\cos\phi$, $x_2 = r\sin\phi$ rewrite these equations in polar coordinates r and ϕ.

(iv) At $t = t_0$ a particle is at the point of the x_1x_2-plane with polar coordinates (r_0, ϕ_0). Find the equations of its trajectory in the parametric form $r = r(t)$, $\phi = \phi(t)$.

(v) You are given that a particle is at distance a from the origin at $t = 0$. Find the moment of time $t_1 > 0$ when the particle is at distance $2a$ from the origin.

3.7: In Cartesian coordinates x, y, where x is the horizontal and y the vertical coordinate, suppose the velocity in a small-amplitude standing surface wave on water of depth h is given by

$$v_x = v_0 \sin(\omega t)\cos(kx)\cosh[k(y + h)],$$
$$v_y = v_0 \sin(\omega t)\sin(kx)\sinh[k(y + h)],$$

where v_0, ω and k are constants. Find the equation of the streamlines written in the form $F(x, y) = \text{const}$.

3.8: Suppose at the initial moment of time, $t = 0$, a continuum occupies a sphere of radius R. It starts to expand, preserving its spherical shape. The velocity of any point at the surface of the sphere is perpendicular to the surface and its magnitude is equal to $a^2V/(a^2 + t^2)$. Inside the sphere the velocity is also in the radial direction and its magnitude is proportional to the distance from the centre of the sphere.

(i) Find the radius of the sphere at time t.

(ii) Find the dependence of the velocity on the distance from the centre of the sphere, r, at time t.

(iii) You are given that the density in the sphere has the form $\rho = r^2 f(t)$, and $f(0) = \rho_0/R^2$. Use the mass conservation equation to determine $f(t)$. Calculate the limiting value of the density as $t \to \infty$.

Chapter 4
Dynamics of Continuum

After describing the kinematics of a mechanical system with a finite number of degrees of freedom, we proceed to its dynamical behaviour and describe its motion under the action of forces applied to this system. If a system consists of a few interacting parts then there is no fundamental difference between the description of the external and internal forces. The situation is different in continuum mechanics. Internal forces play a crucial role in the continuum motion and their description is very much different from that of external forces. We will see below that these forces are completely determined by a so-called stress tensor.

4.1 Forces in Continuum Mechanics

We consider *body* and *surface* forces. A *body force* acts on any particle in a continuum. It is characterised by a vector field $\boldsymbol{b}(\boldsymbol{x}, t)$ which is the force per unit mass of the continuum. If we take an infinitesimal volume dV inside the continuum, then its mass is $\rho\, dV$ and the body force acting on this volume is $\rho \boldsymbol{b}\, dV$. The body force acting on a volume V of a continuum is equal to $\int_V \rho \boldsymbol{b}\, dV$.

A *surface force* acts on the surface of a volume. The surface force acting on an elementary surface dS with the external normal unit vector $\boldsymbol{n}$ is equal to $\boldsymbol{t}(\boldsymbol{x}, t, \boldsymbol{n})\, dS$. We see that the surface force depends on the orientation of the elementary surface dS. The vector $\boldsymbol{t}$ is called the *surface traction* or *stress*. It represents the force per unit area of the surface. The surface force acting on the surface S is equal to $\int_S \boldsymbol{t}\, dS$.

Both body and surface forces can be either *external* or *internal*. Whether the force is external or internal depends on the point of view. For example, the gravity force exerted by one part of the sun on another part is an internal force (the self-gravity force), while the gravity force exerted by the sun on the Earth is external for the Earth if we consider the Earth as a separate system. However, if we consider the whole solar system as one system, then the gravity force exerted by the sun on the Earth

© Springer Nature Switzerland AG 2019

M. S. Ruderman, *Fluid Dynamics and Linear Elasticity*, Springer Undergraduate Mathematics Series, https://doi.org/10.1007/978-3-030-19297-6_4

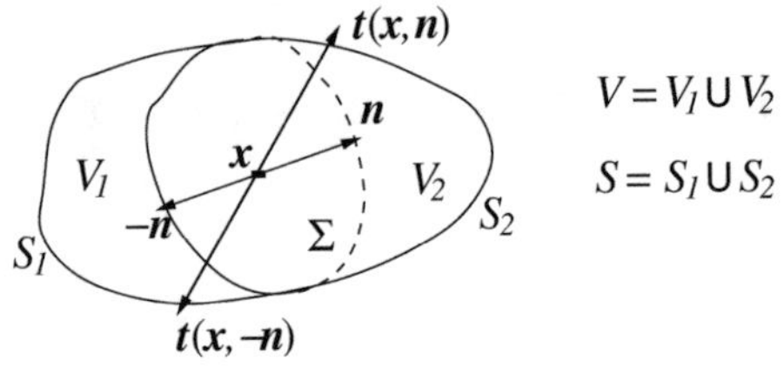

Fig. 4.1 The volume V is cut into two volumes, V_1 and V_2, by the surface Σ

is internal. If we consider a table as a separate system then the pressure exerted by a book lying on the table is an external surface force. However, if we consider the table and the book as two parts of one system, then the pressure exerted by the book on the table is an internal surface force.

The total force exerted on a volume V of a continuum with surface S is equal to

$$\int_S \boldsymbol{t}\, dS + \int_V \rho \boldsymbol{b}\, dV. \tag{4.1.1}$$

The total *torque* on V about the origin is

$$\int_S \boldsymbol{x} \times \boldsymbol{t}\, dS + \int_V \rho \boldsymbol{x} \times \boldsymbol{b}\, dV. \tag{4.1.2}$$

4.2 The Stress Tensor

Consider an arbitrary material volume V of a continuum, i.e. a volume that consists of the same particles of the continuum at any moment of time. The momentum of this volume is equal to $\int_V \rho \boldsymbol{v}\, dV$. Then, in accordance with Newton's second law and Eq. (4.1.1)

$$\frac{d}{dt}\int_V \rho \boldsymbol{v}\, dV = \int_S \boldsymbol{t}\, dS + \int_V \rho \boldsymbol{b}\, dV. \tag{4.2.1}$$

Using Eq. (4.1.2) and the theorem of classical mechanics stating that the rate of change of the *angular momentum* of a mechanical system is equal to the torque of the external forces, we obtain

$$\frac{d}{dt}\int_V \rho \boldsymbol{x} \times \boldsymbol{v}\, dV = \int_S \boldsymbol{x} \times \boldsymbol{t}\, dS + \int_V \rho \boldsymbol{x} \times \boldsymbol{b}\, dV. \tag{4.2.2}$$

Assume now that the volume V is cut by the surface Σ into two parts: V_1 and V_2. If $\boldsymbol{n}$ is the unit normal vector at Σ external with respect to V_1, then the unit normal vector external with respect to V_2 is $-\boldsymbol{n}$ (see Fig. 4.1). Writing Eq. (4.2.1) for V_1 and V_2 yields

$$\frac{d}{dt}\int_{V_1} \rho \boldsymbol{v}\, dV = \int_{S_1} \boldsymbol{t}\, dS + \int_{\Sigma} \boldsymbol{t}(\boldsymbol{n})\, dS + \int_{V_1} \rho \boldsymbol{b}\, dV,$$

$$\frac{d}{dt}\int_{V_2} \rho \boldsymbol{v}\, dV = \int_{S_2} \boldsymbol{t}\, dS + \int_{\Sigma} \boldsymbol{t}(-\boldsymbol{n})\, dS + \int_{V_2} \rho \boldsymbol{b}\, dV.$$

Adding these equations we obtain

$$\frac{d}{dt}\int_{V} \rho \boldsymbol{v}\, dV = \int_{S} \boldsymbol{t}\, dS + \int_{\Sigma} [\boldsymbol{t}(\boldsymbol{n}) + \boldsymbol{t}(-\boldsymbol{n})]\, dS + \int_{V} \rho \boldsymbol{b}\, dV.$$

Subtracting Eq. (4.2.1) from this equation yields

$$\int_{\Sigma} [\boldsymbol{t}(\boldsymbol{n}) + \boldsymbol{t}(-\boldsymbol{n})]\, dS = 0.$$

The volume V and surface Σ are arbitrary. Let the maximum distance between a point inside V and $\boldsymbol{x}$ be dl. Then $\boldsymbol{t}(\boldsymbol{x}', \pm\boldsymbol{n}) = \boldsymbol{t}(\boldsymbol{x}, \pm\boldsymbol{n}) + O(dl)$ and

$$\int_{\Sigma} [\boldsymbol{t}(\boldsymbol{n}) + \boldsymbol{t}(-\boldsymbol{n})]\, dS = [\boldsymbol{t}(\boldsymbol{x}, \boldsymbol{n}) + \boldsymbol{t}(\boldsymbol{x}, -\boldsymbol{n})]\Sigma + O(dl)\Sigma = 0,$$

so that

$$[\boldsymbol{t}(\boldsymbol{x}, \boldsymbol{n}) + \boldsymbol{t}(\boldsymbol{x}, -\boldsymbol{n})] + O(dl) = 0.$$

Taking $dl \to 0$ we eventually obtain

$$\boldsymbol{t}(\boldsymbol{x}, -\boldsymbol{n}) = -\boldsymbol{t}(\boldsymbol{x}, \boldsymbol{n}). \tag{4.2.3}$$

In fact, this is Newton's third law in continuum mechanics.

Let us now consider a tetrahedron with three edges parallel to the coordinate axes (see Fig. 4.2). It is straightforward to obtain the expressions for the areas of triangles ABC, ABD, and BCD in terms of the area of triangle ACD and the components of the unit normal $\boldsymbol{n}$ to this triangle. These expressions are given in Fig. 4.2

Now we again consider the same tetrahedron and assume that it is infinitesimal, so that we can neglect the variation of any quantity inside it and calculate its value at $\boldsymbol{x}$. We take $l_i = \alpha_i \epsilon$ $(i = 1, 2, 3)$ (see Fig. 4.3). Then we obtain

$$S_1 = \tfrac{1}{2}\alpha_2\alpha_3\epsilon^2, \quad S_2 = \tfrac{1}{2}\alpha_1\alpha_3\epsilon^2, \quad S_3 = \tfrac{1}{2}\alpha_1\alpha_2\epsilon^2,$$

$$S = \sqrt{(n_1^2 + n_2^2 + n_3^2)S^2} = \sqrt{S_1^2 + S_2^2 + S_3^2} = \tfrac{1}{2}\epsilon^2\sqrt{\alpha_1^2\alpha_2^2 + \alpha_1^2\alpha_3^2 + \alpha_2^2\alpha_3^2} = \widetilde{S}\epsilon^2,$$

$$V = \tfrac{1}{3}l_3 S_3 = \tfrac{1}{6}\alpha_1\alpha_2\alpha_3\epsilon^3 = \widetilde{V}\epsilon^3.$$

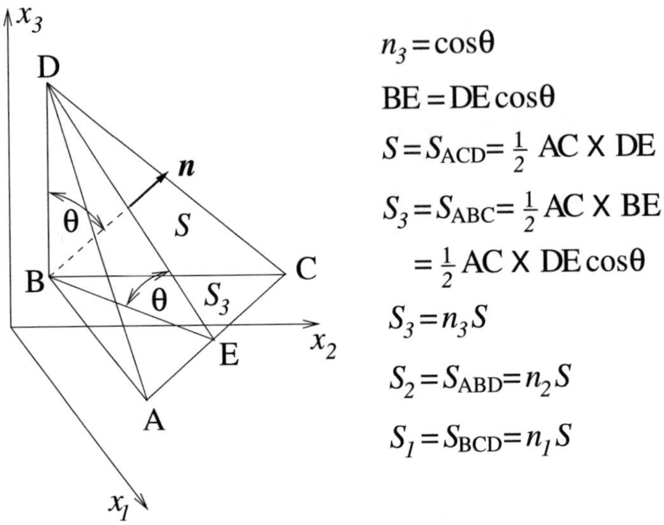

Fig. 4.2 A tetrahedron with edges parallel to the coordinate axis

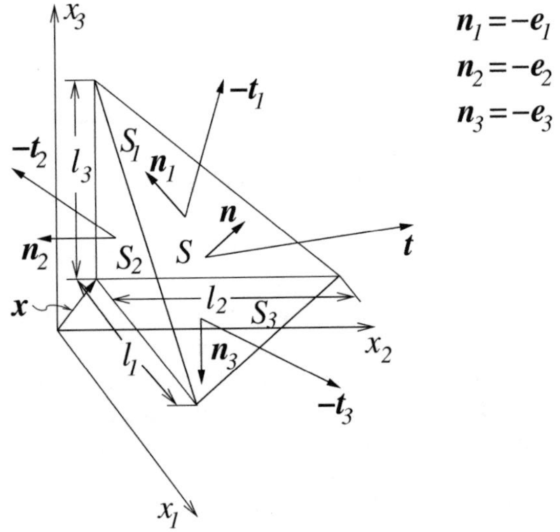

Fig. 4.3 Illustration of the derivation of the expression for the surface traction in terms of the stress tensor. The unit normal at the tetrahedron faces and the surface tractions acting on these faces are shown

The linear momentum of the tetrahedron is equal to $\rho(\boldsymbol{x})\boldsymbol{v}(\boldsymbol{x})V$. The external body force exerted on the tetrahedron is $\rho(\boldsymbol{x})\boldsymbol{b}(\boldsymbol{x})V$. We introduce the notation $\boldsymbol{t}(\boldsymbol{x}) = \boldsymbol{t}(\boldsymbol{x}, \boldsymbol{n})$ and $\boldsymbol{t}_i(\boldsymbol{x}) = \boldsymbol{t}(\boldsymbol{x}, \boldsymbol{e}_i)$ $(i = 1, 2, 3)$, and use the relations $\boldsymbol{n}_i = -\boldsymbol{e}_i$. Then it follows that

$$\boldsymbol{t}(\boldsymbol{x}, \boldsymbol{n}_i) = -\boldsymbol{t}(\boldsymbol{x}, \boldsymbol{e}_i) = -\boldsymbol{t}_i(\boldsymbol{x}),$$

and the surface force exerted on the tetrahedron is

$$t(x)S - t_1(x)S_1 - t_2(x)S_2 - t_3(x)S_3.$$

Hence, the momentum Eq. (4.2.1) for the tetrahedron is

$$\frac{d}{dt}[\rho(x)v(x)V] = \rho(x)b(x)V - t_1(x)S_1 - t_2(x)S_2 - t_3(x)S_3 + t(x)S.$$

We divide this equation by S and use $S_i = n_i S$ to obtain

$$\frac{\epsilon}{\widetilde{S}}\left\{\frac{d}{dt}\left[\rho(x)v(x)\widetilde{V}\right] - \rho(x)b(x)\widetilde{V}\right\} = t(x) - t_1(x)n_1 - t_2(x)n_2 - t_3(x)n_3.$$

Taking $\epsilon \to 0$ yields

$$t(x) = t_1(x)n_1 + t_2(x)n_2 + t_3(x)n_3,$$

or, in components,

$$t_i = t_{i1}n_1 + t_{i2}n_2 + t_{i3}n_3. \tag{4.2.4}$$

Introduce a tensor $\mathbf{T}$ such that in the particular coordinate system that we used to derive Eq. (4.2.4) its components are given by $T_{ij} = t_{ij}$. Then $t_i = T_{ij}n_j$. We have vector quantities on both sides of this relation. Hence, if it is valid in one coordinate system, then it is valid in any. Concluding, we obtain

$$\boxed{t = \mathbf{T} \cdot n.} \tag{4.2.5}$$

The tensor $\mathbf{T}$ is called the *stress tensor*.

Example Suppose the matrix of a stress tensor is given by

$$\hat{\mathbf{T}} = \begin{pmatrix} 1 & 0 & 2 \\ 0 & 1 & 3 \\ 2 & 2 & 1 \end{pmatrix}$$

and the normal vector to a surface is $n = \left(\frac{1}{\sqrt{3}}, \frac{1}{\sqrt{3}}, -\frac{1}{\sqrt{3}}\right)^T$. Then the surface traction at this surface is

$$t = \begin{pmatrix} 1 & 0 & 2 \\ 0 & 1 & 3 \\ 2 & 2 & 1 \end{pmatrix} \begin{pmatrix} 1/\sqrt{3} \\ 1/\sqrt{3} \\ -1/\sqrt{3} \end{pmatrix} = \begin{pmatrix} -1/\sqrt{3} \\ -2/\sqrt{3} \\ \sqrt{3} \end{pmatrix}.$$

4.3 The Main Properties of the Stress Tensor

Now we use the angular momentum equation Eq. (4.2.2). Using Eq. (4.2.5) we can transform the first term on the right-hand side of this equation into

$$\int_S \boldsymbol{x} \times \boldsymbol{t}\, dS = \int_S \boldsymbol{x} \times (\mathbf{T} \cdot \boldsymbol{n})\, dS = \boldsymbol{e}_i \varepsilon_{ijk} \int_S x_j T_{kl} n_l\, dS.$$

Let us introduce a vector $\boldsymbol{a}$ with components $a_l = x_j T_{kl}$, where j and k are fixed. (Note that this relation is only valid in one particular coordinate system. In a new coordinate system, in general, $a'_l \neq x'_j T'_{kl}$.) Then, using Gauss' divergence theorem Eq. (2.1.4), we obtain

$$\begin{aligned}
\int_S \boldsymbol{x} \times \boldsymbol{t}\, dS &= \boldsymbol{e}_i \varepsilon_{ijk} \int_S a_l n_l\, dS = \boldsymbol{e}_i \varepsilon_{ijk} \int_S \boldsymbol{a} \cdot d\boldsymbol{S} \\
&= \boldsymbol{e}_i \varepsilon_{ijk} \int_V \nabla \cdot \boldsymbol{a}\, dV = \boldsymbol{e}_i \varepsilon_{ijk} \int_V \frac{\partial (x_j T_{kl})}{\partial x_l}\, dV \\
&= \boldsymbol{e}_i \varepsilon_{ijk} \int_V \left(\delta_{jl} T_{kl} + x_j \frac{\partial T_{kl}}{\partial x_l} \right) dV = \boldsymbol{e}_i \varepsilon_{ijk} \int_V \left(T_{kj} + x_j \frac{\partial T_{kl}}{\partial x_l} \right) dV.
\end{aligned}$$

Using this result and Eq. (3.9.1) we rewrite Eq. (4.2.2) as

$$\int_V \rho \boldsymbol{x} \times \frac{D\boldsymbol{v}}{Dt}\, dV = \boldsymbol{e}_i \varepsilon_{ijk} \int_V T_{kj}\, dV + \int_V \rho \boldsymbol{x} \times \boldsymbol{c}\, dV + \int_V \rho \boldsymbol{x} \times \boldsymbol{b}\, dV, \qquad (4.3.1)$$

where $c_k = \partial T_{kl}/\partial x_l$. When deriving Eq. (4.3.1) we have used the relation

$$\frac{D}{Dt}(\boldsymbol{x} \times \boldsymbol{v}) = \frac{D\boldsymbol{x}}{Dt} \times \boldsymbol{v} + \boldsymbol{x} \times \frac{D\boldsymbol{v}}{Dt} = \boldsymbol{v} \times \boldsymbol{v} + \boldsymbol{x} \times \frac{D\boldsymbol{v}}{Dt} = \boldsymbol{x} \times \frac{D\boldsymbol{v}}{Dt}.$$

Let us now introduce the size of V, ϵ, which is the maximum distance between two points in V. We also shift the coordinate origin to a point $\boldsymbol{x}_0$ that is inside V, so that $\|\boldsymbol{x}\| \leq \epsilon$. Then the left-hand side of Eq. (4.3.1) as well as the second and third terms on the right-hand side are of the order ϵ^4. The first term on the right-hand side of Eq. (4.3.1) is of the order ϵ^3. Noting that, for $\epsilon \ll 1$, $T_{kj}(\boldsymbol{x}) \approx T_{kj}(\boldsymbol{x}_0)$, dividing Eq. (4.3.1) by ϵ^3, assuming that $\lim_{\epsilon \to 0}(V/\epsilon^3) = \text{const} \neq 0$, and taking $\epsilon \to 0$, we obtain

$$\varepsilon_{ijk} T_{kj}(\boldsymbol{x}_0) = 0. \qquad (4.3.2)$$

Let us take $i = 3$. Then it follows from Eq. (4.3.2) that $T_{12}(\boldsymbol{x}_0) = T_{21}(\boldsymbol{x}_0)$. Since $\boldsymbol{x}_0$ is an arbitrary point, we conclude that $T_{12} = T_{21}$ everywhere. Taking $i = 2$ and $i = 1$ we, in a similar manner, prove that $T_{13} = T_{31}$ and $T_{23} = T_{32}$. Hence, we have proved that the stress tensor $\mathbf{T}$ is a *symmetric* tensor.

As is the case for any symmetric second-order tensor, the stress tensor has three mutually orthogonal principal axes. The directions of these axes are called the *principal stress directions*. The corresponding eigenvalues of the operator $\mathbf{T}$ associated with the stress tensor are called the *principal stresses*.

4.4 Momentum Equation in the Eulerian Description

The momentum equation in the integral form is given by Eq. (4.2.1) with $\boldsymbol{t}$ determined by Eq. (4.2.5). Our aim is now to derive this equation in the differential form assuming that all quantities are continuous together with their first derivatives.

We start by transforming the expression for the surface force. Using Eq. (4.2.5) written in components we obtain

$$\int_S \boldsymbol{t}\, dS = \boldsymbol{e}_i \int_S T_{ij} n_j\, dS.$$

Let us introduce the vector $\boldsymbol{a} = T_{ij}\boldsymbol{e}_j$ (note that $a_j = T_{ij}$ only in one particular coordinate system). Then, using Gauss' divergence theorem, we obtain

$$\int_S \boldsymbol{t}\, dS = \boldsymbol{e}_i \int_S a_j n_j\, dS = \boldsymbol{e}_i \int_S \boldsymbol{a} \cdot \boldsymbol{n}\, dS = \boldsymbol{e}_i \int_S \boldsymbol{a} \cdot d\boldsymbol{S} = \boldsymbol{e}_i \int_V \nabla \cdot \boldsymbol{a}\, dV,$$

and, eventually,

$$\int_S \boldsymbol{t}\, dS = \boldsymbol{e}_i \int_V \frac{\partial T_{ji}}{\partial x_j}\, dV = \int_V \nabla \cdot \mathbf{T}\, dV. \tag{4.4.1}$$

Using Eq. (4.4.1) we rewrite Eq. (4.2.1) as

$$\frac{d}{dt}\int_V \rho \boldsymbol{v}\, dV = \int_V \nabla \cdot \mathbf{T}\, dV + \int_V \rho \boldsymbol{b}\, dV. \tag{4.4.2}$$

Now we use Eq. (3.9.2) with $\boldsymbol{v}$ substituted for f to transform the left-hand side of Eq. (4.4.2) as

$$\frac{d}{dt}\int_V \rho \boldsymbol{v}\, dV = \int_V \rho \frac{D\boldsymbol{v}}{Dt}\, dV.$$

Using this result we rewrite Eq. (4.4.2) as

$$\int_V \left(\rho \frac{D\boldsymbol{v}}{Dt} - \nabla \cdot \mathbf{T} - \rho \boldsymbol{b} \right) dV = 0.$$

Since V is arbitrary, it follows from Lemma 3.1 proved in Sect. 3.8 that

$$\boxed{\rho \frac{D\boldsymbol{v}}{Dt} = \nabla \cdot \mathbf{T} + \rho \boldsymbol{b}.} \tag{4.4.3}$$

This is the momentum equation in the Eulerian variables. Using Eq. (3.1.8) we can rewrite it in an alternative form:

$$\boxed{\frac{\partial \boldsymbol{v}}{\partial t} + (\boldsymbol{v} \cdot \nabla)\boldsymbol{v} = \frac{1}{\rho}\nabla \cdot \mathbf{T} + \boldsymbol{b}.} \tag{4.4.4}$$

We obtain the equation of equilibrium if we take $\boldsymbol{v} = 0$ in Eq. (4.4.4):

$$\boxed{\nabla \cdot \mathbf{T} + \rho \boldsymbol{b} = 0.} \tag{4.4.5}$$

Example Consider a continuum inside an infinite cylinder. In cylindrical coordinates r, ϕ, z with the z-axis coinciding with the cylinder axis suppose the flow velocity is given by $\boldsymbol{v} = \Omega r \boldsymbol{e}_\phi$, where Ω is a constant, and $\rho =$ const, and the stress tensor has the form $\mathbf{T} = -p\mathbf{I}$, $\boldsymbol{b} = 0$, and $p = p_0 =$ const at $r = 0$. Calculate the dependence of p on r.

Solution: First we obtain $\nabla \cdot \mathbf{T} = -\nabla p$. Using Eq. (2.13.10) we obtain from Eq. (4.4.4)

$$\rho \Omega^2 r = \frac{\partial p}{\partial r}, \quad \frac{\partial p}{\partial \phi} = 0, \quad \frac{\partial p}{\partial z} = 0.$$

It follows from these equations that p only depends on r and is given by $p = p_0 + \frac{1}{2}\rho\Omega^2 r^2$.

4.5 Momentum Equation in the Lagrangian Description

Now we derive the momentum equation in the *Lagrangian* variables. With the aid of Eq. (3.7.4) we transform the left-hand side of Eq. (4.4.2) as

$$\frac{d}{dt}\int_V \rho \boldsymbol{v}\, dV = \frac{d}{dt}\int_{V_0} \rho \boldsymbol{v} \mathcal{J}\, dV = \frac{d}{dt}\int_{V_0} \rho_0 \boldsymbol{v}\, dV = \int_{V_0} \rho_0 \frac{D\boldsymbol{v}}{Dt}\, dV. \tag{4.5.1}$$

The second term on the right-hand side of Eq. (4.4.2) can be rewritten as

$$\int_V \rho \boldsymbol{b}\, dV = \int_{V_0} \rho \boldsymbol{b} \mathcal{J}\, dV = \int_{V_0} \rho_0 \boldsymbol{b}\, dV. \tag{4.5.2}$$

Finally, we transform the first term on the right-hand side of Eq. (4.4.2). To do this we recall some facts from matrix algebra. Consider a 3×3 matrix $\hat{\mathbf{A}}$ with entries a_{ij}. The second-order minor M_{ij} of $\hat{\mathbf{A}}$ is the determinant of 2×2 matrix obtained from $\hat{\mathbf{A}}$ by removing the ith row and the jth column:

$$\begin{pmatrix} a_{11} & a_{12} & a_{13} \\ a_{21} & a_{22} & a_{23} \\ a_{31} & a_{32} & a_{33} \end{pmatrix}, \qquad M_{22} = \begin{vmatrix} a_{11} & a_{13} \\ a_{31} & a_{33} \end{vmatrix}.$$

The algebraic adjunct for the entry a_{ij} is $A_{ij} = (-1)^{i+j} M_{ij}$. The determinant of $\hat{\mathbf{A}}$ can be calculated using an expansion with respect to any row:

$$\begin{aligned} \det(\hat{\mathbf{A}}) = |\hat{\mathbf{A}}| &= a_{11}A_{11} + a_{12}A_{12} + a_{13}A_{13} \\ &= a_{21}A_{21} + a_{22}A_{22} + a_{23}A_{23} \\ &= a_{31}A_{31} + a_{32}A_{32} + a_{33}A_{33}. \end{aligned}$$

The entries of the inverse matrix, $\hat{\mathbf{A}}^{-1}$, $\tilde{a}_{ij}$, are given by $\tilde{a}_{ij} = A_{ji}/|\hat{\mathbf{A}}|$, so that

$$\hat{\mathbf{A}}^{-1} = \frac{1}{|\hat{\mathbf{A}}|} \begin{pmatrix} A_{11} & A_{12} & A_{13} \\ A_{21} & A_{22} & A_{23} \\ A_{31} & A_{32} & A_{33} \end{pmatrix}^T.$$

Now we prove the following

Lemma 4.1 *Consider an arbitrary matrix $\hat{\mathbf{A}}$ that depends on a parameter τ. Then*

$$\frac{d|\hat{\mathbf{A}}|}{d\tau} = |\hat{\mathbf{A}}|\, tr\left(\frac{d\hat{\mathbf{A}}}{d\tau}\hat{\mathbf{A}}^{-1}\right). \tag{4.5.3}$$

Proof Calculating the derivative of $\det(\hat{\mathbf{A}}) = |\hat{\mathbf{A}}|$ we obtain

$$\begin{aligned} \frac{d|\hat{\mathbf{A}}|}{d\tau} &= \frac{d}{d\tau}\begin{vmatrix} a_{11} & a_{12} & a_{13} \\ a_{21} & a_{22} & a_{23} \\ a_{31} & a_{32} & a_{33} \end{vmatrix} \\ &= \begin{vmatrix} \frac{da_{11}}{d\tau} & \frac{da_{12}}{d\tau} & \frac{da_{13}}{d\tau} \\ a_{21} & a_{22} & a_{23} \\ a_{31} & a_{32} & a_{33} \end{vmatrix} + \begin{vmatrix} a_{11} & a_{12} & a_{13} \\ \frac{da_{21}}{d\tau} & \frac{da_{22}}{d\tau} & \frac{da_{23}}{d\tau} \\ a_{31} & a_{32} & a_{33} \end{vmatrix} + \begin{vmatrix} a_{11} & a_{12} & a_{13} \\ a_{21} & a_{22} & a_{23} \\ \frac{da_{31}}{d\tau} & \frac{da_{32}}{d\tau} & \frac{da_{33}}{d\tau} \end{vmatrix} \\ &= \frac{da_{11}}{d\tau}A_{11} + \frac{da_{12}}{d\tau}A_{12} + \frac{da_{13}}{d\tau}A_{13} \\ &\quad + \frac{da_{21}}{d\tau}A_{21} + \frac{da_{22}}{d\tau}A_{22} + \frac{da_{23}}{d\tau}A_{23} \\ &\quad + \frac{da_{31}}{d\tau}A_{31} + \frac{da_{32}}{d\tau}A_{32} + \frac{da_{33}}{d\tau}A_{33} \\ &= \frac{da_{ij}}{d\tau}A_{ij} = |\hat{\mathbf{A}}|\frac{da_{ij}}{d\tau}\tilde{a}_{ji} = |\hat{\mathbf{A}}|\,\mathrm{tr}\left(\frac{d\hat{\mathbf{A}}}{d\tau}\hat{\mathbf{A}}^{-1}\right). \end{aligned} \tag{4.5.4}$$

The lemma is proved. □

Lemma 4.2

$$\frac{\partial}{\partial \xi_j}\left(\mathcal{J}\frac{\partial \xi_j}{\partial x_k}\right) = 0. \tag{4.5.5}$$

Proof We recall that $\mathcal{J} = \det(\hat{\mathbf{F}})$, where $\mathbf{F}$ is the deformation-gradient tensor, and introduce $\widetilde{\mathcal{J}} = \det(\hat{\mathbf{F}}^{-1}) = \mathcal{J}^{-1}$. Then

$$\frac{\partial}{\partial \xi_j}\left(\mathcal{J}\frac{\partial \xi_j}{\partial x_k}\right) = \left[\frac{\partial}{\partial x_l}\left(\frac{1}{\widetilde{\mathcal{J}}}\frac{\partial \xi_j}{\partial x_k}\right)\right]\frac{\partial x_l}{\partial \xi_j} = \frac{\partial x_l}{\partial \xi_j}\left(\frac{1}{\widetilde{\mathcal{J}}}\frac{\partial^2 \xi_j}{\partial x_k \partial x_l} - \frac{1}{\widetilde{\mathcal{J}}^2}\frac{\partial \widetilde{\mathcal{J}}}{\partial x_l}\frac{\partial \xi_j}{\partial x_k}\right)$$

$$= \mathcal{J}\frac{\partial x_l}{\partial \xi_j}\left[\frac{\partial^2 \xi_j}{\partial x_k \partial x_l} - \mathcal{J}\frac{\partial \xi_j}{\partial x_k}\widetilde{\mathcal{J}}\,\mathrm{tr}\left(\frac{\partial \hat{\mathbf{F}}^{-1}}{\partial x_l}\hat{\mathbf{F}}\right)\right]$$

$$= \mathcal{J}\frac{\partial^2 \xi_j}{\partial x_k \partial x_l}\frac{\partial x_l}{\partial \xi_j} - \mathcal{J}\frac{\partial x_l}{\partial \xi_j}\frac{\partial \xi_j}{\partial x_k}\left[\frac{\partial}{\partial x_l}\left(\frac{\partial \xi_m}{\partial x_n}\right)\right]\frac{\partial x_n}{\partial \xi_m}$$

$$= \mathcal{J}\frac{\partial^2 \xi_j}{\partial x_k \partial x_l}\frac{\partial x_l}{\partial \xi_j} - \mathcal{J}\delta_{lk}\frac{\partial^2 \xi_m}{\partial x_l \partial x_n}\frac{\partial x_n}{\partial \xi_m} = \mathcal{J}\frac{\partial^2 \xi_j}{\partial x_k \partial x_l}\frac{\partial x_l}{\partial \xi_j} - \mathcal{J}\frac{\partial^2 \xi_j}{\partial x_k \partial x_l}\frac{\partial x_l}{\partial \xi_j} = 0.$$

The lemma is proved. □

Using Lemma 4.2 we can rewrite the first term on the right-hand side of Eq. (4.4.2) in a desirable form. Using Eq. (4.5.5) we obtain

$$\int_V \nabla\cdot\mathbf{T}\,dV = \boldsymbol{e}_i\int_{V_0}\frac{\partial T_{ki}}{\partial x_k}\mathcal{J}\,dV = \boldsymbol{e}_i\int_{V_0}\frac{\partial T_{ki}}{\partial \xi_j}\frac{\partial \xi_j}{\partial x_k}\mathcal{J}\,dV$$

$$= \boldsymbol{e}_i\int_{V_0}\left[\frac{\partial T_{ki}}{\partial \xi_j}\frac{\partial \xi_j}{\partial x_k}\mathcal{J} + T_{ki}\frac{\partial}{\partial \xi_j}\left(\mathcal{J}\frac{\partial \xi_j}{\partial x_k}\right)\right]dV = \boldsymbol{e}_i\int_{V_0}\left[\frac{\partial}{\partial \xi_j}\left(\mathcal{J}T_{ki}\frac{\partial \xi_j}{\partial x_k}\right)\right]dV.$$

Introducing the *nominal stress tensor* $\mathbf{\Pi}$ (also called the *Piola–Kirchhoff stress tensor*) with matrix of coefficients given by $\hat{\mathbf{\Pi}} = \mathcal{J}\hat{\mathbf{F}}^{-1}\hat{\mathbf{T}}$, so that

$$\Pi_{ji} = \mathcal{J}T_{ki}\frac{\partial \xi_j}{\partial x_k}, \tag{4.5.6}$$

and using the notation

$$\nabla_\xi\cdot\mathbf{\Pi} = \boldsymbol{e}_i\frac{\partial \Pi_{ji}}{\partial \xi_j}, \tag{4.5.7}$$

we eventually arrive at

$$\int_V \nabla\cdot\mathbf{T}\,dV = \int_{V_0}\nabla_\xi\cdot\mathbf{\Pi}\,dV. \tag{4.5.8}$$

Using Eqs. (4.5.1), (4.5.2), (4.5.8), and the relation $\boldsymbol{v} = \frac{D\boldsymbol{x}}{Dt} = \frac{D\boldsymbol{u}}{Dt}$ we rewrite Eq. (4.4.2) as

$$\int_{V_0}\left(\rho_0\frac{D^2\boldsymbol{u}}{Dt^2} - \nabla_{\xi}\cdot\boldsymbol{\Pi} - \rho_0\boldsymbol{b}\right)dV = 0.$$

Since V_0 is arbitrary, it follows from this equation that

$$\boxed{\rho_0\frac{D^2\boldsymbol{u}}{Dt^2} = \nabla_{\xi}\cdot\boldsymbol{\Pi} + \rho_0\boldsymbol{b}.} \tag{4.5.9}$$

Taking $\frac{D\boldsymbol{u}}{Dt} = 0$ in Eq. (4.5.9) we obtain the equilibrium equation in Lagrangian variables:

$$\boxed{\nabla_{\xi}\cdot\boldsymbol{\Pi} + \rho_0\boldsymbol{b} = 0.} \tag{4.5.10}$$

Example Suppose the deformation of a continuum is defined by

$$x_1 = \xi_1 + \xi_2\sin\alpha, \quad x_2 = \xi_1\sin\alpha + \xi_2, \quad x_3 = \xi_3 + \frac{1}{3}k^2\xi_3^3,$$

where $0 < \alpha < \pi/2$ and $k > 0$ are positive constants and the components of the stress tensor are given by

$$T_{ij} = \mu\delta_{ij}\frac{\partial u_k}{\partial \xi_k} + \mu\left(\frac{\partial u_i}{\partial \xi_j} + \frac{\partial u_j}{\partial \xi_i}\right),$$

where $\boldsymbol{u} = \boldsymbol{x} - \boldsymbol{\xi}$, and μ is a constant. Suppose also that the continuum density before the deformation is $\rho_0 = \text{const}$. Calculate $\hat{\boldsymbol{\Pi}}$. You are given that the continuum is in equilibrium. Find the body force $\boldsymbol{b}$.

Solution: The matrices $\hat{\mathbf{F}}$ and $\hat{\mathbf{T}}$ are given by

$$\hat{\mathbf{F}} = \begin{pmatrix} 1 & \sin\alpha & 0 \\ \sin\alpha & 1 & 0 \\ 0 & 0 & 1 + k^2\xi_3^2 \end{pmatrix}, \quad \hat{\mathbf{T}} = \mu\begin{pmatrix} k^2\xi_3^2 & 2\sin\alpha & 0 \\ 2\sin\alpha & k^2\xi_3^2 & 0 \\ 0 & 0 & 3k^2\xi_3^2 \end{pmatrix}.$$

Then we obtain

$$J = \cos^2\alpha\left(1 + k^2\xi_3^2\right), \quad \hat{\mathbf{F}}^{-1} = \begin{pmatrix} \sec^2\alpha & -\tan\alpha\sec\alpha & 0 \\ -\tan\alpha\sec\alpha & \sec^2\alpha & 0 \\ 0 & 0 & \left(1 + k^2\xi_3^2\right)^{-1} \end{pmatrix}.$$

Using these results yields

$$\hat{\Pi} = \mu \begin{pmatrix} (1+k^2\xi_3^2)(k^2\xi_3^2 - 2\sin^2\alpha) & (1+k^2\xi_3^2)(2-k^2\xi_3^2)\sin\alpha & 0 \\ (1+k^2\xi_3^2)(2-k^2\xi_3^2)\sin\alpha & (1+k^2\xi_3^2)(k^2\xi_3^2 - 2\sin^2\alpha) & 0 \\ 0 & 0 & 3k^2\xi_3^2\cos^2\alpha \end{pmatrix}.$$

Substituting this expression into Eq. (4.5.10) we finally obtain

$$\boldsymbol{b} = -\frac{6\boldsymbol{e}_3}{\rho_0}\mu k^2\xi_3\cos^2\alpha.$$

4.6 Boundary Conditions

Usually a continuum occupies not the whole space, but only a spatial domain. To solve equations describing the motion of a continuum in this domain we need to define some relations among variables describing the continuum and their derivatives at the boundary of this domain. These relations are called *boundary conditions*. When studying motions of liquids and gases we often consider these motions in domains with rigid boundaries (see Fig. 4.4). Since the liquid cannot penetrate through a rigid boundary, the normal component of the velocity at such a boundary has to vanish:

$$v_n = \boldsymbol{v} \cdot \boldsymbol{n} = 0 \quad \text{for } \forall \boldsymbol{x} \in \Gamma, \tag{4.6.1}$$

where Γ indicates the rigid boundary and $\boldsymbol{n}$ is the unit normal vector at this boundary at point $\boldsymbol{x}$. An ideal fluid can slide along rigid boundaries, so that $\boldsymbol{v}_\tau = \boldsymbol{v} - v_n\boldsymbol{n}$ is arbitrary. A viscous fluid cannot slide along a boundary. This implies that $\boldsymbol{v} = 0$ at Γ.

A more complicated situation arises when the surface of the domain occupied by the continuum is not given and has to be determined from the solution of the problem. At such a surface two boundary conditions have to be satisfied: *kinematic* and *dynamic*.

To obtain the kinematic boundary condition we write the equation of the boundary as $f(x_1, x_2, x_3, t) = 0$. Then we use the condition that if a particle was at the boundary at the initial time, then it remains at the boundary at any time. Let the equations of the trajectory of such a particle be $x_1 = x_1(t, \boldsymbol{\xi})$, $x_2 = x_2(t, \boldsymbol{\xi})$, $x_3 = x_3(t, \boldsymbol{\xi})$, where $\boldsymbol{\xi}$ is the particle's initial position. Then

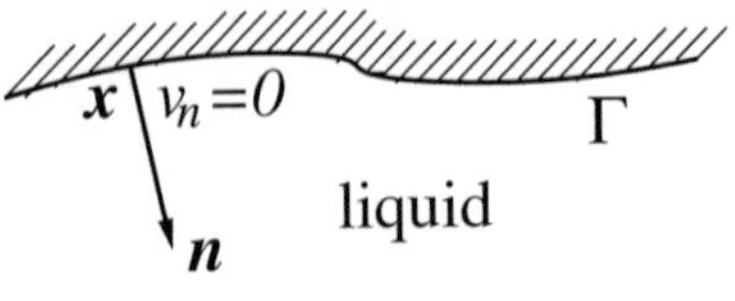

Fig. 4.4 Ideal fluid can slide without friction along a rigid boundary, but it cannot penetrate it

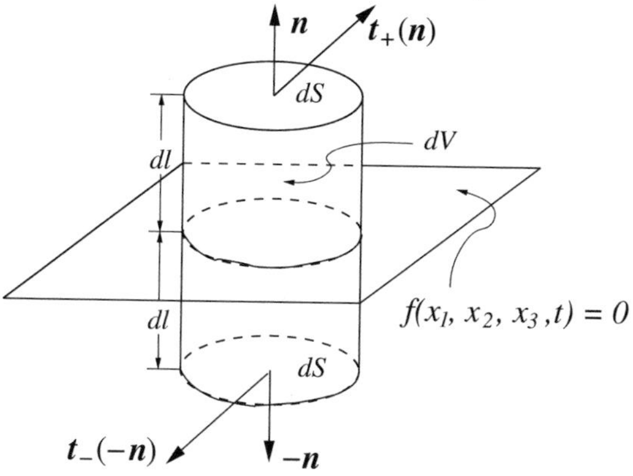

Fig. 4.5 Illustration of the derivation of the dynamic boundary condition. The horizontal plane cuts the cylindrical volume into two equal parts

$$f(x_1(t, \boldsymbol{\xi}), x_2(t, \boldsymbol{\xi}), x_3(t, \boldsymbol{\xi}), t) \equiv 0,$$

where we use the identity sign "≡" to indicate that this equality is valid for any t. Differentiating this identity with respect to t and taking into account that $\dfrac{D\boldsymbol{x}}{Dt} = \boldsymbol{v}$ we obtain the *kinematic* boundary condition

$$\boxed{\frac{\partial f}{\partial t} + \boldsymbol{v} \cdot \nabla f = 0.} \tag{4.6.2}$$

To derive the dynamic boundary condition we consider the cylindrical volume shown in Fig. 4.5. The mass inside the volume dV is $\rho\, dV$. Its momentum is $\rho \boldsymbol{v}\, dV$. The force acting on the side surface of the cylinder is $O(dl)$, so the total surface force on the cylinder is $[\boldsymbol{t}_+(\boldsymbol{n}) + \boldsymbol{t}_-(-\boldsymbol{n})]dS + O(dl)$, and the momentum equation takes the form

$$\frac{1}{dS}\frac{D}{Dt}(\rho \boldsymbol{v}\, dV) = \boldsymbol{t}_+(\boldsymbol{n}) + \boldsymbol{t}_-(-\boldsymbol{n}) + O(dl).$$

Now we take $dl \to 0$. Then it follows that $dV \to 0$, and, using $\boldsymbol{t}_-(-\boldsymbol{n}) = -\boldsymbol{t}_-(\boldsymbol{n})$ and $\boldsymbol{t} = \mathbf{T} \cdot \boldsymbol{n}$, we obtain the *dynamic* boundary condition

$$\boxed{[\![\mathbf{T} \cdot \boldsymbol{n}]\!] = 0,} \tag{4.6.3}$$

where $[\![g]\!]$ indicates the jump of the function g across the surface $f(x_1, x_2, x_3, t) = 0$ defined by

$$[\![g(\boldsymbol{x})]\!] = \lim_{\epsilon \to +0} \{g(\boldsymbol{x} + \epsilon \boldsymbol{n}) - g(\boldsymbol{x} - \epsilon \boldsymbol{n})\}. \tag{4.6.4}$$

Example Suppose a continuum occupies a half-space $x_3 < 0$ and the components of the stress tensor are given by

$$T_{ij} = -p\delta_{ij} + \eta\left(\frac{\partial v_i}{\partial x_j} + \frac{\partial v_j}{\partial x_i} - \frac{2}{3}\delta_{ij}\frac{\partial v_k}{\partial x_k}\right) + \zeta\delta_{ij}\frac{\partial v_k}{\partial x_k},$$

where p is the pressure, and η and ζ are constants. You can assume that the stress is zero above the surface $x_3 = 0$. Write down the dynamic boundary conditions at $x_3 = 0$.

Solution: Since $\mathsf{T} = 0$ for $x_3 > 0$, it follows from Eq. (4.6.3) that $\mathsf{T} \cdot \boldsymbol{n} = 0$ at $x_3 = 0$. Then, taking into account that $\boldsymbol{n} = \boldsymbol{e}_3$ we obtain the boundary conditions that must be satisfied at $x_3 = 0$,

$$\frac{\partial v_1}{\partial x_3} + \frac{\partial v_3}{\partial x_1} = 0, \quad \frac{\partial v_2}{\partial x_3} + \frac{\partial v_3}{\partial x_2} = 0, \quad p = \eta\left(2\frac{\partial v_3}{\partial x_3} - \frac{2}{3}\frac{\partial v_k}{\partial x_k}\right) + \zeta\frac{\partial v_k}{\partial x_k}.$$

Problems

4.1: Considering the equilibrium of an infinitesimal cube with sides parallel to the coordinate axes, derive the equilibrium equation in Cartesian coordinates x_1, x_2, x_3,

$$\frac{\partial T_{ij}}{\partial x_j} + \rho b_i = 0,$$

where T_{ij} are the components of the stress tensor T, ρ is the density, and b_i are the components of the body force $\boldsymbol{b}$. Draw the sketch clearly indicating the forces acting on the cube.

4.2: Suppose the stress tensor at a certain point has components

$$\begin{pmatrix} 3 & 1 & 1 \\ 1 & 0 & 2 \\ 1 & 2 & 0 \end{pmatrix}$$

in a given system of units. Find
(i) the stress vector on the plane normal to the x_1-axis;
(ii) the stress vector acting on an element of the surface with normal in the direction of the vector (0,1,1). Show that the normal stress has magnitude 2 and find the shear component of this stress vector;
(iii) the principal stresses and the principal directions of stress;
(iv) all unit normals $\boldsymbol{n}$ of the form

$$\boldsymbol{n} = (0, n_2, n_3)$$

for which $\boldsymbol{t}(\boldsymbol{n})$ is perpendicular to $\boldsymbol{n}$ (i.e. the stress on elements with normal vector $\boldsymbol{n}$ is purely tangential).

4.3: Suppose the stress components in a plate bounded by $x_1 = \pm L$ and $x_2 = \pm h$ are given by

$$\begin{aligned}T_{11} &= Wm^2\cos(\pi x_1/2L)\sinh(mx_2),\\ T_{22} &= -(W\pi^2/4L^2)\cos(\pi x_1/2L)\sinh(mx_2),\\ T_{12} &= (W\pi m/2L)\sin(\pi x_1/2L)\cosh(mx_2),\\ T_{13} &= T_{23} = T_{33} = 0,\end{aligned}$$

where W and m are constants.
(i) Verify that this stress satisfies the equations of equilibrium with no body force.
(ii) Find the stress on the edges $x_2 = h$ and $x_1 = -L$.

4.4: Suppose the density of a continuous medium is defined by

$$\rho = \rho_0 \exp\left[-\left(x_1^2 + x_2^2 + x_3^2\right)/l^2\right],$$

where ρ_0 and l are constants. The matrix of the stress tensor $\mathbf{T}$ in this medium is given by

$$\hat{\mathbf{T}} = \frac{f}{l^4}\begin{pmatrix} ax_1^2 & 0 & 0 \\ 0 & x_2^2 & 2x_2x_3 \\ 0 & 2x_2x_3 & x_3^2 \end{pmatrix},$$

where f and a is a constant. You are given that the medium is in equilibrium under the action of a body force $\boldsymbol{b}$.
(i) Determine $\boldsymbol{b}$.
(ii) You are given that $\boldsymbol{b}$ is a potential vector meaning that $\nabla \times \boldsymbol{b} = 0$. Determine a.

4.5: A continuous medium occupies the exterior of a solid sphere of radius R. The sphere creates a gravity field with acceleration due to gravity directed to the centre of the sphere and inversely proportional to the square of the distance from it, i.e. in spherical coordinates r, θ, φ with the origin at the centre of the sphere the body force is $\boldsymbol{b} = (-gR^2/r^2, 0, 0)$. You are also given that the stress tensor has the form $\mathbf{T} = -p\mathbf{I}$, where $\mathbf{I}$ is the unit tensor and p the pressure. The pressure p is proportional to the continuum density ρ, $p = \alpha\rho$, and $p = p_0 = \text{const}$ at $r = R$. Determine the dependence of p on r for $r > R$.

4.6: Suppose a charged medium with density $\rho = \text{const}$ and electric charge density $\rho_e = \text{const}$ is in equilibrium in the gravity field with acceleration due to gravity $\boldsymbol{g} = \text{const}$ and electric field $\boldsymbol{E} = \text{const}$. You are given that $\boldsymbol{g}$ is antiparallel to the x_3-axis and $\boldsymbol{E}$ is parallel to the x_2-axis. The body force imposed on the medium by the electric field is equal to $(\rho_e/\rho)\boldsymbol{E}$. You are also given that the matrix of the stress tensor has the form

$$\hat{\mathbf{T}} = \begin{pmatrix} T_1 & 0 & 0 \\ 0 & T_1 & T_3 \\ 0 & T_3 & T_2 \end{pmatrix},$$

$\mathbf{T} = 0$ at $x_3 = 0$, and T_1 has the form $T_1 = qx_2^2x_3$, where q is a given constant. Calculate T_2 and T_3.

Chapter 5
Ideal Incompressible Fluids

5.1 Main Assumptions and Governing Equations

In what follows we apply the general theory developed in the previous chapters to the study of particular models of continuum mechanics. In the previous sections we obtained two equations that are valid for any continuum medium. They are the mass conservation equation and the momentum equation (either Eqs. (3.7.4) and (4.5.9) in the Lagrangian description, or Eqs. (3.8.3) and (4.4.4) in the Eulerian description). These two equations do not constitute a full system of equations for the variables describing continuum motions. Since the momentum equation is a vector equation, we have only four scalar equations. However, there are ten variables describing continuum motion. They are the density ρ, six components of the stress tensor T_{ij}, and either three components of the displacement u_i in the Lagrangian description, or three components of the velocity v_i in the Eulerian description.

To obtain a closed system of equations describing continuum motion we need additional equations. One such equation is the energy equation. However, it does not solve the problem because we still have more variables than equations. We will not use this equation until Chap. 8. Another class of equations that are used to obtain a closed system of equations describing continuum motion are the so-called *constitutive equations*. These equations are related to the constitution of a medium and vary from one medium to another. In this chapter we consider the first example of using constitution equations, which is the model of an *ideal incompressible fluid*.

When an external pressure is applied to a continuum occupying a finite domain its volume decreases. However, for many continuum media the volume decrease is extremely small and can be neglected. Then it is a very good approximation to consider such a medium as incompressible. One such example is water.

In general, when two layers of a fluid slide against each other there is friction between them. This is called viscosity. However, viscosity is only important when the fluid velocity varies on a relatively small spatial scale, otherwise it can be neglected. The exact criterion showing when viscosity can be neglected will be given in Chap. 7. There are also fluids where there is no viscosity at all. They are called superfluids.

© Springer Nature Switzerland AG 2019

M. S. Ruderman, *Fluid Dynamics and Linear Elasticity*, Springer Undergraduate Mathematics Series, https://doi.org/10.1007/978-3-030-19297-6_5

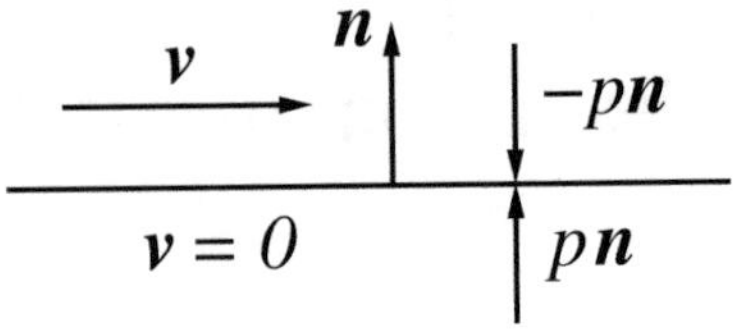

Fig. 5.1 Stresses at a velocity discontinuity in an ideal fluid

Superfluidity occurs in two isotopes of helium (helium-3 and helium-4) when they are liquified by cooling to cryogenic temperatures. A fluid without viscosity is often called ideal.

The model of an *ideal incompressible fluid* is also interesting from a mathematical point of view because it is the simplest model of continuum mechanics. It allows us to use advanced mathematical methods to develop a very elegant description of its motion. The first constitutive equation in the model of an ideal incompressible fluid is the expression for the stress tensor. It is given by

$$\mathbf{T} = -p\,\mathbf{I}. \tag{5.1.1}$$

The quantity p is called the pressure. It follows from Eq. (5.1.1) that $\boldsymbol{t} = \mathbf{T}\cdot\boldsymbol{n} = -p\boldsymbol{n}$, so that the stress in this fluid reduces to pure pressure p. There is no tangential stress: $\boldsymbol{t}_\tau = 0$ (Fig. 5.1). The absence of tangential stresses means that the layers of an ideal fluid can freely slide against each other without friction. Using (5.1.1), we obtain

$$\nabla\cdot\mathbf{T} = \boldsymbol{e}_i\frac{\partial T_{ij}}{\partial x_j} = -\boldsymbol{e}_i\delta_{ij}\frac{\partial p}{\partial x_j} = -\boldsymbol{e}_i\frac{\partial p}{\partial x_i} = -\nabla p.$$

Substituting this result into (4.4.3), we obtain *Euler's equation* (after Swiss scientist Leonard Euler):

$$\boxed{\rho\frac{D\boldsymbol{v}}{Dt} = -\nabla p + \rho\boldsymbol{b}.} \tag{5.1.2}$$

This equation can also be written in the form

$$\boxed{\frac{\partial\boldsymbol{v}}{\partial t} + (\boldsymbol{v}\cdot\nabla)\boldsymbol{v} = -\frac{1}{\rho}\nabla p + \boldsymbol{b}.} \tag{5.1.3}$$

The word "*incompressible*" means that the volume of the material does not change, $V(t) = V(0) = V_0$, meaning that

$$\frac{dV}{dt} = \frac{d}{dt}\int_{V_0}\mathcal{J}\,dV = \int_{V_0}\frac{D\mathcal{J}}{Dt}\,dV = 0.$$

Since this equation is valid for any V_0, it follows that $D\mathcal{J}/Dt = 0$. Then, using Eq. (3.7.4), we obtain

$$\boxed{\frac{D\rho}{Dt} = 0.} \tag{5.1.4}$$

Now it follows from (3.8.3) that

$$\boxed{\nabla \cdot \boldsymbol{v} = 0,} \tag{5.1.5}$$

which is the second constitutive equation. Equation (5.1.4) implies that the density of a small material particle does not change with time. Still ρ can depend on $\boldsymbol{\xi}$. If it depends on $\boldsymbol{\xi}$, then the fluid is called *stratified*. Otherwise it is called *homogeneous*.

In what follows we only consider homogeneous fluids, so that $\rho = \text{const}$. Then Eq. (5.1.2) (or Eqs. (5.1.3)) and (5.1.5) constitute the full system of equations describing the motion of a homogeneous ideal incompressible fluid: there is one vector equation and one scalar equation for one vector variable, $\boldsymbol{v}$, and one scalar variable, p.

Since an ideal fluid can slide along a rigid boundary, only its normal component must vanish at such a boundary, $v_n = 0$ at Γ. With the use of (5.1.1) the dynamic boundary condition (4.6.3) reduces to

$$[\![p]\!] = 0. \tag{5.1.6}$$

5.2 Equilibrium in a Gravitational Field

Let us consider the equilibrium ($\boldsymbol{v} = 0$) of a homogeneous ideal incompressible fluid in a gravitational field with constant acceleration due to gravity $\boldsymbol{g}$, so that $\boldsymbol{b} = \boldsymbol{g}$. We introduce Cartesian coordinates x_1, x_2, x_3 with the x_3-axis antiparallel to $\boldsymbol{g}$, so that $\boldsymbol{g} = -g\boldsymbol{e}_3$, $g = \text{const}$. Then it follows from (5.1.2) that the equation of equilibrium is

$$\nabla p = -\rho g \boldsymbol{e}_3.$$

In particular, this equation implies $\dfrac{\partial p}{\partial x_1} = \dfrac{\partial p}{\partial x_2} = 0$, so that p depends only on x_3. In what follows we use $z = x_3$.

Now we rewrite the equilibrium equation as

$$\frac{dp}{dz} = -\rho g. \tag{5.2.1}$$

We assume that there is a surface separating water from the air, and the air pressure at this surface is $p_a = \text{const}$. In accordance with Eq. (5.1.6) the water pressure is also constant at this surface. Equation (5.2.1) implies that this is only possible if the surface is horizontal. We take this surface as the zero level for z, so that its equation is $z = 0$, and the water occupies the half-space $z < 0$. The dynamic boundary condition is

Fig. 5.2 A submarine. U.S. Navy photo by General Dynamics Electric Boat

$$p = p_a \quad \text{at} \quad z = 0. \tag{5.2.2}$$

The solution to (5.2.1) with the boundary condition (5.2.2) is straightforward:

$$p = p_a - \rho g z, \quad z < 0. \tag{5.2.3}$$

The atmospheric pressure is $p_a \approx 10^5\ \mathrm{Nm}^{-2} = 10^5$ Pa (Pa = Pascal after French scientist Blaise Pascal). Other units used to measure pressure are 1 bar $= 10^5\ \mathrm{Nm}^{-2}$ and 1 millibar $= 10^2\ \mathrm{Nm}^{-2}$ used by meteorologists.

Since $\rho \approx 10^3\ \mathrm{kg\,m}^{-3}$ and $g \approx 10\ \mathrm{ms}^{-2}$,

$$p \approx p_a(1 + 0.1|z|)\ \mathrm{Nm}^{-2}, \tag{5.2.4}$$

where $|z|$ is the depth measured in meters.

Example The force acting on a square meter of the surface of a submarine (see Fig. 5.2) moving 300 m below the sea surface is $31 p_a \approx 300$ ton. In 1960 the US Navy sent the *bathyscaphe Trieste* (a *submersible* – a mini-submarine designed to operate at great depths) down into the Marianas trench, the deepest place in the world. They touched the bottom at 10,915 m. That means, at this deepest point, there was almost 11 km of water over their heads! The pressure at this depth is $p \approx 1.1 \times 10^3 p_a$, so the force acting on each m^2 of the bathyscaphe surface was about 11,000 ton.

5.3 Bernoulli's Integral

We start by proving the identity

$$(\boldsymbol{v}\cdot\nabla)\boldsymbol{v} = \frac{1}{2}\nabla\big(\|\boldsymbol{v}\|^2\big) + (\nabla\times\boldsymbol{v})\times\boldsymbol{v}. \tag{5.3.1}$$

Using the identity (2.9.6) we obtain

$$\frac{1}{2}\nabla\big(\|\boldsymbol{v}\|^2\big) + (\nabla\times\boldsymbol{v})\times\boldsymbol{v} = \boldsymbol{e}_i\left\{\frac{1}{2}\frac{\partial(v_j v_j)}{\partial x_i} + \varepsilon_{ijk}\left(\varepsilon_{jlm}\frac{\partial v_m}{\partial x_l}\right)v_k\right\}$$
$$= \boldsymbol{e}_i\left(v_j\frac{\partial v_j}{\partial x_i} - \varepsilon_{ikj}\varepsilon_{lmj}v_k\frac{\partial v_m}{\partial x_l}\right) = \boldsymbol{e}_i\left(v_j\frac{\partial v_j}{\partial x_i} - (\delta_{il}\delta_{km} - \delta_{im}\delta_{kl})v_k\frac{\partial v_m}{\partial x_l}\right)$$
$$= \boldsymbol{e}_i\left(v_j\frac{\partial v_j}{\partial x_i} - v_k\frac{\partial v_k}{\partial x_i} + v_k\frac{\partial v_i}{\partial x_k}\right) = \boldsymbol{e}_i v_k\frac{\partial v_i}{\partial x_k} = (\boldsymbol{v}\cdot\nabla)\boldsymbol{v}.$$

Using (5.3.1) and assuming $\boldsymbol{b} = -\nabla\varphi$, where φ is the gravity potential, we rewrite Euler's equation (5.1.3) in the *Gromeka–Lamb* form

$$\frac{\partial\boldsymbol{v}}{\partial t} + (\nabla\times\boldsymbol{v})\times\boldsymbol{v} = -\nabla\left(\frac{p}{\rho} + \frac{1}{2}\|\boldsymbol{v}\|^2 + \varphi\right). \tag{5.3.2}$$

We assume that the motion is stationary, i.e. $\dfrac{\partial\boldsymbol{v}}{\partial t} = 0$, and integrate Eq. (5.3.2) along a curve C connecting two points, A and B:

$$\int_C [(\nabla\times\boldsymbol{v})\times\boldsymbol{v}]\cdot d\boldsymbol{l} = -\int_C\left[\nabla\left(\frac{p}{\rho} + \frac{1}{2}\|\boldsymbol{v}\|^2 + \varphi\right)\right]\cdot d\boldsymbol{l}$$
$$= -\int_C \frac{d}{dl}\left(\frac{p}{\rho} + \frac{1}{2}\|\boldsymbol{v}\|^2 + \varphi\right)dl = \left(\frac{p}{\rho} + \frac{1}{2}\|\boldsymbol{v}\|^2 + \varphi\right)\Bigg|_{\mathrm{B}}^{\mathrm{A}}.$$

Now we assume that C is a streamline. Then $d\boldsymbol{l}\parallel\boldsymbol{v}$, $[(\nabla\times\boldsymbol{v})\times\boldsymbol{v}]\cdot d\boldsymbol{l} = 0$, and it follows that

$$\left(\frac{p}{\rho} + \frac{1}{2}\|\boldsymbol{v}\|^2 + \varphi\right)\Bigg|_{\mathrm{A}} = \left(\frac{p}{\rho} + \frac{1}{2}\|\boldsymbol{v}\|^2 + \varphi\right)\Bigg|_{\mathrm{B}}.$$

We have obtained *Bernoulli's integral* (after Swiss scientist Daniel Bernoulli): on any *streamline*

$$\boxed{p + \frac{\rho}{2}\|\boldsymbol{v}\|^2 + \rho\varphi = \text{const.}} \tag{5.3.3}$$

We have to specially emphasise that the right-hand side of Eq. (5.3.3) is constant only on a fixed streamline. In general, it takes different values on different streamlines.

Using Eq. (5.3.3) we can easily re-derive Eq. (5.2.3). Since now $\boldsymbol{v} = 0$, the left-hand side of Eq. (5.3.2) is zero, and we do not even have to assume that $\mathcal{C}$ is a streamline. Then we immediately obtain Eq. (5.3.3) with $\boldsymbol{v} = 0$. Note that in this case the constant on the right-hand side of Eq. (5.3.3) is universal. It is the same in the whole domain occupied by the fluid. If we use the coordinate system described in Sect. 5.2, then $\boldsymbol{g} = -\nabla\varphi$, where $\varphi = gx_3 = gz$, and Eq. (5.3.3) gives $p + g\rho z =$ const. Using the boundary condition $p = p_a$ at $z = 0$, we obtain const $= p_a$ and eventually arrive at Eq. (5.2.3).

Example Suppose there is a straight tube of length L with a circular cross-section. In cylindrical coordinates with the z-axis coinciding with the tube axis the tube cross-section radius is $R(z)$. There is a flow of an ideal incompressible fluid through the tube. The pressure and speed at the tube entrance are p_1 and v_1. The tube is long so that its length is much larger than max $R(z)$. In addition its radius varies slowly, which implies that R' is of the order of $R/L \ll 1$, where the prime indicates the derivative. Calculate the pressure at the tube exit. You can assume that the pressure in the tube is independent of r and ϕ.

Solution: Since there is no body force Bernoulli's integral is written as

$$p + \tfrac{1}{2}\rho v^2 = p_1 + \tfrac{1}{2}\rho v_1^2. \tag{5.3.4}$$

Since p is independent of r and ϕ it follows from this equation that v is also independent of r and ϕ. We assume that the motion is axisymmetric, that is, it is independent of ϕ and $v_\phi = 0$. Then $v_z = v + \mathcal{O}(R^2/L^2)$, which implies that we can neglect the dependence of v_z on r. Now, using Eqs. (2.13.8) and (5.1.5) yields

$$\frac{1}{r}\frac{\partial(rv_r)}{\partial r} + \frac{dv_z}{dz} = 0.$$

Multiplying this equation by r and then integrating with respect to r we obtain

$$Rv_r\big|_{r=R} + \frac{1}{2}R^2\frac{dv_z}{dz} = 0. \tag{5.3.5}$$

The vector $\boldsymbol{e}_r - R'\boldsymbol{e}_z$ is orthogonal to the tube axis. Since the normal component of the velocity must be zero at the tube boundary it follows that the scalar product of this vector and the velocity is zero. This gives $v_r\big|_{r=R} = R'v_z$. Using this result we transform Eq. (5.3.5) into

$$RR'v_z + \frac{1}{2}R^2\frac{dv_z}{dz} = 0. \tag{5.3.6}$$

Integrating this equation, taking $v \approx v_z$, and using the boundary conditions at the tube entrance we obtain

$$R^2 v = R_1^2 v_1, \tag{5.3.7}$$

where R_1 is the cross-section radius at the tube entrance. This equation has a simple physical meaning. The quantity $R^2 v$ is approximately equal to the fluid flux through a tube cross-section orthogonal to its axis. Since the fluid is incompressible this flux is conserved. Substituting Eq. (5.3.7) into Eq. (5.3.4) finally gives

$$p_2 = p_1 + \frac{\rho v_1^2}{2}\left[1 - \left(\frac{R_1}{R_2}\right)^4\right], \tag{5.3.8}$$

where R_2 is the radius of the tube cross-section at its exit. We note that $p_2 > p_1$ when $R_2 > R_1$, while $p_2 < p_1$ when $R_2 < R_1$. This result puts on a rigorous mathematical footing the qualitative analysis of aneurysm development given in Sect. 1.1.

5.4 D'Alembert's Paradox and the Lifting Force

Bernoulli's integral (5.3.3) has been used to explain many important and interesting phenomena. We apply it to explain the *dry leaf free kick* in football, and the *lifting force* keeping airplanes in the air.

The dry leaf free kick is especially impressive when a player manages to score a goal from a corner. The first player who managed to do this was the Argentinian player Cesáreo Onzari in a game against Uruguay in 1924. It is well known that if we throw a stone at some angle with respect to the horizon then it will move in a vertical plane along a parabolic trajectory under the action of gravity. The projection of its trajectory onto the horizontal plane will be a straight line. However, the ball in a dry leaf kick from the corner does not do this. Rather the projection of its trajectory onto the horizontal plane resembles an arc of a circle with the starting point at the corner and the end point somewhere in the goal (see Fig. 5.3). Below we will try to find out what force deviates the ball from the vertical plane.

The description of the motion of a kicked ball is an extremely difficult problem that can only be solved numerically using powerful computers. One of the fundamental approaches of applied mathematics is the following. We are trying to simplify a problem describing a particular phenomenon as much as possible, however we still want to keep its important properties. In line with this approach we first neglect viscosity and consider the air as an inviscid fluid. The second much less obvious simplification is to reduce the three-dimensional problem describing the motion of

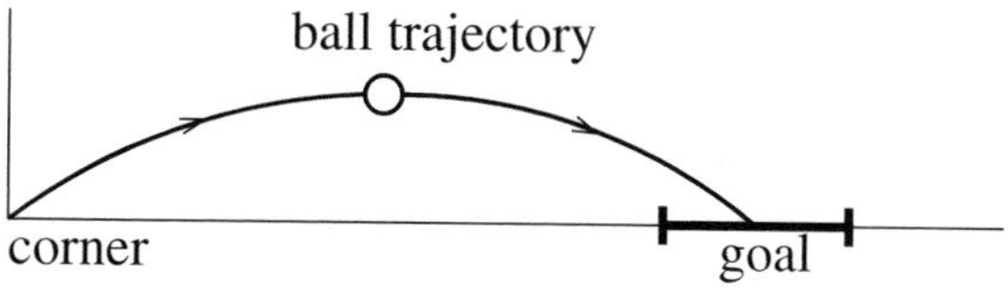

Fig. 5.3 Projection on the horizontal plane of the trajectory of a ball in a dry leaf kick from a corner

a ball to a two-dimensional problem. Rather than studying the motion of a ball we consider the motion of an infinite cylinder.

Hence, the mathematical formulation of the problem that we shall study is the following. Consider a stationary potential flow of an incompressible fluid near a rigid cylinder. The cylinder surface is defined by the equation $x^2 + y^2 = a^2$ in Cartesian coordinates x, y, z. You are given that the flow is invariant in the z-direction, i.e. it is independent of z, and symmetric with respect to the x-axis. You are also given that the flow velocity has magnitude V and it is in the negative x-direction far from the cylinder. Below we use cylindrical coordinates r, ϕ, z with the origin at the centre of the cylinder and the z-axis coinciding with the same axis in Cartesian coordinates.

The normal component of the velocity at the cylinder's surface is

$$v_r = \frac{\partial \Phi}{\partial r},$$

where Φ is the velocity potential. It must be zero at the surface. As a result, we have the boundary condition

$$\frac{\partial \Phi}{\partial r} = 0 \quad \text{at} \quad r = a. \tag{5.4.1}$$

Substituting $\boldsymbol{v} = \nabla\Phi$ into $\nabla \cdot \boldsymbol{v} = 0$ we obtain $\nabla^2\Phi = 0$. Using (2.13.1) and (2.13.8) yields

$$\begin{aligned} \nabla^2\Phi = \nabla \cdot \nabla\Phi &= \frac{1}{r}\frac{\partial}{\partial r} r \frac{\partial \Phi}{\partial r} + \frac{1}{r}\frac{\partial}{\partial \phi}\frac{1}{r}\frac{\partial \Phi}{\partial \phi} + \frac{\partial}{\partial z}\frac{\partial \Phi}{\partial z} \\ &= \frac{1}{r}\frac{\partial}{\partial r} r \frac{\partial \Phi}{\partial r} + \frac{1}{r^2}\frac{\partial^2 \Phi}{\partial \phi^2} + \frac{\partial^2 \Phi}{\partial z^2}. \end{aligned} \tag{5.4.2}$$

Taking into account that Φ is independent of z we obtain

$$r\frac{\partial}{\partial r} r \frac{\partial \Phi}{\partial r} + \frac{\partial^2 \Phi}{\partial \phi^2} = 0. \tag{5.4.3}$$

We now expand Φ in a Fourier series. The condition that the flow is symmetric with respect to the x-axis implies that Φ is an even function of ϕ. Hence,

$$\Phi = \sum_{n=0}^{\infty} \Phi_n(r)\cos(n\phi). \tag{5.4.4}$$

Substituting this expression into (5.4.3) yields

$$\sum_{n=0}^{\infty}[r(r\Phi_n')' - n^2\Phi_n]\cos(n\phi) \quad \Longrightarrow \quad r(r\Phi_n')' - n^2\Phi_n = 0, \tag{5.4.5}$$

Fig. 5.4 Calculation of components of an infinitesimal vector $\boldsymbol{l}$

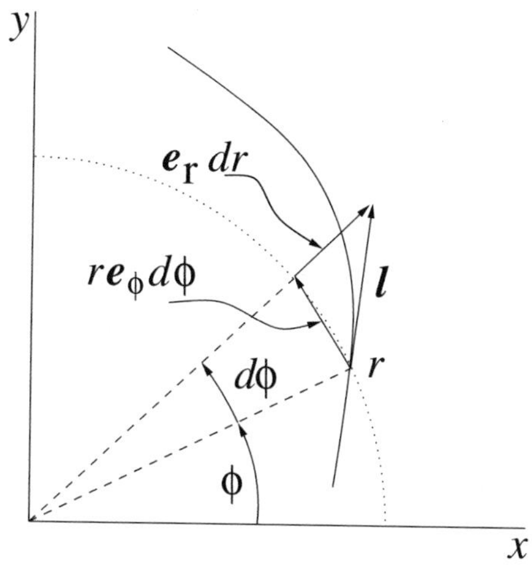

where the prime indicates the derivative with respect to r. The solution is

$$\Phi_0 = A_0 + B_0 \ln r, \quad \Phi_n = A_n r^n + B_n r^{-n}, \quad n = 1, 2, \ldots, \tag{5.4.6}$$

where A_0, B_0, A_n, and B_n are constants. Substituting this result into (5.4.1) we obtain $B_0 = 0$ and

$$n\big(A_n a^{n-1} - B_n a^{-n-1}\big) = 0 \implies B_n = a^{2n} A_n, \quad n = 1, 2, \ldots \tag{5.4.7}$$

Now we use the condition at $x \to -\infty$. This condition can be written as

$$\boldsymbol{v} \to -V\boldsymbol{e}_r \quad \text{as} \quad r \to \infty \quad \text{and} \quad \phi = 0. \tag{5.4.8}$$

Using (5.4.4), (5.4.6), and (5.4.7) yields

$$\boldsymbol{v} = \boldsymbol{e}_r \sum_{n=1}^{\infty} nA_n\big(r^{n-1} - a^{2n}r^{-n-1}\big)\cos(n\phi) - \boldsymbol{e}_\phi \sum_{n=1}^{\infty} nA_n\big(r^{n-1} + a^{2n}r^{-n-1}\big)\sin(n\phi). \tag{5.4.9}$$

Then it follows from (5.4.8) that $A_1 = -V$, $A_n = 0$, $n = 2, 3, \ldots$ The potential Φ is defined up to an additive constant, so we can take $A_0 = 0$. Hence, eventually,

$$\begin{aligned} \Phi &= -V\left(r + \frac{a^2}{r}\right)\cos\phi, \\ \boldsymbol{v} &= -\boldsymbol{e}_r V\left(1 - \frac{a^2}{r^2}\right)\cos\phi + \boldsymbol{e}_\phi V\left(1 + \frac{a^2}{r^2}\right)\sin\phi. \end{aligned} \tag{5.4.10}$$

Example Find the equation of the streamlines of the flow with velocity given by Eq. (5.4.10).

Solution: A tangential vector to a curve in polar coordinates r, ϕ is $\boldsymbol{e}_r\,dr + \boldsymbol{e}_\phi\,r\,d\phi$ (see Fig 5.4). Then the equation determining the streamlines is

$$\frac{dr}{v_r} = \frac{r\,d\phi}{v_\phi}. \tag{5.4.11}$$

Using (5.4.10) we rewrite this equation as

$$\frac{dr}{d\phi} = -r\,\frac{r^2 - a^2}{r^2 + a^2}\cot\phi.$$

Separating variables yields

$$\frac{r^2 + a^2}{r^2 - a^2}\frac{dr}{r} = -\frac{\cos\phi}{\sin\phi}d\phi. \tag{5.4.12}$$

It is straightforward to obtain

$$\int \frac{r^2 + a^2}{r^2 - a^2}\frac{dr}{r} = \ln\frac{r^2 - a^2}{r} + \text{const}, \quad \int \frac{\cos\phi}{\sin\phi}d\phi = \ln|\sin\phi| + \text{const}. \tag{5.4.13}$$

Using these results we obtain from (5.4.12)

$$\left(\frac{r}{a}\right)^2 - 2C\left(\frac{r}{a}\right)|\sin\phi|^{-1} - 1 = 0,$$

where C is a positive constant. The positive solution to this quadratic equation is

$$\frac{r}{a} = C|\sin\phi|^{-1} + \sqrt{1 + C^2\sin^{-2}\phi}, \quad C > 0. \tag{5.4.14}$$

The streamlines described by this equation are shown in Fig. 5.5.

Definition Let C be a simple closed contour. The quantity $\Gamma = \oint_C \boldsymbol{v}\cdot d\boldsymbol{l}$ is called the *velocity circulation* around the contour C.

It can be shown (see Problem **5.2**) that Γ is the same for any contour enclosing the cylinder. Hence, we can take C to be a circle $r = a$. Then $d\boldsymbol{l} = a\boldsymbol{e}_\phi\,d\phi$ and

$$\Gamma = \int_0^{2\pi} a v_\phi\,d\phi = 2aV\int_0^{2\pi}\sin\phi\,d\phi = -2aV\cos\phi\Big|_0^{2\pi} = 0.$$

Now we calculate the force acting on the cylinder per unit length along its axis. It follows from Bernoulli's integral with $\varphi = 0$ that

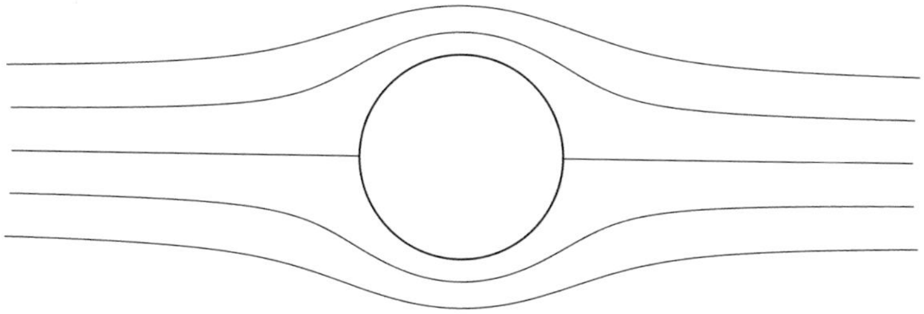

Fig. 5.5 Streamlines of a potential flow near a cylinder. They are described by Eq. (5.4.14)

$$p + \frac{\rho}{2}\|\boldsymbol{v}\|^2 = p_0 + \frac{\rho}{2}V^2, \tag{5.4.15}$$

where we have assumed that $p \to p_0$ as $r \to \infty$. Using Eq. (5.4.10) we obtain

$$\|\boldsymbol{v}\|^2 = V^2\left[\left(1 - \frac{a^2}{r^2}\right)^2 \cos^2\phi + \left(1 + \frac{a^2}{r^2}\right)^2 \sin^2\phi\right] = V^2\left(1 + \frac{a^4}{r^4} - \frac{2a^2}{r^2}\cos 2\phi\right).$$

Then at the cylinder's surface ($r = a$)

$$p = p_0 + \frac{\rho}{2}V^2 - \rho V^2(1 - \cos 2\phi) = p_0 - \frac{\rho}{2}V^2(1 - 2\cos 2\phi).$$

The force acting on an elementary surface of the cylinder between ϕ and $\phi + d\phi$ is $d\boldsymbol{f} = -ap\boldsymbol{e}_r\,d\phi$, so $df_x = -ap\cos\phi\,d\phi$ and $df_y = -ap\sin\phi\,d\phi$. Then the x-component of the total force acting on the cylinder is

$$f_x = -a\int_0^{2\pi} p\cos\phi\,d\phi = -a\int_0^{2\pi}\left[\left(p_0 - \frac{\rho}{2}V^2\right)\cos\phi + \rho V^2\cos\phi\cos 2\phi\right]d\phi.$$

Using $2\cos\alpha\cos\beta = \cos(\alpha+\beta) + \cos(\alpha-\beta)$ we eventually obtain

$$f_x = -a\int_0^{2\pi}\left(p_0\cos\phi + \frac{\rho}{2}V^2\cos 3\phi\right)d\phi = -a\left(p_0\sin\phi + \frac{\rho}{6}V^2\sin 3\phi\right)\Big|_0^{2\pi} = 0.$$

In a similar way it can be shown that $f_y = 0$. Hence, the force acting on the cylinder is zero. This result obviously does not match our everyday experience. We know that when swimming or rowing we need to make great effort to keep moving forward. In these examples we still might think that our energy is only being spent to create surface waves. But it is also well known that a submarine completely immersed in water needs a powerful engine to more forward. If the engine stops then in a short time the submarine will come to rest. Hence, in the real world a body moving in water experiences resistance. This is why the result that a cylinder moving with a constant velocity in an ideal fluid does not experience any resistance is called *d'Alembert's paradox*.

In fact, the resistance that appears when a body is moving in water is related to viscosity. Does this mean that we need to take viscosity into account, or it is still possible to explain the dry leaf phenomenon using the model of an ideal fluid? Let us consider what we missed when formulating the problem. A thorough examination of films showing a dry leaf kick reveals that the player kicks a ball not at its centre but on the side, making it rotate. Hence, could we succeed in explaining the dry leaf phenomenon if we take this rotation into account? If we simply assume that the cylinder is rotating then nothing will change. Indeed, an ideal fluid can slide across the cylinder's surface without friction, so it will simply not "notice" that the cylinder is rotating. However, in reality air is viscous, meaning that the air velocity at the surface of the cylinder must coincide with the surface velocity. Hence, the rotating ball will involve the nearby air in the rotation.

To mimic the air rotation related to viscosity but still enjoy the simplicity of the model of an ideal fluid we assume that the water rotates from the very beginning. To add water rotation to our original model we consider another solution to Eq. (5.4.3): $\Phi = C\phi$, where C is a constant. This expression can also be considered as a potential of a flow. Note that it is multivalued. The corresponding velocity is $\boldsymbol{v} = (C/r)\boldsymbol{e}_\phi$. We have $\Gamma = 2\pi a(C/a) = 2\pi C$. Hence $\Phi = (\Gamma/2\pi)\phi$. Equation (5.4.3) is linear, meaning, in particular, that the sum of two solutions is again a solution. Hence we can take the potential

$$\Phi = -V\left(r + \frac{a^2}{r}\right)\cos\phi + \frac{\Gamma\phi}{2\pi}. \tag{5.4.16}$$

It corresponds to the velocity

$$\boldsymbol{v} = -\boldsymbol{e}_r V\left(1 - \frac{a^2}{r^2}\right)\cos\phi + \boldsymbol{e}_\phi\left[V\left(1 + \frac{a^2}{r^2}\right)\sin\phi + \frac{\Gamma}{2\pi r}\right]. \tag{5.4.17}$$

Example Find the equation of the streamlines of the flow with velocity given by Eq. (5.4.17).

Solution: Using Eq. (5.4.11) we obtain that the equation of the streamlines is

$$\frac{d\phi}{dr} = -\frac{qr + (r^2 + a^2)\sin\phi}{r(r^2 - a^2)\cos\phi}, \quad q = \frac{\Gamma}{2\pi V}.$$

The variable substitution $s = \sin\phi$ reduces this equation to

$$\frac{ds}{dr} + \frac{(r^2 + a^2)s}{r(r^2 - a^2)} = -\frac{q}{r^2 - a^2}. \tag{5.4.18}$$

Using Eq. (5.4.13) we obtain that the integration factor for this equation is $\left(r^2 - a^2\right)/r$. Multiplying Eq. (5.4.18) by the integration factor we reduce it to

$$\frac{d}{dr}\left(\frac{r^2 - a^2}{r}s\right) = -\frac{q}{r}.$$

It follows from this equation that the equation of the streamlines is

$$\sin\phi = \frac{Cr - qr\ln(r/a)}{r^2 - a^2}. \tag{5.4.19}$$

The streamlines defined by Eq. (5.4.19) are shown in Fig. 5.6.

Let us determine the coordinates of the two points on the cylinder surface where $\boldsymbol{v} = 0$. They are on the two streamlines that end at the cylinder surface. The equations of these streamlines are given by Eq. (5.4.19). Since at these points $r = a$, it follows that the numerator in Eq. (5.4.19) is zero at $r = a$, meaning that $C = 0$. Then, using L'Hôpital's rule, we obtain

$$\sin\phi = -q\lim_{r\to a}\frac{r\ln(r/a)}{r^2 - a^2} = -q\lim_{r\to a}\frac{\ln(r/a) + 1}{2r} = -\frac{q}{2a}.$$

Hence, $\phi = -\alpha$ or $\phi = \pi + \alpha$, where $\alpha = \arcsin(q/2a)$. We see that these points only exist if $q < 2a$. In this case the streamlines are shown in the upper panel in Fig. 5.6. When $q = 2a$ they merge into one point. In this case the streamlines are shown in the middle panel in Fig. 5.6. And finally, if $q > 2a$ then there are no points on the cylinder surface where $\boldsymbol{v} = 0$. In this case the streamlines are shown in the lower panel in Fig. 5.6.

We now calculate the force acting on the cylinder. It follows from Eq. (5.4.17) that $\|\boldsymbol{v}\| \to V$ as $r \to \infty$. This implies that Eq. (5.4.15) remains valid. Using Eq. (5.4.17) we obtain

$$\begin{aligned}
\|\boldsymbol{v}\|^2 &= V^2\left(1 - \frac{a^2}{r^2}\right)^2\cos^2\phi + \left[V\left(1 + \frac{a^2}{r^2}\right)\sin\phi + \frac{\Gamma}{2\pi r}\right]^2 \\
&= V^2\left[\left(1 + \frac{a^4}{r^4} - \frac{2a^2}{r^2}\cos 2\phi\right) + \frac{2q}{r}\left(1 + \frac{a^2}{r^2}\right)\sin\phi + \frac{q^2}{r^2}\right].
\end{aligned}$$

Then it follows from Eq. (5.4.15) that at the surface of the cylinder ($r = a$)

$$p = p_0 - \rho V^2\left(\frac{1}{2} + \frac{q^2}{2a^2} - \cos 2\phi + \frac{2q}{a}\sin\phi\right).$$

Again the force acting on an elementary surface of the cylinder between ϕ and $\phi + d\phi$ is $d\boldsymbol{f} = -ap\boldsymbol{e}_r\,d\phi$, so $df_x = -ap\cos\phi\,d\phi$ and $df_y = -ap\sin\phi\,d\phi$. Then, similar to the case without circulation, we obtain that the x-component of the total force acting on the cylinder is $f_x = 0$.

For the y-component of the force we obtain

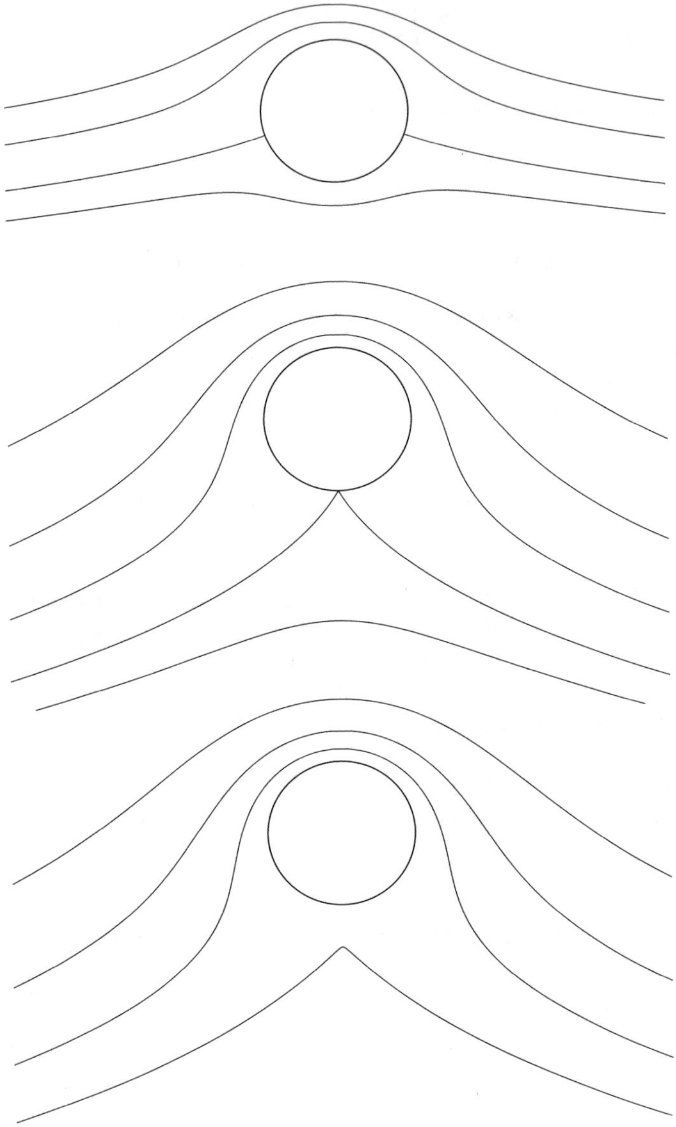

Fig. 5.6 Streamlines of potential flow with circulation near a cylinder defined by Eq. (5.4.19). The upper, middle, and lower panels correspond to $q < 2a$, $q = 2a$, and $q > 2a$, respectively

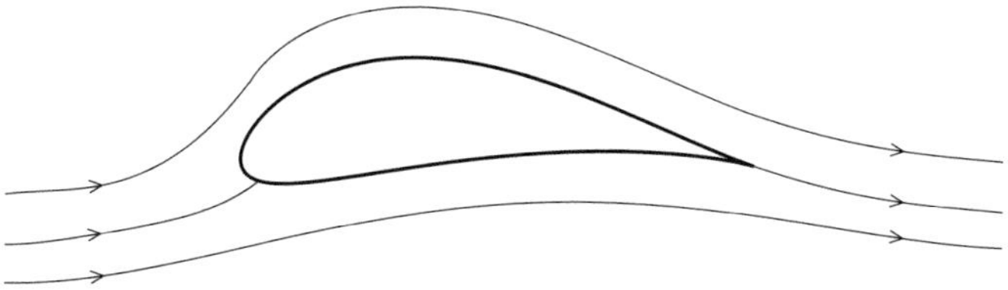

Fig. 5.7 Streamlines of air flow near an airplane wing

$$
\begin{aligned}
f_y &= -a\int_0^{2\pi} p\sin\phi\,d\phi = -a\int_0^{2\pi}\left[\left(p_0 - \frac{\rho}{2}V^2 - \frac{\rho q^2}{2a^2}V^2\right)\sin\phi\right.\\
&\quad \left. + \rho V^2\sin\phi\cos 2\phi - \frac{2\rho q}{a}V^2\sin^2\phi\right]d\phi\\
&= a\int_0^{2\pi}\left[\frac{\rho q}{a}V^2 - \left(p_0 - \rho V^2 - \frac{\rho q^2}{2a^2}V^2\right)\sin\phi - \rho V^2\left(\frac{1}{2}\sin 3\phi + \frac{q}{a}\cos 2\phi\right)\right]d\phi\\
&= a\left[\frac{\rho q}{a}V^2\phi + \left(p_0 - \rho V^2 - \frac{\rho q^2}{2a^2}V^2\right)\cos\phi + \frac{\rho}{6}V^2\cos 3\phi - \frac{\rho q}{2a}V^2\sin 2\phi\right]_0^{2\pi}\\
&= 2\pi\rho q V^2 = \rho V\Gamma.
\end{aligned}
\tag{5.4.20}
$$

We see that there is a net force acting in the direction perpendicular to the cylinder velocity. This force deviates the cylinder from its path along a straight line. Note that, since the force is orthogonal to the cylinder velocity, its work on the cylinder is zero and, while it changes the direction of the cylinder velocity, it cannot decrease its speed. Hence, there is still no water resistance to the cylinder motion and we still have d'Alembert's paradox. However, our result explains the phenomenon of the dry leaf free kick. The force acting in the direction perpendicular to the velocity deviates the cylinder trajectory from a straight line. The same occurs with a real ball and, as a result, the projection of its trajectory onto the horizontal plane becomes curved.

Now we can explain why airplanes fly. In Fig. 5.7 you see a typical airfoil of an old propeller-type airplane together with the streamlines of the air flow near it.

The flow is stationary in the reference frame moving together with the wing, so we can use Bernoulli's integral. Now we assume that far to the left of the wing the flow is not perturbed and has constant horizontal velocity with magnitude v_0 and constant pressure p_0. Then Eq. (5.4.15) is valid. Calculations show that the flow speed near the upper wing boundary is larger than that near the lower boundary. As a result, the total pressure force exerted on the upper boundary and directed downwards is smaller than the total pressure force exerted on the lower boundary and directed upward. Hence, the net force exerted on the wing, which is the difference between these two forces, is directed upward. This force is the *lifting force* supporting an airplane when it flies.

In general the flow near the wing can only be calculated numerically, even when we use the approximation of an ideal incompressible fluid. However there is a two-parameter family of airfoils when the flow can be calculated analytically using the theory of functions of complex variables. These particular airfoils are called *Joukowsky airfoils* after the Russian scientist Nikolai Zhukovsky who first suggested them. When considering the flow near an infinite cylinder we chose the circulation Γ arbitrarily. However, we cannot choose it arbitrarily when studying the flow near a Joukowsky airfoil. Zhukovsky introduced a natural postulate that the flow velocity must be finite everywhere. It is now called the *Joukowsky postulate*. A Joukowsky airfoil has a cusp at its trailing edge. It turns out that there is exactly one value of the circulation Γ such that the velocity at the cusp point is finite, while it is infinite for any other values of Γ. As a result, for any Joukowsky airfoil we can uniquely determine the lifting force.

Problems

5.1: Show that for a potential flow the circulation $\Gamma = \oint_C \boldsymbol{v} \cdot d\boldsymbol{l}$ is the same for any contour enclosing the cylinder.

5.2: Show that $f_y = 0$ in the potential flow without circulation near the cylinder. [Hint: Use $2 \sin\alpha \cos\beta = \sin(\alpha + \beta) + \sin(\alpha - \beta)$.]

5.3: You are given that the fluid motion is potential, $\boldsymbol{v} = \nabla\Phi$. Use Euler's equation written in the Gromeka–Lamb form to derive the Cauchy–Lagrange integral

$$\frac{\partial \Phi}{\partial t} + \frac{p}{\rho} + \frac{1}{2}\|\nabla\Phi\|^2 + \varphi = C(t),$$

where φ is the body force potential and $C(t)$ is an arbitrary function.

5.4: A bathysphere (a manned submersible attached to a ship by a steel rope, designed for deep sea observation) has the shape of a sphere of radius $r = 1$ m. It is immersed at a depth of $h = 2$ km. What is the pressure force exerted on the bathysphere?

5.5: Consider a tank in the form of cube with side $H = 1$ m. The tank is opened from above and filled with water up to the top. There is a circular hole of radius $r = 1$ cm at the bottom of the tank, so that the water leaks through this hole.

(i) Considering the flow as approximately stationary and water as an ideal incompressible fluid, show that

$$v^2 = 2gh,$$

where $v = \|\boldsymbol{v}\|$ is the water speed at the hole, and h is the water depth in the tank. Derive this formula in two different ways: (a) Using Bernoulli's integral; (b) using the Lagrange–Cauchy integral.

(ii) Determine the period of time after which only half of the initial volume of water remains in the tank.

5.6: Prove Archimedes' law: the pressure force exerted on the surface of a body immersed in water is in the vertical direction, and its magnitude is equal to the weight of water displaced by the body.

5.7: A bathysphere has the shape of a sphere of radius $r = 1$ m. It is completely immersed in water and attached to a ship by a steel rope. The weight of the bathysphere is 6×10^4 N. What is the tension of the rope?

5.8: A fluid motion is called planar when, in Cartesian coordinates $x,\ y,\ z$, the z-component of the fluid velocity is zero while the x and y-component are independent of z.

(i) Show that the velocity of a planar motion of an incompressible fluid can be expressed in terms of the flux function Ψ as

$$v_x = \frac{\partial \Psi}{\partial y}, \quad v_y = -\frac{\partial \Psi}{\partial x}.$$

(ii) Show that streamlines of a planar motion of an incompressible fluid are defined by the equation

$$\Psi(x, y) = \text{const.}$$

(iii) A fluid motion is called potential when $\boldsymbol{v} = \nabla\Phi$, where Φ is called the potential. Show that the potential of a planar motion of an ideal incompressible fluid satisfies the Laplace equation

$$\nabla^2\Phi = 0.$$

5.9: Consider a potential flow of an incompressible fluid near a rigid sphere. The surface of the sphere is defined by the equation $x^2 + y^2 + z^2 = a^2$ in Cartesian coordinates $x,\ y,\ z$. You are given that the flow is axisymmetric, i.e. it is independent of ϕ in spherical coordinates $r,\ \theta,\ \phi$ with the angle θ measured from the positive z-axis. You are also given that the flow velocity has magnitude V and is in the negative z-direction far from the sphere.

(i) Write down the boundary condition for the velocity at the surface of the sphere in terms of the velocity potential Φ.

(ii) Find the expression for the velocity potential Φ in spherical coordinates. [Hint: Look for the solution in the form $\Phi = R(r)\cos\theta$, where $R(r)$ is the function to be determined.]

(iii) Use the Bernoulli integral to calculate the pressure at the surface of the sphere. Calculate the total force acting on the sphere. Thus prove d'Alembert's paradox: The drag force acting on the sphere is zero.

5.10: (Landau and Lifshitz (1987), see Further Reading). A rigid sphere is immersed in an ideal incompressible fluid of density ρ. At the initial time it starts to expand, so that its radius is $R(t)$. At a large distance r from the sphere the water pressure $p_0 = \text{const}$. There is no body force. You can assume that the water motion caused by the sphere expansion is potential and axisymmetric, and the water speed is zero at a large distance from the sphere.

(i) Use the mass conservation equation in Eulerian variables to calculate the potential of the water velocity.

(ii) Use the Lagrange–Cauchy integral derived in problem **5.3** to calculate the force acting on the surface of the sphere.

5.11: Consider a spherical shell made of an elastic material. The shell is filled with air at a pressure of $p_0 = 6 \times 10^5$ N/m^2. The radius of the shell is $r_0 = 0.2$ m. There is a weight of mass $m = 30$ kg attached to the shell. At the initial time the shell is anchored at the bottom of a water basin of depth $h_0 = 40$ m. Then it is released and starts to move upward under the action of Archimedes' force. You are given that the volume of the shell is proportional to the difference between the internal pressure of the air in the shell and the external pressure, and the pressure of the air inside the shell is inversely proportional to its volume.

(i) Find the radius of the shell at depth h. (You can assume that the motion is quasi-static and the shell preserves its spherical shape during the motion. You can also take $g = 10$ m/s^2, the water density $\rho = 10^3$ kg/m^3, and the atmospheric pressure $p_a = 10^5$ N/m^2.)

(ii) Derive the equation for the depth $h(t)$ at which the shell will be at time t (but do not try to solve it). Thus calculate the acceleration of the shell at the moment when it reaches the water surface. (You can neglect the weight of the spherical shell, and the water resistance to its motion.)

5.12: A sphere of radius R and mass M is immersed in an ideal incompressible fluid of density ρ. It is moving with constant acceleration $\boldsymbol{a} = a\boldsymbol{e}_z$ under the action of a constant force $\boldsymbol{F} = F\boldsymbol{e}_z$, where $\boldsymbol{e}_z$ is the unit vector of the z-axis of Cartesian coordinates x, y, z. The quantity $F/a - M$ is called the *added mass*. Calculate the added mass assuming that the fluid flow is potential.

Chapter 6
Linear Elasticity

6.1 Main Assumptions and Governing Equations

The next model that we consider is *linear elasticity*. A material is called *elastic* if the stress tensor is a function of the deformation-gradient tensor (which, for brevity, we will later call simply the deformation tensor) and, possibly, such parameters as the temperature, concentration of different admixtures and so on. In what follows we only consider the simplest case where the stress tensor is only a function of the deformation tensor, $\mathbf{T} = \mathbf{T}(\mathbf{F})$, or, in components, $T_{ij} = T_{ij}(F_{kl})$. In particular, this implies that when external forces are removed, the deformed solid completely regains its initial shape. Of course, this is an approximation. Real solid materials are not exactly elastic. Usually, after forces causing deformation are removed some small residual deformations are left. However, they are often so small that they can be neglected.

In the *linear theory of elasticity* it is assumed that $\mathbf{T}$ only depends on the *symmetric part* of $\mathbf{F}$, which is $\frac{1}{2}(\mathbf{F} + \mathbf{F}^T)$. Since $\frac{1}{2}(\mathbf{F} + \mathbf{F}^T) = \mathbf{I} + \mathbf{E}$, this assumption is equivalent to the assumption that $\mathbf{T}$ is a function of $\mathbf{E}$, $\mathbf{T} = \mathbf{T}(\mathbf{E})$, or, in components, $T_{ij} = T_{ij}(E_{kl})$ (recall that $\mathbf{E}$ is the infinitesimal strain tensor, see (3.4.5)). It is also assumed that there are no stresses when $\mathbf{E} = 0$, so that $\mathbf{T}(0) = 0$. Let us use this condition and expand the functions $T_{ij}(E_{kl})$ in a Taylor series:

$$T_{ij} = A_{ijkl}E_{kl} + B_{ijklmn}E_{kl}E_{mn} + \dots \tag{6.1.1}$$

where

$$A_{ijkl} = \frac{\partial T_{ij}}{\partial E_{kl}}, \quad B_{ijklmn} = \frac{\partial^2 T_{ij}}{\partial E_{kl}\partial E_{mn}}, \quad \dots$$

The last assumption made in linear elasticity is that the quantities $\partial u_i/\partial \xi_j$ are small, which implies that $\mathbf{E}$ is small and we can only retain the first term in (6.1.1). Then we arrive at *Hooke's law*:

$$T_{ij} = A_{ijkl}E_{kl}. \tag{6.1.2}$$

© Springer Nature Switzerland AG 2019
M. S. Ruderman, *Fluid Dynamics and Linear Elasticity*, Springer Undergraduate Mathematics Series, https://doi.org/10.1007/978-3-030-19297-6_6

In the new coordinates

$$A'_{ijkl} = \frac{\partial T'_{ij}}{\partial E'_{kl}} = a_{im}a_{jn}\frac{\partial T_{mn}}{\partial E_{pq}}\frac{\partial E_{pq}}{\partial E'_{kl}},$$

$E_{pq} = \tilde{a}_{pu}\tilde{a}_{qv}E'_{uv} = a_{up}a_{vq}E'_{uv}$, where $\tilde{a}_{ij}$ are the entries of the matrix inverse to the transformation matrix from the old to new coordinates. Then

$$\frac{\partial E_{pq}}{\partial E'_{kl}} = a_{up}a_{vq}\delta_{ku}\delta_{lv} = a_{kp}a_{lq},$$

so, eventually,

$$A'_{ijkl} = a_{im}a_{jn}a_{kp}a_{lq}\frac{\partial T_{mn}}{\partial E_{pq}} = a_{im}a_{jn}a_{kp}a_{lq}A_{mnpq}. \tag{6.1.3}$$

Let us introduce the fourth-order tensor **A** with components A_{ijkl} in the old coordinates. Then Eq. (6.1.3) implies that A'_{ijkl} are components of **A** in any new coordinates. Hence, our final conclusion is that A_{ijkl} in (6.1.2) are the components of a fourth-order tensor **A**. Equation (6.1.2) is the general form of a constitutive equation in linear elasticity.

Substituting Eq. (6.1.2) into Eq. (4.5.9) we obtain that Eqs. (3.7.4) and (4.5.9) constitute a closed system of equations for ρ and $\boldsymbol{u}$.

6.2 Hooke's Law for Isotropic Materials

In accordance with Eq. (6.1.2) the relation between tensors **E** and **T** involves the fourth-order tensor **A**. Hence, in general it is defined by the 81 components of the tensor **A**, i.e. it contains 81 parameters. It is then not surprising that, in general, it is extremely complex. We now consider isotropic materials, which are materials with the same properties in any direction. This, in particular, implies that the tensor **A** has the same components in any Cartesian coordinates. This condition imposes a very severe restriction on **A**. We will see in what follows that, as a result, the 81 components of **A** are expressed in term of only two scalar quantities.

To obtain the general expression for **A** for isotropic materials we, first of all, choose the coordinate system with axes in the principal directions of the tensor **E**. Then

$$T_{ij} = A_{ij11}E_1 + A_{ij22}E_2 + A_{ij33}E_3, \tag{6.2.1}$$

where $E_1 = E_{11}$, $E_2 = E_{22}$, and $E_3 = E_{33}$, while $E_{ij} = 0$ for $i \neq j$.

Consider the coordinate system rotated about the x_3-axis by the angle ϕ (see Fig. 6.1). The coordinate transformation is given by

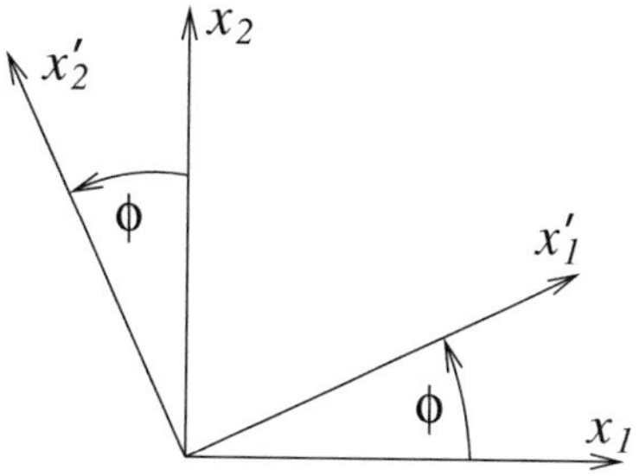

Fig. 6.1 Rotation of the coordinate system about the x_3-axis

$$x_1' = x_1 \cos\phi + x_2 \sin\phi$$
$$x_2' = -x_1 \sin\phi + x_2 \cos\phi. \tag{6.2.2}$$

The third coordinate remains the same, $x_3' = x_3$. Let us take $\phi = \pi$, so that $x_1' = -x_1$ and $x_2' = -x_2$. This implies that $a_{11} = a_{22} = -1, a_{33} = 1$, and $a_{ij} = 0\ (i \neq j)$. Then it follows from (6.1.3) that

$$A_{31[k][k]} = a_{3m}a_{1n}a_{[k]p}a_{[k]q}A_{mnpq} = a_{33}a_{11}a_{[k][k]}a_{[k][k]}A_{31[k][k]} = -A_{31[k][k]}.$$

This result implies that $A_{31[k][k]} = 0,\ k = 1, 2, 3$ (recall that the square brackets at a repeating index indicate that there is no summation with respect to this index). We wrote $A_{31[k][k]}$ instead of $A'_{31[k][k]}$ on the left-hand of this equation because, due to isotropy, $A'_{31[k][k]} = A_{31[k][k]}$. Similarly

$$A_{32[k][k]} = a_{33}a_{22}a_{[k][k]}a_{[k][k]}A_{32[k][k]} = -A_{32[k][k]},$$

implying that $A_{32[k][k]} = 0,\ k = 1, 2, 3$.

In the same way, we use the rotation by $\phi = \pi$ about the x_2-axis to prove that $A_{21[k][k]} = A_{23[k][k]} = 0$, and the rotation by $\phi = \pi$ about the x_1-axis to prove that $A_{12[k][k]} = A_{13[k][k]} = 0$ for $k = 1, 2, 3$. Summarising, we obtain that

$$A_{ij[k][k]} = 0 \quad \text{for} \quad i \neq j \quad \text{and} \quad k = 1, 2, 3. \tag{6.2.3}$$

Now we rotate the coordinate system by the angle $\phi = \pi/2$ about the x_3-axis. As a result we obtain $x_1' = x_2$, $x_2' = -x_1$, and $x_3' = x_3$, so that $a_{12} = 1$, $a_{21} = -1$, and $a_{33} = 1$, while all other entries of the transformation matrix are equal to zero. Then it follows from (6.1.3) that

$$\begin{cases} A_{1111} = a_{1m}a_{1n}a_{1p}a_{1q}A_{mnpq} = a_{12}^4 A_{2222} = A_{2222}, \\ A_{1122} = a_{1m}a_{1n}a_{2p}a_{2q}A_{mnpq} = a_{12}^2 a_{21}^2 A_{2211} = A_{2211}, \\ A_{1133} = a_{1m}a_{1n}a_{3p}a_{3q}A_{mnpq} = a_{12}^2 a_{33}^2 A_{2233} = A_{2233}, \\ A_{3311} = a_{3m}a_{3n}a_{1p}a_{1q}A_{mnpq} = a_{33}^2 a_{12}^2 A_{3322} = A_{3322}. \end{cases} \tag{6.2.4}$$

Finally, we rotate by $\pi/2$ about the x_2-axis. As a result, we obtain $x_1' = -x_3$, $x_2' = x_2$, and $x_3' = x_1$, so that $a_{13} = -1$, $a_{22} = 1$, $a_{31} = 1$, while all other entries of the transformation matrix are equal to zero. Then it follows from (6.1.3) that

$$\begin{cases} A_{1111} = a_{1m}a_{1n}a_{1p}a_{1q}A_{mnpq} = a_{13}^4 A_{3333} = A_{3333}, \\ A_{1122} = a_{1m}a_{1n}a_{2p}a_{2q}A_{mnpq} = a_{13}^2 a_{22}^2 A_{3322} = A_{3322}, \\ A_{1133} = a_{1m}a_{1n}a_{3p}a_{3q}A_{mnpq} = a_{13}^2 a_{31}^2 A_{3311} = A_{3311}. \end{cases} \tag{6.2.5}$$

It follows from (6.2.4) and (6.2.5) that

$$A_{1111} = A_{2222} = A_{3333} = 2\mu + \lambda, \tag{6.2.6}$$

$$A_{1122} = A_{2211} = A_{1133} = A_{3311} = A_{2233} = A_{3322} = \lambda. \tag{6.2.7}$$

The quantities μ and λ are called the *Lamé constants* after the outstanding French mathematician Gabriel Lamé who first introduced them. We see that the 81 components of the tensor $\mathbf{A}$ are expressed in terms of only two quantities, μ and λ.

Other important parameters of an elastic material are *Poisson's ratio* ν and *Young's modulus of elasticity* E introduced by two renowned scientists, French mathematician, engineer, and physicist Siméon Denis Poisson, and British mathematician and physicist Thomas Young. They are given by

$$\nu = \frac{\lambda}{2(\lambda + \mu)}, \quad E = \mu\frac{3\lambda + 2\mu}{\lambda + \mu}. \tag{6.2.8}$$

These two parameters have a simple physical interpretation. If we compress an elastic cylinder in the axial direction, then the ratio of its relative expansion in the radial direction to its relative contraction in the axial direction is equal to ν. If we extend an elastic cylinder in the axial direction, then the ratio of its relative contraction in the radial direction to its relative expansion in the axial direction is also equal to ν. If the stress at the flat surface of the cylinder is equally distributed then in both cases of extension or contraction its ratio to the relative change of the cylinder length is equal to E (see problems 6.5 and 6.6).

Using Eqs. (6.2.3), (6.2.6), and (6.2.7) we obtain from Eq. (6.2.1) that

$$T_{ij} = 0 \text{ for } i \neq j, \tag{6.2.9}$$

$$\begin{cases} T_{11} = (2\mu + \lambda)E_1 + \lambda E_2 + \lambda E_3, \\ T_{22} = \lambda E_1 + (2\mu + \lambda)E_2 + \lambda E_3, \\ T_{33} = \lambda E_1 + \lambda E_2 + (2\mu + \lambda)E_3. \end{cases} \tag{6.2.10}$$

The last equation can be rewritten as

$$T_{[i][i]} = \lambda I_1(\mathbf{E}) + 2\mu E_i,$$

where $I_1(\mathbf{E})$ is the first invariant of the tensor $\mathbf{E}$ given by $I_1(\mathbf{E}) = E_1 + E_2 + E_3$. Using this equation and Eq. (6.2.9) we eventually obtain

$$T_{ij} = \lambda I_1(\mathbf{E})\delta_{ij} + 2\mu E_{ij} \tag{6.2.11}$$

(recall that $E_{ij} = 0$ for $i \neq j$). In particular, it follows from Eqs. (6.2.9) and (6.2.10) that the principal axes of the tensor $\mathbf{T}$ coincide with the principal axes of the tensor $\mathbf{E}$.

We derived Eq. (6.2.11) in a special coordinate system with the axes in the principal directions of the tensor $\mathbf{E}$. However, now we see that both sides of Eq. (6.2.11) are components of second-order tensors. If their components coincide in one coordinate system, then they coincide in any coordinate system. Hence, Eq. (6.2.11) holds in any coordinate system. It can be rewritten as the relation between tensors:

$$\boxed{\mathbf{T} = \lambda I_1(\mathbf{E})\mathbf{I} + 2\mu\mathbf{E}.} \tag{6.2.12}$$

Using the expression for $\mathbf{E}$ (see (3.4.2) and (3.4.5)) we also can rewrite (6.2.11) as

$$\boxed{T_{ij} = \lambda\delta_{ij}\frac{\partial u_k}{\partial \xi_k} + \mu\left(\frac{\partial u_i}{\partial \xi_j} + \frac{\partial u_j}{\partial \xi_i}\right).} \tag{6.2.13}$$

6.3 The Momentum Equation for Isotropic Elastic Materials

Now we can write down the closed system of equations for linear elasticity. But before doing this we will make further simplifications. The assumption that $\partial u_i/\partial \xi_j$ is small can be written as $|\partial u_i/\partial \xi_j| = \mathcal{O}(\epsilon)$, $\epsilon \ll 1$. In what follows we retain only terms of the order of ϵ, while we neglect the terms of the order of ϵ^2 and of higher orders. Then

$$\mathcal{J} = \begin{vmatrix} 1 + \dfrac{\partial u_1}{\partial \xi_1} & \dfrac{\partial u_1}{\partial \xi_2} & \dfrac{\partial u_1}{\partial \xi_3} \\ \dfrac{\partial u_2}{\partial \xi_1} & 1 + \dfrac{\partial u_2}{\partial \xi_2} & \dfrac{\partial u_2}{\partial \xi_3} \\ \dfrac{\partial u_3}{\partial \xi_1} & \dfrac{\partial u_3}{\partial \xi_2} & 1 + \dfrac{\partial u_3}{\partial \xi_3} \end{vmatrix} = \frac{\partial u_1}{\partial \xi_3}\begin{vmatrix} \dfrac{\partial u_2}{\partial \xi_1} & 1 + \dfrac{\partial u_2}{\partial \xi_2} \\ \dfrac{\partial u_3}{\partial \xi_1} & \dfrac{\partial u_3}{\partial \xi_2} \end{vmatrix}$$

$$- \frac{\partial u_1}{\partial \xi_2}\begin{vmatrix} \dfrac{\partial u_2}{\partial \xi_1} & \dfrac{\partial u_2}{\partial \xi_3} \\ \dfrac{\partial u_3}{\partial \xi_1} & 1 + \dfrac{\partial u_3}{\partial \xi_3} \end{vmatrix} + \left(1 + \frac{\partial u_1}{\partial \xi_1}\right)\begin{vmatrix} 1 + \dfrac{\partial u_2}{\partial \xi_2} & \dfrac{\partial u_2}{\partial \xi_3} \\ \dfrac{\partial u_3}{\partial \xi_2} & 1 + \dfrac{\partial u_3}{\partial \xi_3} \end{vmatrix}$$

$$= \left(1 + \frac{\partial u_1}{\partial \xi_1}\right)\left(1 + \frac{\partial u_2}{\partial \xi_2}\right)\left(1 + \frac{\partial u_3}{\partial \xi_3}\right) + \mathcal{O}(\epsilon^2)$$

$$= 1 + \frac{\partial u_1}{\partial \xi_1} + \frac{\partial u_2}{\partial \xi_2} + \frac{\partial u_3}{\partial \xi_3} + \mathcal{O}(\epsilon^2),$$

so eventually

$$\mathcal{J} = 1 + \nabla_\xi \cdot \boldsymbol{u} + \mathcal{O}(\epsilon^2) = 1 + \mathcal{O}(\epsilon). \tag{6.3.1}$$

Using (3.7.4) and (6.3.1) we obtain

$$\rho = \frac{\rho_0}{\mathcal{J}} = \frac{\rho_0}{1 + \nabla_\xi \cdot \boldsymbol{u} + \mathcal{O}(\epsilon^2)} = \rho_0[1 + \mathcal{O}(\epsilon)]. \tag{6.3.2}$$

In accordance with (3.4.1) $\mathbf{F} = \mathbf{I} + \mathbf{H}$. It follows from the assumption $|\partial u_i/\partial \xi_j| = \mathcal{O}(\epsilon)$ that $\mathbf{H} = \mathcal{O}(\epsilon)$, so that $\mathbf{F} = \mathbf{I} + \mathcal{O}(\epsilon)$. Then we obtain

$$\mathbf{F}^{-1} = \mathbf{I} + \mathcal{O}(\epsilon). \tag{6.3.3}$$

Now we note that the assumption $|\partial u_i/\partial \xi_j| = \mathcal{O}(\epsilon)$ leads to $\mathbf{E} = \mathcal{O}(\epsilon)$, and, due to (6.2.12), to $\mathbf{T} = \mathcal{O}(\epsilon)$. Then, using (6.3.1) and (6.3.3), we obtain for the nominal stress tensor:

$$\mathbf{\Pi} = \mathcal{J}\mathbf{T}\mathbf{F}^{-1} = [1 + \mathcal{O}(\epsilon)]\mathbf{T}[\mathbf{I} + \mathcal{O}(\epsilon)] = \mathbf{T} + \mathcal{O}(\epsilon^2), \tag{6.3.4}$$

so that we can substitute $\mathbf{T}$ for $\mathbf{\Pi}$ in the momentum equation in Lagrangian variables (4.5.9). Using (6.2.13) we obtain

$$\begin{aligned}
\nabla_\xi \cdot \mathbf{T} &= \boldsymbol{e}_i \frac{\partial}{\partial \xi_j}\left\{\lambda \delta_{ij} \frac{\partial u_k}{\partial \xi_k} + \mu\left(\frac{\partial u_i}{\partial \xi_j} + \frac{\partial u_j}{\partial \xi_i}\right)\right\} \\
&= \boldsymbol{e}_i \left\{\lambda \frac{\partial}{\partial \xi_i}\left(\frac{\partial u_k}{\partial \xi_k}\right) + \mu \frac{\partial^2 u_i}{\partial \xi_j \partial \xi_j} + \mu \frac{\partial}{\partial \xi_i}\left(\frac{\partial u_j}{\partial \xi_j}\right)\right\} \\
&= (\lambda + \mu)\nabla_\xi(\nabla_\xi \cdot \boldsymbol{u}) + \mu \nabla_\xi^2 \boldsymbol{u}.
\end{aligned}$$

With the aid of this result we transform (4.5.9) into

$$\rho_0 \frac{D^2 \boldsymbol{u}}{Dt^2} = (\lambda + \mu)\nabla_\xi(\nabla_\xi \cdot \boldsymbol{u}) + \mu \nabla_\xi^2 \boldsymbol{u} + \rho_0 \boldsymbol{b}. \tag{6.3.5}$$

The equilibrium equation ($\partial \boldsymbol{u}/\partial t = 0$) is

$$(\lambda + \mu)\nabla_\xi(\nabla_\xi \cdot \boldsymbol{u}) + \mu \nabla_\xi^2 \boldsymbol{u} + \rho_0 \boldsymbol{b} = 0. \tag{6.3.6}$$

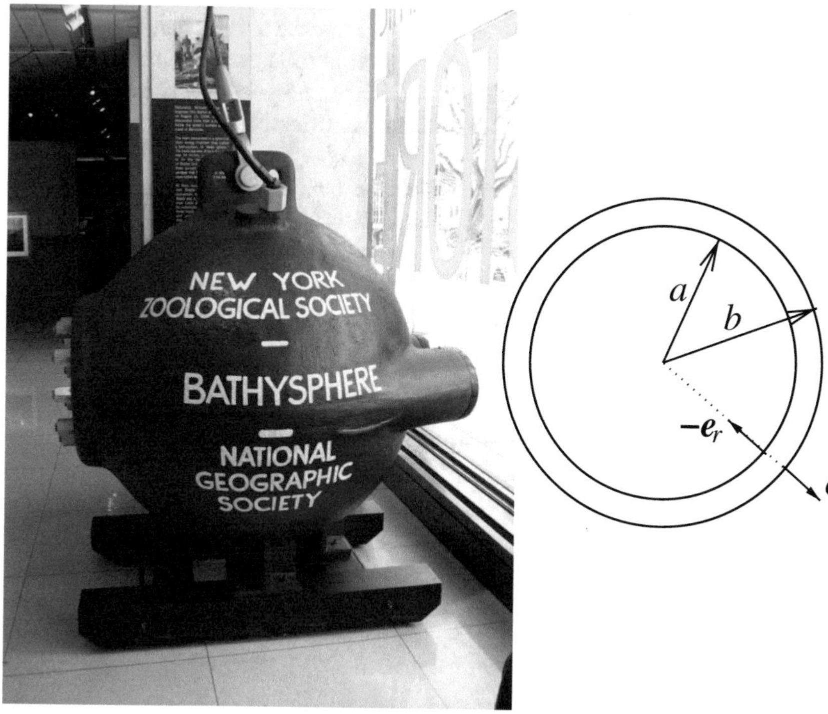

Fig. 6.2 Left: A bathysphere; photograph by Mike Cole, distributed under a CC-BY 2.0 license. Right: Cross-section of a bathysphere

Now we note that, in linear elasticity there is no difference between the Lagrangian and Eulerian description. Indeed,

$$\frac{\partial u_i}{\partial \xi_j} = \frac{\partial u_i}{\partial x_k}\frac{\partial x_k}{\partial \xi_j} = \frac{\partial u_i}{\partial x_k}\left(\delta_{jk} + \frac{\partial u_k}{\partial \xi_j}\right) = \frac{\partial u_i}{\partial x_j} + \mathcal{O}(\epsilon^2).$$

Also the difference between $D\boldsymbol{u}/Dt$ and $\partial \boldsymbol{u}/\partial t$ is of the order of ϵ^2. Since in linear elasticity we neglect terms of the order of ϵ^2, we can rewrite Eqs. (6.3.5) and (6.3.6) as

$$\boxed{\rho_0 \frac{\partial^2 \boldsymbol{u}}{\partial t^2} = (\lambda + \mu)\nabla(\nabla \cdot \boldsymbol{u}) + \mu \nabla^2 \boldsymbol{u} + \rho_0 \boldsymbol{b},} \tag{6.3.7}$$

$$\boxed{(\lambda + \mu)\nabla(\nabla \cdot \boldsymbol{u}) + \mu \nabla^2 \boldsymbol{u} + \rho_0 \boldsymbol{b} = 0.} \tag{6.3.8}$$

Example A bathysphere is a manned submersible that is attached to a ship by a steel rope, designed for deep sea observations (see Fig. 6.2). It is a spherical shell made of steel, with internal radius $a = 1$ m and external radius $b = 1.1$ m. Its cross-section is shown in Fig. 6.2. The Lamé constants of the steel that the bathysphere is made of are approximately equal, $\lambda \approx \mu$. The bathysphere would collapse if the stress magnitude at least at one point inside the shell and at least in one direction is larger than the critical value $t_{\rm cr} = 10^8$ N m^{-2}. What is the maximum depth at which the bathysphere can be immersed?

Solution: The maximum depth at which the bathysphere can be immersed is much greater than the bathysphere's diameter. This implies that we can neglect the variation of pressure on the bathysphere's external surface and take the pressure to be the same at any point of this surface. This pressure is

$$p = p_a + \rho g h, \tag{6.3.9}$$

where $p_a \approx 10^5$ N m^{-2} is the atmospheric pressure, $\rho \approx 10^3$ kg m^{-3} the water density, and h the depth the bathysphere is immersed at. The pressure inside the bathysphere is p_a. The body force exerted on the shell is the force of gravity, so that $\boldsymbol{b} = \boldsymbol{g}$. Then the last term on the left-hand side of the equilibrium Eq. (6.3.8) is $\rho_s \boldsymbol{g}$, where $\rho_s \approx 8 \times 10^3$ kg m^{-3} is the steel density. We have the estimate $\|\rho_0 \boldsymbol{b}\| = \|\rho_s \boldsymbol{g}\| \approx 8 \times 10^4$ N m^{-3}. The first term on the left-hand side of Eq. (6.3.8) is of the order of $(p - p_a)/(b - a) \sim \rho g h/(b - a)$. If we take $h \sim 1$ km then $(p - p_a)/(b - a) \sim 10^8$ N m^{-3} $\gg \|\rho_0 \boldsymbol{b}\|$. This means that we can neglect $\rho_0 \boldsymbol{b}$ in Eq. (6.3.8) and write the equilibrium equation as

$$(\lambda + \mu)\nabla(\nabla \cdot \boldsymbol{u}) + \mu \nabla^2 \boldsymbol{u} = 0. \tag{6.3.10}$$

In what follows we use spherical coordinates, so that $x_1 = r$, $x_2 = \theta$, and $x_3 = \phi$. The condition that the surface traction must be continuous at any surface of discontinuity gives

$$\begin{aligned} \boldsymbol{t}_1 &= p_a \boldsymbol{e}_r \quad \text{at } r = a, \\ \boldsymbol{t}_2 &= -p \boldsymbol{e}_r \quad \text{at } r = b, \end{aligned} \tag{6.3.11}$$

where $\boldsymbol{t}_1 = -\mathsf{T} \cdot \boldsymbol{e}_r$ and $\boldsymbol{t}_2 = \mathsf{T} \cdot \boldsymbol{e}_r$ ($-\boldsymbol{e}_r$ and $\boldsymbol{e}_r$ are the unit normal vectors at the inner and outer surfaces of the shell, see Fig. 6.2).

The external pressure causes the displacement of points of the shell in the radial direction, $\boldsymbol{u} = u \boldsymbol{e}_r$. The configuration is spherically symmetric, which implies that $u = u(r)$. To simplify Eq. (6.3.10) we prove the identity

$$\nabla^2 \boldsymbol{a} = \nabla(\nabla \cdot \boldsymbol{a}) - \nabla \times \nabla \times \boldsymbol{a}, \tag{6.3.12}$$

where $\boldsymbol{a}$ is an arbitrary vector. It is enough to only prove it in Cartesian coordinates. Then it will be valid in any curvilinear coordinates.

$$\nabla(\nabla \cdot \boldsymbol{a}) - \nabla \times \nabla \times \boldsymbol{a} = \boldsymbol{e}_i \frac{\partial}{\partial x_i}\left(\frac{\partial a_j}{\partial x_j}\right) - \boldsymbol{e}_i \epsilon_{ijk} \frac{\partial}{\partial x_j}(\nabla \times \boldsymbol{a})_k$$

$$= \boldsymbol{e}_i \frac{\partial^2 a_j}{\partial x_i \partial x_j} - \boldsymbol{e}_i \epsilon_{ijk}\epsilon_{klm} \frac{\partial}{\partial x_j}\left(\frac{\partial a_m}{\partial x_l}\right) = \boldsymbol{e}_i \left(\frac{\partial^2 a_j}{\partial x_i \partial x_j} - \epsilon_{ijk}\epsilon_{lmk} \frac{\partial^2 a_m}{\partial x_j \partial x_l}\right)$$

$$= \boldsymbol{e}_i \left(\frac{\partial^2 a_j}{\partial x_i \partial x_j} - (\delta_{il}\delta_{jm} - \delta_{im}\delta_{jl}) \frac{\partial^2 a_m}{\partial x_j \partial x_l}\right)$$

$$= \boldsymbol{e}_i \left(\frac{\partial^2 a_j}{\partial x_i \partial x_j} - \frac{\partial^2 a_j}{\partial x_j \partial x_i} + \frac{\partial^2 a_i}{\partial x_j \partial x_j}\right) = \boldsymbol{e}_i \frac{\partial^2 a_i}{\partial x_j \partial x_j} = \nabla^2 \boldsymbol{a}.$$

Using Eq. (2.13.18) we obtain $\nabla \times \boldsymbol{u} = 0$, so that

$$\nabla^2 \boldsymbol{u} = \nabla(\nabla \cdot \boldsymbol{u}) = \boldsymbol{e}_r \frac{d}{dr}\left[\frac{1}{r^2}\frac{d(r^2 u)}{dr}\right],$$

and Eq. (6.3.10) reduces to

$$\frac{d}{dr}\left[\frac{1}{r^2}\frac{d(r^2 u)}{dr}\right] = 0. \tag{6.3.13}$$

Integrating we obtain

$$\frac{d(r^2 u)}{dr} = 3Ar^2, \tag{6.3.14}$$

where A is a constant. Integrating again we get

$$u = Ar + \frac{B}{r^2}, \tag{6.3.15}$$

where B is a constant. In accordance with Eq. (6.2.12)

$$\mathbf{T} = \lambda \mathbf{I} \nabla \cdot \boldsymbol{u} + \mu\left[\nabla \boldsymbol{u} + (\nabla \boldsymbol{u})^T\right]. \tag{6.3.16}$$

Using Eq. (2.13.15) we obtain

$$\nabla \boldsymbol{u} = \frac{du}{dr}\boldsymbol{e}_r\boldsymbol{e}_r + \frac{u}{r}(\boldsymbol{e}_\theta\boldsymbol{e}_\theta + \boldsymbol{e}_\phi\boldsymbol{e}_\phi). \tag{6.3.17}$$

We see that $\nabla \boldsymbol{u}$ is symmetric, $\nabla \boldsymbol{u} = (\nabla \boldsymbol{u})^T$. Substituting the expression for $\nabla \boldsymbol{u}$ into Eq. (6.3.16) and using $\mathbf{I} = \boldsymbol{e}_r\boldsymbol{e}_r + \boldsymbol{e}_\theta\boldsymbol{e}_\theta + \boldsymbol{e}_\phi\boldsymbol{e}_\phi$ yields

$$\mathbf{T} = \left(\lambda \nabla \cdot \boldsymbol{u} + 2\mu \frac{du}{dr}\right)\boldsymbol{e}_r\boldsymbol{e}_r + \left(\lambda \nabla \cdot \boldsymbol{u} + 2\mu\,\frac{u}{r}\right)(\boldsymbol{e}_\theta\boldsymbol{e}_\theta + \boldsymbol{e}_\phi\boldsymbol{e}_\phi). \tag{6.3.18}$$

The stress at the outer boundary is

$$t_2 = \mathbf{T} \cdot \boldsymbol{e}_r\big|_{r=b} = \left(\lambda \nabla \cdot \boldsymbol{u} + 2\mu \frac{du}{dr}\right)\boldsymbol{e}_r\bigg|_{r=b}$$
$$= \left(\frac{\lambda}{r^2}\frac{d(r^2 u)}{dr}\bigg|_{r=b} + 2\mu \frac{du}{dr}\bigg|_{r=b}\right)\boldsymbol{e}_r = \left[(3\lambda + 2\mu)A - 4\mu B b^{-3}\right]\boldsymbol{e}_r. \tag{6.3.19}$$

Similarly,

$$\boldsymbol{t}_1 = -\mathbf{T} \cdot \boldsymbol{e}_r\big|_{r=a} = \left[-(3\lambda + 2\mu)A + 4\mu B a^{-3}\right]\boldsymbol{e}_r. \tag{6.3.20}$$

Using Eqs. (6.3.19) and (6.3.20) we reduce Eq. (6.3.11) to

$$\begin{aligned} (3\lambda + 2\mu)A - 4\mu B a^{-3} &= -p_a, \\ (3\lambda + 2\mu)A - 4\mu B b^{-3} &= -p. \end{aligned} \tag{6.3.21}$$

Then, with the aid of Eq. (6.3.9), we find

$$A = -\frac{(b^3 - a^3)p_a + \rho g h b^3}{(3\lambda + 2\mu)(b^3 - a^3)}, \quad B = -\frac{\rho g h a^3 b^3}{4\mu(b^3 - a^3)}. \tag{6.3.22}$$

Using Eq. (6.3.18) we find that the surface traction inside the shell at the surface with unit normal vector $\boldsymbol{n}$ is given by

$$\boldsymbol{t} = \mathbf{T} \cdot \boldsymbol{n} = \left(\lambda \nabla \cdot \boldsymbol{u} + 2\mu \frac{du}{dr}\right) n_r \boldsymbol{e}_r + \left(\lambda \nabla \cdot \boldsymbol{u} + 2\mu \frac{u}{r}\right)(n_\theta \boldsymbol{e}_\theta + n_\phi \boldsymbol{e}_\phi). \tag{6.3.23}$$

With the aid of Eqs. (6.3.14) and (6.3.15) we rewrite this expression as

$$\boldsymbol{t} = \left[(3\lambda + 2\mu)A - 4\mu B r^{-3}\right] n_r \boldsymbol{e}_r + \left[(3\lambda + 2\mu)A + 2\mu B r^{-3}\right](n_\theta \boldsymbol{e}_\theta + n_\phi \boldsymbol{e}_\phi). \tag{6.3.24}$$

Then, using the identity $n_r^2 + n_\theta^2 + n_\phi^2 = 1$, we obtain for the square of the stress magnitude

$$\|\boldsymbol{t}\|^2 = \left[(3\lambda + 2\mu)A + 2\mu B r^{-3}\right]^2 - 12\mu B r^{-3}\left[(3\lambda + 2\mu)A - \mu B r^{-3}\right] n_r^2. \tag{6.3.25}$$

Using the first equation in (6.3.21) and taking into account that, in accordance with (6.3.22), $B < 0$, we rewrite the second term in (6.3.25) as

$$-12\mu|B|r^{-3}\left[\mu|B|\left(4a^{-3} - r^{-3}\right) + p_a\right] n_r^2.$$

It is obvious that this expression is negative unless $n_r = 0$, which implies that, for fixed r, $\|\boldsymbol{t}\|$ takes its maximum value, $\tilde{t}(r)$, when $n_r = 0$, i.e. when the normal vector to the surface is orthogonal to the radial direction. This maximum value is given by

$$\tilde{t}(r) = (3\lambda + 2\mu)|A| + 2\mu|B|r^{-3},$$

where we have taken into account that $A < 0$ and $B < 0$. It is obvious that $\tilde{t}(r)$ is a monotonically decreasing function, so it takes its maximum value, $t_{\max}$, at $r = a$. Hence, the maximum stress in the shell is

$$t_{\max} = (3\lambda + 2\mu)|A| + 2\mu|B|a^{-3}. \tag{6.3.26}$$

Substituting the numerical values into (6.3.26) and using (6.3.22) we obtain

$$t_{\max} \approx (10 + 5.89h) \times 10^4 \text{ N m}^{-2}.$$

The bathysphere can only endure the water pressure if $t_{\max} \leq t_{\text{cr}}$, which gives $h_{\max} \approx$ 1700 m.

6.4 Elastic Body Waves

Let us assume that a homogeneous elastic material fills the whole space and there is no body force: $\boldsymbol{b} = 0$. The motion of this material is described by Eq. (6.3.7). Consider one-dimensional motions where all variables only depend on $x = x_1$ in Cartesian coordinates x_1, x_2, x_3. Then the three components of Eq. (6.3.7) are written as

$$\frac{\partial^2 u_1}{\partial t^2} = c_p^2 \frac{\partial^2 u_1}{\partial x^2}, \tag{6.4.1}$$

$$\frac{\partial^2 u_2}{\partial t^2} = c_s^2 \frac{\partial^2 u_2}{\partial x^2}, \quad \frac{\partial^2 u_3}{\partial t^2} = c_s^2 \frac{\partial^2 u_3}{\partial x^2}, \tag{6.4.2}$$

where

$$c_p = \sqrt{\frac{\lambda + 2\mu}{\rho_0}}, \quad c_s = \sqrt{\frac{\mu}{\rho_0}}. \tag{6.4.3}$$

Introducing the vector $\boldsymbol{w} = u_2\boldsymbol{e}_2 + u_3\boldsymbol{e}_3$, multiplying the first equation in (6.4.2) by $\boldsymbol{e}_2$, the second equations in (6.4.2) by $\boldsymbol{e}_3$, and adding the results, we unite the two equations in (6.4.2) into one vector equation

$$\frac{\partial^2 \boldsymbol{w}}{\partial t^2} = c_s^2 \frac{\partial^2 \boldsymbol{w}}{\partial x^2}. \tag{6.4.4}$$

In the rest of this subsection we will write u instead of u_1.

Let us make the variable substitution in Eq. (6.4.1):

$$X = t - x/c_p, \quad Y = t + x/c_p.$$

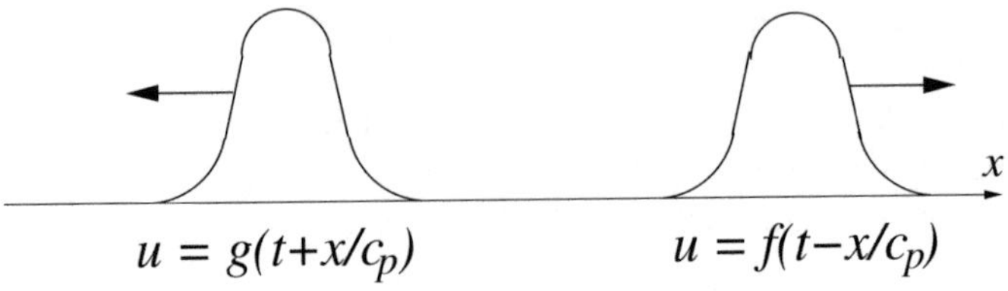

Fig. 6.3 Perturbations propagating in the positive and negative x-directions

Then

$$\frac{\partial^2 u}{\partial t^2} = \frac{\partial^2 u}{\partial X^2} + 2\frac{\partial^2 u}{\partial X \partial Y} + \frac{\partial^2 u}{\partial Y^2},$$

$$\frac{\partial^2 u}{\partial x^2} = \frac{1}{c_p^2}\left(\frac{\partial^2 u}{\partial X^2} - 2\frac{\partial^2 u}{\partial X \partial Y} + \frac{\partial^2 u}{\partial Y^2}\right),$$

and Eq. (6.4.1) is transformed into

$$\frac{\partial^2 u}{\partial X \partial Y} = 0.$$

It follows from this equation that

$$\frac{\partial u}{\partial X} = \tilde{f}(X),$$

where $\tilde{f}(X)$ is an arbitrary function. Integrating this equation with respect to X we obtain

$$u = f(X) + g(Y),$$

where $g(Y)$ is an arbitrary function, and $f(X) = \int \tilde{f}(X)\,dX$. Returning to the initial variables we eventually arrive at

$$\boxed{u = f(t - x/c_p) + g(t + x/c_p).} \tag{6.4.5}$$

This expression is called the *d'Alembert solution.*

We take $g = 0$, so that $u = f(t - x/c_p)$. Then the value of u at the spatial position $x + c_p t$ at time t is the same as it was at $t = 0$ at the spatial position x. Hence, u describes a perturbation of a permanent shape propagating with speed c_p in the positive x-direction (see Fig. 6.3).

If we take $f = 0$, so that $u = g(t + x/c_p)$, then the value of u at the spatial position $x - c_p t$ at time t is the same as it was at $t = 0$ at the spatial position x. Hence, u describes a perturbation of a permanent shape propagating with speed c_p in the negative x-direction (again see Fig. 6.3).

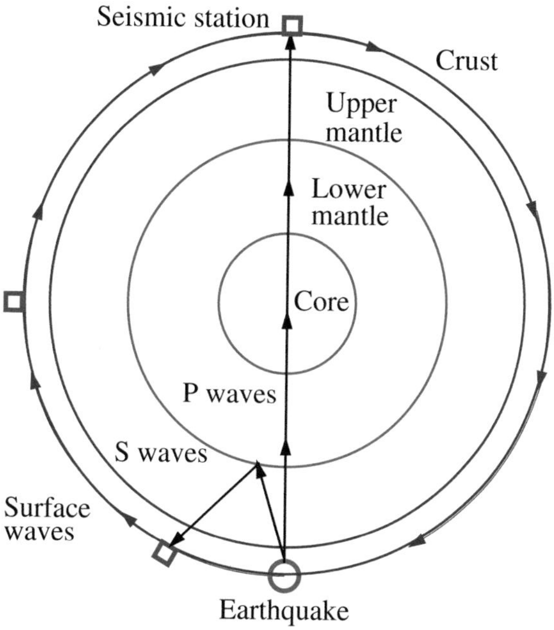

Fig. 6.4 Schematic picture of wave propagation in the Earth's interior

The waves that are described by Eq. (6.4.1) and propagate with speed c_p are called *longitudinal* waves because in these waves $\boldsymbol{u}$ is in the direction of wave propagation. The speed of the longitudinal waves is quite high. For example, in iron $c_p \approx 5.8\ \mathrm{km\,s^{-1}}$.

Equation (6.4.4) has the same form as Eq. (6.4.1), so that its general solution is also given by the d'Alembert solution:

$$\boldsymbol{w} = \boldsymbol{f}(t - x/c_s) + \boldsymbol{g}(t + x/c_s), \tag{6.4.6}$$

where now $\boldsymbol{f}$ and $\boldsymbol{g}$ are two-dimensional vector functions with the x-component equal to zero. The waves described by Eq. (6.4.4) are called *transverse* waves because in these waves $\boldsymbol{u}$ is perpendicular (transverse) to the direction of wave propagation. The propagation speed of these waves is c_s. Although, as follows from Eq. (6.4.3), $c_s < c_p$, it is also quite high. For example, in iron $c_s \approx 3.14\ \mathrm{km\,s^{-1}}$.

If in a transverse wave $\boldsymbol{w} = w\boldsymbol{n}$, where $\boldsymbol{n}$ is a constant unit vector perpendicular to the direction of wave propagation, then the wave is called *plane polarised*. The direction of the vector $\boldsymbol{n}$ is called the direction of *wave polarisation*.

The theory of wave propagation in elastic media is the basis of *seismology*. In seismology the longitudinal waves are called P-waves, and the transversal waves are called S-waves. P-waves can propagate both in solids and liquids, while S-waves can only propagate in solids.

In Fig. 6.4 a schematic picture of the Earth's interior is shown. While the P-waves can propagate through any part of the interior, the S-waves cannot propagate through

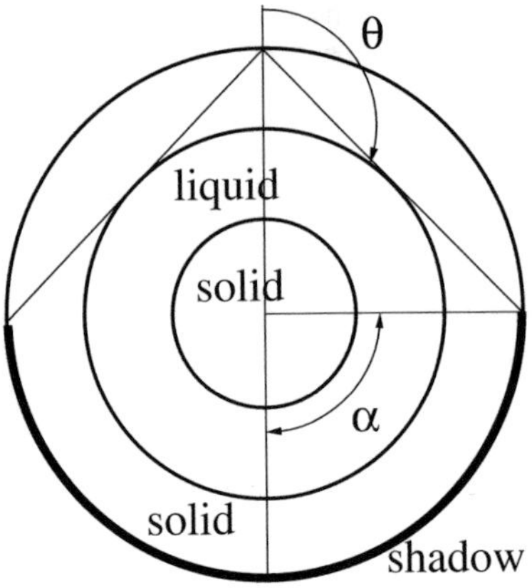

Fig. 6.5 Illustration of the solution to the Example

the lower mantle because it is liquid. As a result, if an earthquake occurred at the south pole then the P-waves created by this earthquake can be registered at any seismic station, even by one at the north pole. On the other hand, the S-waves will only be registered at seismic stations that are not very far from the south pole. This picture is clarified in the example given below.

Example The Earth's radius is $R \approx 6400$ km. Suppose the source of perturbations is at the north pole. There is the shadow for S-waves defined by $\pi - \alpha/2 \le \theta \le \pi$ in spherical coordinates $r,\ \theta,\ \phi$ (see Fig. 6.5). Calculate the radius r of the liquid core if $\alpha = \pi/2$.

Solution: The angle between the ray that touches the outer surface of the liquid core and the vertical direction is $\alpha/2$ (see Fig. 6.5). Then the radius of the liquid core is $R\sin(\alpha/2) = R\sqrt{2}/2 \approx 4525$ km.

6.5 Rayleigh Waves

Let us now assume that an elastic material occupies the region $x_3 \ge 0$, and that the boundary $x_3 = 0$ is stress-free. In what follows we use the notation $x = x_1$, $y = x_2$, and $z = x_3$. We consider a two-dimensional motion of the elastic material, i.e. we assume that $\boldsymbol{u}$ is independent of y and $u_2 = 0$. Then $\nabla \cdot \boldsymbol{u} = \partial u_1/\partial x + \partial u_3/\partial z$. We also assume that there is no body force, $\boldsymbol{b} = 0$. Then we obtain from Eq. (6.3.7)

$$\rho_0 \frac{\partial^2 u_1}{\partial t^2} = (\lambda + 2\mu)\frac{\partial^2 u_1}{\partial x^2} + (\lambda + \mu)\frac{\partial^2 u_3}{\partial x \partial z} + \mu \frac{\partial^2 u_1}{\partial z^2}, \tag{6.5.1}$$

$$\rho_0 \frac{\partial^2 u_3}{\partial t^2} = \mu \frac{\partial^2 u_3}{\partial x^2} + (\lambda + \mu)\frac{\partial^2 u_1}{\partial x \partial z} + (\lambda + 2\mu)\frac{\partial^2 u_3}{\partial z^2}. \tag{6.5.2}$$

The unit normal vector at the surface $z = 0$ is $\boldsymbol{e}_3$. Then, using (6.3.16), we obtain for the surface traction at the positive side of this surface

$$\boldsymbol{t} = \mathbf{T} \cdot \boldsymbol{e}_3 = \lambda \boldsymbol{e}_3 \nabla \cdot \boldsymbol{u} + \mu \nabla u_3 + \mu \frac{\partial \boldsymbol{u}}{\partial z}.$$

Since the surface traction is continuous at any surface, and the surface traction at the negative side of the surface $z = 0$ is zero, we conclude that $\boldsymbol{t} = 0$. Then, using the expression for $\nabla \cdot \boldsymbol{u}$, we obtain two boundary conditions valid at $z = 0$:

$$\frac{\partial u_1}{\partial z} + \frac{\partial u_3}{\partial x} = 0, \quad \lambda \frac{\partial u_1}{\partial x} + (\lambda + 2\mu) \frac{\partial u_3}{\partial z} = 0. \tag{6.5.3}$$

Let us look for the solution to Eqs. (6.5.1) and (6.5.2) with the boundary conditions (6.5.3) in the form

$$u_1 = U(z) \exp[i\omega(t - x/V)], \quad u_3 = W(z) \exp[i\omega(t - x/V)].$$

Substituting these expressions into Eqs. (6.5.1)–(6.5.3) we obtain the system of equations,

$$\mu \frac{d^2 U}{dz^2} - (\lambda + \mu) \frac{i\omega}{V} \frac{dW}{dz} + \omega^2 \left(\rho_0 - \frac{\lambda + 2\mu}{V^2} \right) U = 0, \tag{6.5.4}$$

$$(\lambda + 2\mu) \frac{d^2 W}{dz^2} - (\lambda + \mu) \frac{i\omega}{V} \frac{dU}{dz} + \omega^2 \left(\rho_0 - \frac{\mu}{V^2} \right) W = 0, \tag{6.5.5}$$

and the boundary conditions at $z = 0$,

$$\frac{dU}{dz} - \frac{i\omega}{V} W = 0, \quad (\lambda + 2\mu) \frac{dW}{dz} - \frac{i\lambda\omega}{V} U = 0. \tag{6.5.6}$$

We are looking for the solution to the system of Eqs. (6.5.4) and (6.5.5) with the boundary conditions (6.5.6) in the form $U = \widetilde{U} e^{\kappa z}$, $W = \widetilde{W} e^{\kappa z}$, where $\widetilde{U}$ and $\widetilde{W}$ are constants. Substituting these expressions into Eqs. (6.5.4) and (6.5.5) and using Eq. (6.4.3) to express λ and μ in terms of c_p and c_s, we obtain

$$\begin{aligned} &\left(1 + \frac{c_s^2 \kappa^2}{\omega^2} - \frac{c_p^2}{V^2} \right) \widetilde{U} - \left(c_p^2 - c_s^2 \right) \frac{i\kappa}{\omega V} \widetilde{W} = 0, \\ &\left(c_p^2 - c_s^2 \right) \frac{i\kappa}{\omega V} \widetilde{U} - \left(1 + \frac{c_p^2 \kappa^2}{\omega^2} - \frac{c_s^2}{V^2} \right) \widetilde{W} = 0. \end{aligned} \tag{6.5.7}$$

This is the system of two linear homogeneous algebraic equations for two variables, $\widetilde{U}$ and $\widetilde{W}$. It has a non-trivial solution only if its determinant is zero. This gives the equation determining κ:

$$\left(1+\frac{c_s^2\kappa^2}{\omega^2}-\frac{c_p^2}{V^2}\right)\left(1+\frac{c_p^2\kappa^2}{\omega^2}-\frac{c_s^2}{V^2}\right)+\frac{\kappa^2\left(c_p^2-c_s^2\right)^2}{\omega^2V^2}=0.$$

This equation can be rewritten as

$$c_p^2c_s^2V^4\kappa^4+\kappa^2\omega^2V^2\left[\left(c_p^2+c_s^2\right)V^2-2c_p^2c_s^2\right]+\omega^4\left(c_p^2-V^2\right)\left(c_s^2-V^2\right)=0. \tag{6.5.8}$$

Considering this equation as a quadratic equation in κ^2, we easily obtain that its two roots are given by

$$\kappa_1^2=\frac{\omega^2}{V^2}\left(1-\frac{V^2}{c_p^2}\right),\quad \kappa_2^2=\frac{\omega^2}{V^2}\left(1-\frac{V^2}{c_s^2}\right). \tag{6.5.9}$$

Then the four roots of the biquadratic equation (6.5.8) are $\pm\kappa_1$ and $\pm\kappa_2$.

In what follows we are only interested in a solution describing a surface wave, i.e. the wave where all perturbations vanish as $z\to\infty$. If $V>c_p$, then $\kappa_1^2<0$ and $\kappa_2^2<0$, which implies that κ_1 and κ_2 are purely imaginary. Then $\exp(\pm\kappa_1 z)$ and $\exp(\pm\kappa_2 z)$ are oscillating functions not tending to zero as $z\to\infty$. This means that there is no solution describing a surface wave with $V>c_p$.

If $c_s<V<c_p$, then $\kappa_1^2>0$ and $\exp(-\kappa_1 z)\to 0$ as $z\to\infty$. On the other hand, we still have $\kappa_2^2<0$, so that $\exp(\pm\kappa_2 z)\nrightarrow 0$ as $z\to\infty$. Hence, we have only one solution of the system of Eqs. (6.5.4) and (6.5.5) that tends to zero as $z\to\infty$, namely, the solution proportional to $\exp(-\kappa_1 z)$. It is not enough to satisfy the two boundary conditions (6.5.6), so that there is also no solution describing a surface wave when $c_s<V<c_p$.

Summarising our analysis, we conclude that a solution describing a surface wave can only exist if $V<c_s$. In this case there are two linearly independent solutions to the system of Eqs. (6.5.4) and (6.5.5) tending to zero at infinity, namely the solutions proportional to $\exp(-\kappa_1 z)$ and $\exp(-\kappa_2 z)$. Substituting the expression for κ into any of the two equations in (6.5.7), we easily obtain the relation between $\widetilde{U}$ and $\widetilde{W}$ in each of these two solutions, so that eventually we obtain that the two linearly independent solutions to the system of Eqs. (6.5.4) and (6.5.5) vanishing at infinity are

$$\begin{pmatrix}U\\W\end{pmatrix}=\begin{pmatrix}-\omega\\ iV\kappa_1\end{pmatrix}e^{-\kappa_1 z},\quad \begin{pmatrix}U\\W\end{pmatrix}=\begin{pmatrix}iV\kappa_2\\ \omega\end{pmatrix}e^{-\kappa_2 z}.$$

The general solution to the system of Eqs. (6.5.4) and (6.5.5) vanishing at infinity is a linear combination of these two solutions, so that

$$\begin{aligned}U&=-A\omega e^{-\kappa_1 z}+iBV\kappa_2 e^{-\kappa_2 z},\\ W&=iAV\kappa_1 e^{-\kappa_1 z}+B\omega e^{-\kappa_2 z},\end{aligned} \tag{6.5.10}$$

where A and B are arbitrary constants. Substituting this solution into the boundary conditions (6.5.6) and using (6.4.3) we obtain

$$2\omega\kappa_1 V A - i\left(\omega^2 + \kappa_2^2 V^2\right)B = 0,$$
$$\left[c_p^2\kappa_1^2 V^2 - \omega^2\left(c_p^2 - 2c_s^2\right)\right]A - 2i\omega c_s^2\kappa_2 V B = 0. \tag{6.5.11}$$

Again we have obtained a system of two linear homogeneous algebraic equations for two variables, A and B. It only has a non-trivial solution if its determinant is zero. This condition gives the equation determining V:

$$\left(\omega^2 + \kappa_2^2 V^2\right)\left[c_p^2\kappa_1^2 V^2 - \omega^2\left(c_p^2 - 2c_s^2\right)\right] - 4\omega^2 c_s^2\kappa_1\kappa_2 V^2 = 0.$$

Using (6.5.9) we rewrite this equation as

$$\left(2 - \frac{V^2}{c_s^2}\right)^2 = 4\left(1 - \frac{V^2}{c_s^2}\right)^{1/2}\left(1 - \frac{V^2}{c_p^2}\right)^{1/2}.$$

Taking the square of this equation and introducing the notation $\chi = V^2/c_s^2$ and $q = c_s/c_p < 1$ we eventually arrive at

$$\chi\left\{\chi^3 - 8\chi^2 + 8\left(3 - 2q^2\right)\chi - 16\left(1 - q^2\right)\right\} = 0.$$

Hence, apart from the root $\chi = 0$ corresponding to the trivial solution $V = 0$, the wave propagation speed is determined by

$$f(\chi) \equiv \chi^3 - 8\chi^2 + 8\left(3 - 2q^2\right)\chi - 16\left(1 - q^2\right) = 0. \tag{6.5.12}$$

Since $0 < V < c_s$, we are looking for the root to this equation satisfying $0 < \chi < 1$. Since $f(0) = -16(1 - q^2) < 0$ and $f(1) = 1$, it follows that there is at least one root to (6.5.12) in the interval $(0, 1)$. There are two possibilities: either there is one root, or there are three roots in $(0, 1)$. If there are three roots, then $f'(\chi)$ has two zeros in $(0, 1)$, and $f''(\chi)$ has one zero in $(0, 1)$. However, $f''(\chi) = 2(3\chi - 8) < 0$ in $(0, 1)$, so that there is exactly one root in $(0, 1)$. This root gives the phase speed of the surface wave, which is called the *Rayleigh wave* after the famous English scientist Lord Rayleigh who first described this wave theoretically. Rayleigh waves are very important in seismology because they are observed propagating along the Earth's surface after earthquakes.

The parameter q can be expressed in terms of Poisson's ratio ν defined by Eq. (6.2.8), $q^2 = \frac{1}{2}(1 - 2\nu)/(1 - \nu)$, and V/c_s is a function of ν only. Obviously $\nu \in (0, 1/2)$. Figure 6.6 shows the dependence of $\chi = V/c_s$ on ν.

Example In steel $\lambda \approx \mu \approx 8 \times 10^{10}$ N m^{-2} and $v_s \approx 3.14$ km s^{-1}. Calculate the speed of the Rayleigh wave on the surface of steel.

Solution: Using Eq. (6.2.8) we obtain $\nu \approx 1/4$. Then it follows from Fig. 6.6 that $\chi \approx 0.85$. Then the speed of the Rayleigh wave is $V = v_s\sqrt{\chi} \approx 2.9$ km s^{-1}.

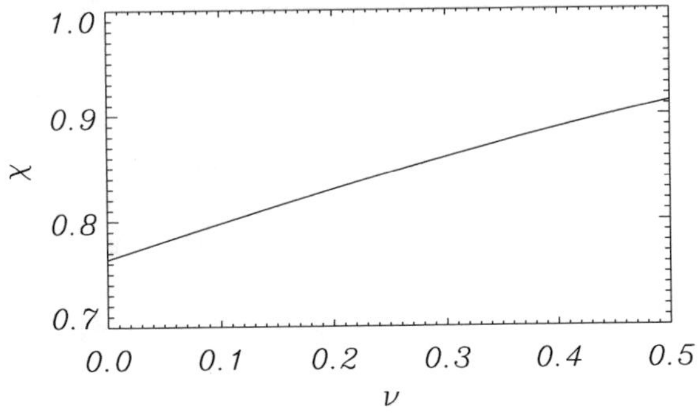

Fig. 6.6 Dependence of the solution to Eq. (6.5.12) on ν

6.6 Viscoelasticity

The assumption that the stress tensor is uniquely defined by the infinitesimal strain tensor implies that the deformation of a solid is reversible, that is, when an external force is removed the solid returns to the state that it had before the force was applied. Moreover, this return occurs instantaneously. If, conversely, we deform a continuum, and then let the continuum return to its initial state where there were no stresses, then the stresses will disappear instantaneously. Of course, this is an idealisation. In reality a body only returns to its initial state after a relaxation time. Hence, the elasticity only properly describes the deformation of solids if the characteristic time of variation of external forces or the continuum deformation is much greater than the relaxation time. Otherwise there is a time delay between the variation of external forces and the change of state of solids. This time delay is taken into account in a branch of continuum mechanics called *viscoelasticity*.

One possibility to account for the relaxation time is to assume that the stress tensor depends not only on the infinitesimal strain tensor, but also on its time derivative. Then in the linear viscoelasticity of isotropic materials Eq. (6.2.12) is generalised to

$$\boxed{\mathbf{T} = \lambda \mathbf{I} I_1 \left(\mathbf{E} + \theta_\lambda \frac{\partial \mathbf{E}}{\partial t}\right) + 2\mu \left(\mathbf{E} + \theta_\mu \frac{\partial \mathbf{E}}{\partial t}\right),} \tag{6.6.1}$$

where θ_λ and θ_μ are constants characterising the viscosity of the material. We obtain the *Kelvin–Voigt* model of a viscoelastic medium. Substituting Eq. (6.6.1) into Eq. (4.5.9) and recalling that in the linear theory we can substitute $\mathbf{T}$ for $\mathbf{\Pi}$ and we do not have to distinguish between the Eulerian and Lagrangian description we obtain

$$\boxed{\rho_0 \frac{\partial^2 \boldsymbol{u}}{\partial t^2} = (\lambda + \mu)\nabla\nabla \cdot \left(\boldsymbol{u} + \theta_\lambda \frac{\partial \boldsymbol{u}}{\partial t}\right) + \mu \nabla^2 \left(\boldsymbol{u} + \theta_\mu \frac{\partial \boldsymbol{u}}{\partial t}\right) + \rho_0 \boldsymbol{b}.} \tag{6.6.2}$$

To clarify the process of relaxation in a viscoelastic material we consider one

Example Suppose there is a slab of elastic material of thickness H. It lays on an absolutely hard surface. In Cartesian coordinates $x,\ y,\ z$ the equation of this surface is $z = 0$. The slab dimensions in the x and y-directions are much larger than H, so that we can neglect boundary effects and consider the slab as infinite in the x and y-directions. In the initial state there are no stresses inside the continuum. Then the slab is slowly compressed so that its thickness varies as $H - h(1 - e^{-t/t_0})$, where $h \ll H$. Describe the time dependence of the displacement in the continuum assuming that there is no body force, $\boldsymbol{b} = 0$.

Solution: We can consider the problem as one-dimensional and take $\boldsymbol{u} = u(z)\boldsymbol{e}_z$. Then Eq. (6.6.2) reduces to

$$\rho_0 \frac{\partial^2 u}{\partial t^2} = (\lambda + 2\mu)\frac{\partial^2}{\partial z^2}\left(u + \varsigma \frac{\partial u}{\partial t}\right), \tag{6.6.3}$$

where $\varsigma = (\lambda + 2\mu)^{-1}[(\lambda + \mu)\theta_\lambda + \mu\theta_\mu]$. Since the displacement is given at $z = 0$ and $z = H$, we have the boundary conditions

$$u = 0 \quad \text{at} \quad z = 0, \quad u = -h(1 - e^{-t/t_0}) \quad \text{at} \quad z = H. \tag{6.6.4}$$

We imposed the second boundary condition at the unperturbed upper boundary because we are considering a linear problem. We also need to impose the initial conditions. Obviously the displacement is zero everywhere at the initial time. On the other hand, at the initial time the time derivative of the displacement is not zero at the upper boundary. However, the perturbation of the upper boundary cannot spread instantaneously through the slab, rather it propagates with a finite speed. Hence, we take that the time derivative of the displacement is also zero at the initial time for $0 \le z < H$. Summarising, the initial conditions are

$$u = \frac{\partial u}{\partial t} = 0 \quad \text{at} \quad t = 0. \tag{6.6.5}$$

Now we make the variable substitution

$$w = u + \frac{hz}{H}(1 - e^{-t/t_0}). \tag{6.6.6}$$

Substituting this expression into Eq. (6.6.3) yields

$$\frac{\partial^2 w}{\partial t^2} - c_p^2 \frac{\partial^2}{\partial z^2}\left(w + \varsigma \frac{\partial w}{\partial t}\right) = -\frac{hz}{t_0^2 H} e^{-t/t_0}. \tag{6.6.7}$$

The boundary condition Eq. (6.6.4) and initial conditions (6.6.5) reduce to

$$w = 0 \quad \text{at} \quad z = 0,\ H, \tag{6.6.8}$$

$$w = 0, \quad \frac{\partial w}{\partial t} = \frac{hz}{t_0 H} \quad \text{at} \quad t = 0. \tag{6.6.9}$$

The boundary conditions (6.6.8) imply that w can be expanded in the Fourier series

$$w(t, z) = \sum_{n=1}^{\infty} w_n(t) \sin \frac{\pi n z}{H}. \tag{6.6.10}$$

Then, using the identity

$$z = \frac{2H}{\pi} \sum_{n=1}^{\infty} \frac{(-1)^{n+1}}{n} \sin \frac{\pi n z}{H}, \tag{6.6.11}$$

valid for $z \in (0, H)$, we obtain from Eq. (6.6.7)

$$\frac{d^2 w_n}{dt^2} + \frac{\pi^2 c_p^2 n^2}{H^2} \left(w_n + \varsigma \frac{dw_n}{dt} \right) = \frac{2h(-1)^n e^{-t/t_0}}{\pi n t_0^2}. \tag{6.6.12}$$

Using Eqs. (6.6.9) and (6.6.11) we obtain that w_n must satisfy the initial conditions

$$w_n = 0, \quad \frac{dw_n}{dt} = \frac{2h(-1)^{n+1}}{\pi n t_0} \quad \text{at} \quad t = 0. \tag{6.6.13}$$

The solution to Eq. (6.6.12) satisfying the initial condition (6.6.13) is given by

$$w_n = \Upsilon_n \left[e^{-t/t_0} - e^{-\Gamma_n t} \cos(t\Omega_n) \right] + \left[\pi n \Upsilon_n (1 - t_0 \Gamma_n) - 2h(-1)^n \right] \frac{e^{-\Gamma_n t} \sin(t\Omega_n)}{\pi n t_0 \Omega_n} \tag{6.6.14}$$

when $n \leq N$, and by

$$w_n = \Upsilon_n \left[e^{-t/t_0} - e^{-\Gamma_n t} \cosh(t|\Omega_n|) \right] + \left[\pi n \Upsilon_n (1 - t_0 \Gamma_n) - 2h(-1)^n \right] \frac{e^{-\Gamma_n t} \sinh(t|\Omega_n|)}{\pi n t_0 |\Omega_n|} \tag{6.6.15}$$

when $n > N$. In these equations

$$\Upsilon_n = \frac{2hH^2(-1)^n}{\pi n [H^2 + \pi^2 n^2 c_p^2 t_0 (t_0 - \varsigma)]}, \tag{6.6.16}$$

$$\Gamma_n = -\frac{\varsigma \pi^2 n^2 c_p^2}{2H^2}, \quad \Omega_n = \frac{\pi n c_p}{2H^2} \sqrt{4H^2 - \varsigma^2 \pi^2 n^2 c_p^2}. \tag{6.6.17}$$

We recall that $c_p^2 = (\lambda + 2\mu)/\rho_0$. The integer N is defined by the condition that $\Omega_n^2 > 0$ when $n \le N$, and $\Omega_n^2 < 0$ when $n > N$.

The problem contains three characteristic times: t_0, ς, and H/c_p, which is the time needed for the longitudinal wave to travel from the lower to the upper slab boundary. First we assume the slab thickness changes very slowly, so that $\max(\varsigma, H/c_p)/t_0 = \epsilon \ll 1$. It is straightforward to see that, in this case, $w_n = \mathcal{O}(\epsilon)$, which implies that

$$u \approx -\frac{hz}{H}(1 - e^{-t/t_0}). \tag{6.6.18}$$

We obtain exactly the same solution if we assume that the variation of the slab thickness is quasi-static and the displacement in the slab is described by the equilibrium Eq. (6.3.8) in linear elasticity where the time is only present as a parameter. On the basis of this analysis we can conclude that the solution to the example in Sect. 6.3 is only valid when the bathysphere is immersed into the water very slowly.

Next, we consider the case where there is no viscosity ($\varsigma = 0$), while $t_0 \le H/c_p$. In this case $u(H) \to -h$ as $t \to \infty$. However, the displacement of the upper boundary drives an undamped standing wave in the slab with amplitude of the order of h and the period of the fundamental mode equal to $2H/c_p$. If $\varsigma \ne 0$ then $w \to 0$ and $u \to -hz/H$ as $t \to \infty$. When $\varsigma\pi c_p < 2H$, w has the form of a standing wave with an exponentially decaying amplitude. However, when $\varsigma\pi c_p > 2H$, w describes an aperiodically decaying perturbation.

In linear elasticity there is only one model. It is based on Hooke's law. For isotropic materials this law is given by Eq. (6.2.12). It is more complex for anisotropic materials, however it is essentially the same Hooke's law. The situation is different in the case of viscoelasticity. There are a few different models. The next popular model is the Maxwell model. In this model the right-hand side of the relation between $\mathbf{T}$ and $\mathbf{E}$ is only expressed in terms of the time derivative of $\mathbf{E}$, while on the left-hand side there is a linear combination of $\mathbf{T}$ and its time derivative. In the Alfrey–Burgers model there is a linear combination of the first and second derivative of $\mathbf{E}$ on the right-hand side, and a linear combination of $\mathbf{T}$ and its first and second derivatives on the left-hand side. There is also the generalised Maxwell model, also known as the Maxwell–Wiechert model, which is considered to be the most general model of viscoelasticity. We do not consider any of these models here but instead refer the interested reader to the suggested literature for further reading. The fact that there is no unique generally accepted model of viscoelasticity has prompted some authors to call this theory non-classical.

6.7 Plasticity

We consider an extension of a metallic rod with constant cross-section radius R and length L. The rod is extended by two forces of magnitude F applied to the flat surfaces of the rod. The forces have opposite directions. We assume that the

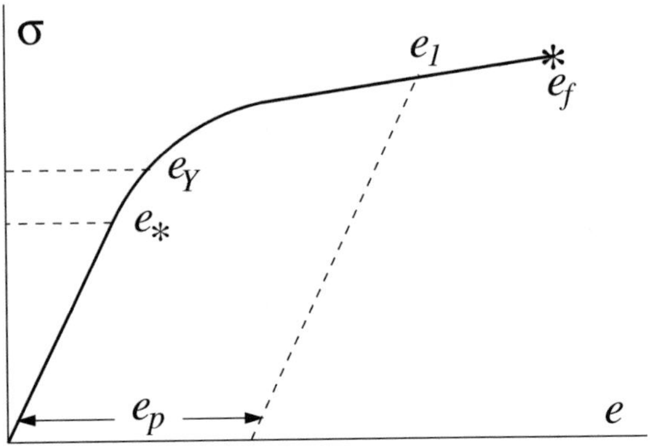

Fig. 6.7 Typical dependence of σ on e for metals

forces are homogeneously distributed and introduce the notation $\sigma = F/\pi R^2$. We also introduce the relative rod extension $e = l/L$, also called the strain, where l is the absolute rod extension. Then in the framework of linear elasticity we obtain $\sigma = Ee$ (see problem **6.6**). However, this relation is only valid for $e \le e_*$. In the case of metals usually $10^{-3} \le e_* \le 5 \times 10^{-3}$. A typical dependence of σ on e is shown in Fig. 6.7. When $e_* \le e \le e_Y$ the dependence of σ on e is nonlinear. However, if e is in this interval, then the rod length will return to its initial value after we remove the forces extending it. For such values of e the deformation of the rod is described by nonlinear elasticity. When $e > e_Y$ the behaviour of the rod is qualitatively different. If the strain exceeds e_Y then after removing the force applied to the rod a residual strain remains. In accordance with this e_Y is called the *yield strain*, and $\sigma_Y = \sigma(e_Y)$ the *yield stress*. For example, if after the rod extension the strain is $e_1 > e_Y$, then the plastic strain is $e_p = e_1 - \sigma(e_1)/E$. This quantity is positive because $\sigma''(e) < 0$ for $e > e_*$, where the prime indicates the derivative. This behaviour of a solid is called *yielding*. Finally, if e increases further the rod will completely fail at $e = e_f$. The curve showing the dependence of σ on e in the case of compression ($e < 0$) is symmetric to that shown in Fig. 6.7, that is, $\sigma(-e) = -\sigma(e)$. In a simplified version of the theory of plasticity the dependence of σ on e is approximated by the curve shown in Fig. 6.8. In this model the rod deformation is described by linear elasticity for $\sigma \le \sigma_Y$, while the stress cannot exceed σ_Y because the material starts to behave like a viscous fluid when $\sigma \ge \sigma_Y$. A solid where the relation between σ and e is described by Fig. 6.8 is called an *elastic-perfectly plastic solid*.

Below we consider the model for elastic-perfectly plastic solids. In the one-dimensional case the condition of yielding in this model is very simple: $|\sigma| = \sigma_Y$. We write the absolute value because $\sigma(e) < 0$ when $e < 0$. It is much more difficult to obtain the yielding condition in the case of three-dimensional deformation. It is natural to assume that it should be written as the condition that a function of the stress tensor is equal to σ_Y. Experiments show that the yielding condition is independent of the external pressure. This means that it must remain the same if we add $p\mathbf{I}$ to the stress tensor, where p is an arbitrary constant. This condition implies that the

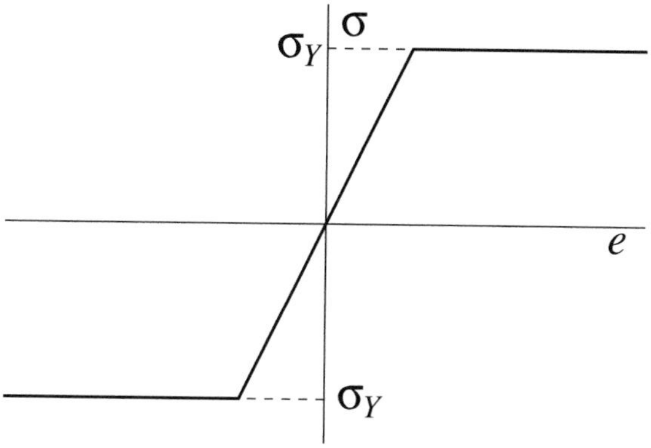

Fig. 6.8 Stress-strain curve for an elastic-perfectly plastic solid

function determining the yielding condition must depend not on the stress tensor itself, but on its deviator $\mathbf{T}' = \mathbf{T} - \frac{1}{3}I_1(\mathbf{T})\mathbf{I}$, where $I_1(\mathbf{T})$ is the first invariant of the stress tensor. We note that the first invariant of the deviator of any tensor is zero. There are a few different expressions for the function of stress tensor determining the yield condition. The two most popular expressions are those given by von Mises and by Tresla. Below we will only consider the *von Mises criterion*. It is written as

$$\sqrt{-3I_2(\mathbf{T}')} \equiv \sqrt{\frac{3}{2}T'_{ij}T'_{ij}} = \sigma_Y, \tag{6.7.1}$$

where $I_2(\mathbf{T}')$ is the second invariant of the deviator of the stress tensor. It can be written in terms of principal stresses as

$$\boxed{\sqrt{(T_1 - T_2)^2 + (T_2 - T_3)^2 + (T_3 - T_1)^2} = \sqrt{2}\,Y,} \tag{6.7.2}$$

where, for simplicity, we use the notation $Y = \sigma_Y$. This equation defines a surface in a three-dimensional space with Cartesian coordinates T_1, T_2, T_3. We make the orthogonal coordinate transformation in this space defined by

$$\begin{pmatrix} X_1 \\ X_2 \\ X_3 \end{pmatrix} = \begin{pmatrix} \frac{1}{\sqrt{3}} & \frac{1}{\sqrt{3}} & \frac{1}{\sqrt{3}} \\ -\frac{1}{\sqrt{2}} & \frac{1}{\sqrt{2}} & 0 \\ -\frac{1}{\sqrt{6}} & -\frac{1}{\sqrt{6}} & \frac{2}{\sqrt{6}} \end{pmatrix} \begin{pmatrix} T_1 \\ T_2 \\ T_3 \end{pmatrix}. \tag{6.7.3}$$

It is straightforward to verify that the matrix in this equation is proper orthogonal. Using this coordinate transformation we transform Eq. (6.7.3) into

$$\boxed{X_2^2 + X_3^2 = \frac{2}{3}Y^2.} \tag{6.7.4}$$

This is the equation of a circular cylinder with cross-section radius $Y\sqrt{2/3}$ and axis coinciding with the X_1-axis. In the space of principal stresses the axis of this cylinder is parallel to the vector (1, 1, 1).

Equation (6.7.1) provides one constitutive equation for an elastic-perfectly plastic solid. In order to completely describe the behaviour of this solid we need five more equations. Below we only consider small displacements that can be described by the linear theory. It is obvious that these equations cannot relate the stress and strain tensor because in the model that we consider in the one-dimensional case σ remains constant when $e > e_Y$. Hence, it looks like a viable assumption that these equations relate the stress tensor and the infinitesimal increment of the strain tensor. Since the plastic behaviour is independent of the external pressure, this relation must only contain the deviator of the stress tensor. It is natural to assume that, in the linear theory, an infinitesimal increment of the strain tensor is a linear homogeneous function of the deviator of the stress tensor. Then similar to the derivation of Eq. (6.2.12) we obtain that for an isotropic material

$$\boxed{d\mathbf{E} = \mathbf{T}' d\Lambda,} \tag{6.7.5}$$

where we have taken into account that $I_1(\mathbf{T}') = 0$. In this equation $\Lambda \geq 0$ is a function of time and spatial variables that must be determined from the solution of a problem. Equation (6.7.5) is a particular form of the *Levy–von Mises equation*. We emphasise again that it is only valid for small displacements when the linear description can be used. Equations (6.7.1) and (6.7.5) are the constitutive equations for plastic deformation. We now consider an example to show how this theory works.

Example We consider the same problem of bathysphere collapse that is studied in Sect. 6.3. The yield stress for the material that the bathysphere is made of is $Y = 10^8$ N m^{-2}. The depth that the bathysphere is immersed in is slowly increasing. Determine the depth h_Y at which plastic deformation will be initiated in the bathysphere material. Also calculate the minimum depth at which the plasticity region will spread through the whole shell.

Solution: Since the water pressure at the typical depth that we consider is much higher than the atmospheric pressure, to simplify the analysis we take $p_a = 0$. It is shown in Sect. 6.3 that $\mathbf{T}$ is given by Eq. (6.3.18). It follows from this expression that the principal stresses are given by

$$T_1 = \lambda \nabla \cdot \boldsymbol{u} + 2\mu \frac{du}{dr}, \quad T_2 = T_3 = \lambda \nabla \cdot \boldsymbol{u} + 2\mu \frac{u}{r}, \tag{6.7.6}$$

where

$$\boldsymbol{u} = u(r)\boldsymbol{e}_r, \quad u = Ar + \frac{B}{r^2}, \tag{6.7.7}$$

and the constants A and B are given by Eq. (6.3.22). Equation (6.7.2) reduces to $|T_1 - T_2| = Y$. Using Eq. (6.3.22), (6.7.6), and (6.7.7) we rewrite this equation as

$$\frac{3\rho g h a^3 b^3}{2r^3(b^3 - a^3)} = Y. \tag{6.7.8}$$

The left-hand side of this equation takes a maximum at $r = a$. This implies that plastic deformation will be initiated at $r = a$. Substituting $a = 1$ m, $b = 1.1$ m, $g = 10\,\mathrm{m\,s^{-2}}$, and $\rho = 10^3\,\mathrm{kg\,m^{-3}}$ we obtain that plastic deformation will be initiated at the depth

$$h_Y = \frac{2Y(b^3 - a^3)}{3\rho g b^3} \approx 1660\,\mathrm{m}. \tag{6.7.9}$$

When the depth increases beyond h_Y the yield spreads outside the inner boundary of the shell. While the expression for u given by Eq. (6.7.7) is not valid in the plasticity region, we still have $\boldsymbol{u} = u(r)\boldsymbol{e}_r$ because of the problem symmetry. This implies that in the plasticity region the expression for $\nabla\boldsymbol{u}$ given by Eq. (6.3.17) remains valid. Since the tensor $\nabla\boldsymbol{u}$ is symmetric, it follows that $\mathsf{E} = \nabla\boldsymbol{u}$. Substituting this expression into Eq. (6.7.5) we obtain that $\boldsymbol{e}_r$, $\boldsymbol{e}_\theta$, $\boldsymbol{e}_\phi$ are the unit vectors in the principal directions of the tensor T, and $T_2 = T_3$. Then the yield Eq. (6.7.2) reduces to

$$|T_1 - T_2| = Y. \tag{6.7.10}$$

To obtain the second equation for T_1 and T_2 we use the momentum equation. We neglect the body force and assume that the depth increase occurs so slowly that we can also neglect the acceleration term. Then the momentum equation reduces to the equilibrium Eq. (4.4.5) with $\boldsymbol{b} = 0$, that is, $\nabla \cdot \mathsf{T} = 0$. After a long but straightforward calculation with the use of Eq. (2.13.13) and the expression

$$\mathsf{T} = T_1\boldsymbol{e}_r\boldsymbol{e}_r + T_2(\boldsymbol{e}_\theta\boldsymbol{e}_\theta + \boldsymbol{e}_\phi\boldsymbol{e}_\phi) \tag{6.7.11}$$

we obtain that the equilibrium equation reduces to

$$r\frac{dT_1}{dr} + 2(T_1 - T_2) = 0. \tag{6.7.12}$$

It follows from Eqs. (6.7.10) and (6.7.12) that

$$r\frac{dT_1}{dr} = \pm 2Y. \tag{6.7.13}$$

At the internal boundary the pressure is zero. Then we have the boundary condition $T_1 = 0$. Integrating Eq. (6.7.13) and using this boundary condition we obtain

$$T_1 = \pm 2Y \ln\frac{r}{a}. \tag{6.7.14}$$

The medium behaviour is elastic in the region $r_0 < r \le b$, and plastic in $a \le r \le r_0$. The surface traction must be continuous at $r = r_0$. Using Eqs. (6.7.6), (6.7.7), and

(6.7.14) we obtain that this boundary condition reads

$$\pm 2Y \ln\frac{r_0}{a} = (3\lambda + 2\mu)A - \frac{4\mu B}{r_0^3}. \tag{6.7.15}$$

It is worth noting that Eqs. (6.7.6) and (6.7.7) remain valid in the part of the shell where the deformations are elastic even when there is a plasticity region, however Eq. (6.3.22) determining A and B cannot be used. The second boundary condition must be satisfied at the external boundary, where the surface traction must be continuous. Using Eqs. (6.7.6) and (6.7.7) we write it as

$$(3\lambda + 2\mu)A - 4\mu B b^{-3} = -\rho g h. \tag{6.7.16}$$

Finally, the yield condition is satisfied at $r = r_0$. Using Eqs. (6.7.2), (6.7.6), and (6.7.7) we obtain

$$6\mu|B| = r_0^3 Y. \tag{6.7.17}$$

When $h = h_Y$ this condition is satisfied at $r_0 = a$, and in this case $B < 0$. When h increases A and B must vary continuously. It follows from Eq. (6.7.17) that $B \neq 0$. This implies that B remains negative and we must substitute $|B| = -B$ into Eq. (6.7.17). Equations (6.7.15)–(6.7.17) constitute a system of equations determining three unknown quantities A, B, and r_0. It follows from Eqs. (6.7.16) and (6.7.17) that

$$A = -\frac{3\rho g h b^3 + 2r_0^3 Y}{3b^3(3\lambda + 2\mu)}, \quad B = -\frac{r_0^3 Y}{6\mu}. \tag{6.7.18}$$

Substituting these expressions into Eq. (6.7.15) and using Eq. (6.7.9) we obtain

$$\pm 2Y \ln\frac{r_0}{a} = \rho g(h - h_Y) + \frac{2Y(r_0^3 - a^3)}{3b^3}. \tag{6.7.19}$$

The right-hand side of this equation is positive when $h > h_Y$ and $r_0 > a$. This implies that the left-hand side also must be positive and we must choose the sign plus. We rewrite Eq. (6.7.19) as

$$f(x) = y, \quad f(x) = \frac{2Y}{\rho g}\left(\ln x - \frac{a^3(x^3 - 1)}{3b^3}\right), \tag{6.7.20}$$

where $x = r_0/a$ and $y = h - h_Y$. It follows from Eq. (6.7.9) that $f(1) = 0$. It is easy to show that $f'(x) > 0$ for $x < b/a$, which implies that Eq. (6.7.20) has a unique solution in the interval $x \in [1, b/a]$ when $0 \le y \le f(b/a)$. In particular, it follows that the yield spreads through the whole shell when

$$h = h_Y + f(b/a) = \frac{2Y}{\rho g}\ln\frac{b}{a} \approx 1906\,\text{m}. \tag{6.7.21}$$

When the bathysphere is immersed at this depth the shell will start to shrink quickly and collapse. In fact, it can collapse at a shallower depth because of the instability that results when the shell loses its spherical shape.

Problems

6.1: Consider an infinite elastic tube of internal radius a and external radius b. At the initial time the pressure inside and outside the tube is equal to the atmospheric pressure p_a. Then the pressure inside the tube is increased to $p_0 > p_a$.

(i) Introduce cylindrical coordinates r, ϕ, z with the z-axis coinciding with the tube axis. Neglecting the body force ($\boldsymbol{b} = 0$), and assuming that the displacement of the tube material is in the radial direction ($\boldsymbol{u} = u\boldsymbol{e}_r$, where $\boldsymbol{e}_r$ is the unit vector in the radial direction), and depends on r only ($u = u(r)$), show that the equilibrium equation of linear elasticity for isotropic material (Eq. (6.3.8)) reduces to

$$\frac{d}{dr}\left(\frac{1}{r}\frac{d(ru)}{dr}\right) = 0.$$

Then show that the displacement is given by

$$u = Ar + \frac{B}{r},$$

where A and B are constants.

(ii) Show that the surface traction at the tube boundaries is given by

$$\boldsymbol{t}(r_0) = \left(\frac{\lambda}{r}\frac{d(ru)}{dr}\bigg|_{r=r_0} + 2\mu\frac{du}{dr}\bigg|_{r=r_0}\right)\boldsymbol{n},$$

where $r_0 = a$ and $\boldsymbol{n} = -\boldsymbol{e}_r$ at the internal boundary, and $r_0 = b$ and $\boldsymbol{n} = \boldsymbol{e}_r$ at the external boundary. Use the boundary conditions, $\boldsymbol{t} = p_0\boldsymbol{e}_r$ at $r = a$ and $\boldsymbol{t} = -p_a\boldsymbol{e}_r$ at $r = b$, to determine the increase in the internal radius of the tube δ. Calculate the numerical value of δ if $a = 11$ cm, $b = 12$ cm, $p_a = 10^5\ \mathrm{N/m^2}$, $p_0 = 15p_a$, $\lambda = 2\mu$, and $\mu = 10^8\ \mathrm{N/m^2}$.

6.2: Suppose there is an elastic spherical shell of internal radius a and external radius b. The shell is filled with water at 0° C through a small hole, and then the hole is tightly sealed. Then the water in the shell is heated to a temperature of 80° C. As a result its density decreases and its volume increases. You are given that the water density at 0° C is $\rho_1 \approx 1000\ \mathrm{kg\,m^{-3}}$, while it is $\rho_2 \approx 985\ \mathrm{kg\,m^{-3}}$ at 80° C.

(i) Calculate the internal radius of the shell after the water has been heated.

(ii) Assuming that the displacement is small everywhere, find the dependence of the displacement on the radial coordinate r in a spherical coordinate system with the origin at the shell's centre.

(iii) Find the expression for the stress tensor in terms of a, b, r, and the Lamé constants λ and μ.

6.3: The shell described in the previous problem collapses if the stress magnitude in at least one point inside the shell and at least in one direction is larger than the critical value $t_{\rm cr} = 10^8$ N m^{-2}. You are given that $a = 5$ cm, $b = 7$ cm, $\lambda = \mu$, and $p_a = 10^5$ N m^{-2}. What is the maximum value of μ for which the shell still endures the pressure imposed by the heated water that is inside it? Does it endure this pressure if it is made of cast iron ($\mu = 4 \times 10^{10}$ N m^{-2})?

6.4: Suppose there is an elastic rod of radius R and length L. The rod is made of an isotropic material. There is no body force, $\boldsymbol{b} = 0$. The rod axis coincides with the z-axis of cylindrical coordinates. Its ends are fixed at the planes $z = 0$ and $z = L$. Initially there are no stresses in the rod. Then the upper plane is rotated by a small angle α, so the displacement at $z = L$ is $\boldsymbol{u} = (0, \alpha r, 0)$. The lower plane does not move, so the displacement is zero at $z = 0$. You can assume that the displacement in the rod has the form $\boldsymbol{u} = (0, u(r, z), 0)$.

(i) Looking for a solution to the equation describing the equilibrium of an elastic isotropic medium in the form $u = rf(z)$, determine the function $f(z)$.

(ii) You are given that the momentum of the force applied to the upper plane is M. Determine the angle α.

6.5: Suppose there is a cylinder of radius R and height H made of an elastic isotropic material. The cylinder is put on a horizontal surface. The surface is smooth, so that the base of the cylinder can slide without friction against the horizontal surface. The deformation of the surface can be neglected. There is a force of magnitude F applied to the flat top of the cylinder. The force is directed downward and evenly distributed, meaning that the pressure is the same at all points of the flat top.

(i) You can assume that, in cylindrical coordinates r, ϕ, z with the z-axis directed upward and coinciding with the cylinder axis, the displacement in the cylinder has the form $\boldsymbol{u} = u_r(r)\boldsymbol{e}_r + u_z(z)\boldsymbol{e}_z$, where $\boldsymbol{e}_r$ and $\boldsymbol{e}_z$ are the unit vectors in the r and z-direction. Show that the equilibrium equation for the cylinder reduces to the system of two equations,

$$\frac{d}{dr}\left(\frac{1}{r}\frac{d(ru_r)}{dr}\right) = 0, \quad \frac{d^2u_z}{dz^2} = 0.$$

Assuming that the origin of the cylindrical coordinates is on the horizontal plane, show that the general solution to this system of equations, regular at $r = 0$, is given by $u_r = Ar$ and $u_z = Bz$, where A and B are constants.
(ii) Use Eq. (2.13.7) to show that

$$\nabla\boldsymbol{u} = A\boldsymbol{e}_r\boldsymbol{e}_r + A\boldsymbol{e}_\phi\boldsymbol{e}_\phi + B\boldsymbol{e}_z\boldsymbol{e}_z,$$

where $\boldsymbol{e}_\phi$ is the unit vector in the ϕ-direction. Then use Eq. (6.3.16) to find the expression for the stress tensor $\mathbf{T}$.

(iii) Obtain the expressions for the surface tractions at the flat top and the side surface of the cylinder. Use the continuity of surface traction to express the constants A and B in terms of R, H, F, and the Lamé constants λ and μ. Then show that the ratio of the relative increase in the cylinder radius and the relative decrease in its height is equal to Poisson's ratio, and the ratio of the stress magnitude at the top of the cylinder to the absolute value of the relative change of its length is equal to Young's modulus E.

(iv) You are given that $H = R = 1$ m, $F = 10^8$ N, and $\lambda = \mu = 10^{10}$ N m^{-2}. Calculate the decrease in the cylinder's height and the increase in its radius.

6.6: Suppose there is a cylindrical rod of radius R and length L made of an elastic isotropic material. There are two forces of magnitude F stretching the rod that are applied to the flat boundaries of the rod. They are parallel to the rod axis and act in opposite directions. The forces are evenly distributed on the flat surfaces, meaning that the surface traction is the same at all points of the flat surfaces. You are given that the rod extension is l.

(i) You can assume that, in cylindrical coordinates r, ϕ, z with the z-axis coinciding with the cylinder axis and the origin at the centre of the rod, the displacement in the cylinder has the form $\boldsymbol{u} = u_r(r)\boldsymbol{e}_r + u_z(z)\boldsymbol{e}_z$, where $\boldsymbol{e}_r$ and $\boldsymbol{e}_z$ are the unit vectors in the r and z-directions. Show that, for this particular form of $\boldsymbol{u}$, Eq. (6.3.8) with $\boldsymbol{b} = 0$ reduces to the system of two equations,

$$\frac{\partial}{\partial r}\left(\frac{1}{r}\frac{\partial(ru_r)}{\partial r}\right) = 0, \qquad \frac{\partial^2 u_z}{\partial z^2} = 0.$$

Assuming that the displacement at the coordinate origin is zero show that the general solution to this system of equations, regular at $r = 0$, is given by $u_r = Ar$, $u_z = Bz$, where A and B are constants.

(ii) Using Eq. (6.3.16) show that

$$\mathsf{T} = [2A(\lambda + \mu) + \lambda B](\boldsymbol{e}_r\boldsymbol{e}_r + \boldsymbol{e}_\phi\boldsymbol{e}_\phi) + [2A\lambda + B(\lambda + 2\mu)]\boldsymbol{e}_z\boldsymbol{e}_z,$$

where $\boldsymbol{e}_\phi$ is the unit vector in the ϕ-direction. Use this expression to calculate the surface traction at the flat surfaces of the rod and at its side surface. Then, using the condition of continuity of the surface traction at these two surfaces and neglecting the air pressure determine the constants A and B, and then express the force F acting on the flat surfaces in terms of R, L, l, and the Lamé constants λ and μ.

(iii) Calculate the decrease in the cylinder's radius. Show that the ratio of the relative decrease in the cylinder's radius and the relative increase in its length is equal to Poisson's ratio. Also show that the ratio of the stress magnitude at the rod's flat end to the rod's relative extension is equal to Young's modulus E.

6.7: Suppose there is an elastic cylindrical tube of internal radius R and external radius $R + l$. You are given that the displacement in the tube is in the radial direction and only depends on the distance from the tube axis. Hence, in the cylindrical coordinates, $\boldsymbol{u} = u(t, r)\boldsymbol{e}_r$, where $\boldsymbol{e}_r$ is the unit vector in the radial direction.

(i) Show that, in this case, the equation of motion (6.3.7) reduces to

$$\frac{\partial^2 u}{\partial t^2} = c_p^2 \left(\frac{\partial^2 u}{\partial r^2} + \frac{1}{r}\frac{\partial u}{\partial r} - \frac{u}{r^2} \right). \qquad (*)$$

(ii) Looking for a solution to equation $(*)$ in the form $u = U(r)\sin(\omega t)$ derive the equation determining ω. You can neglect the air pressure when writing the boundary conditions.

(iii) You are now given that $\epsilon = l/R \ll 1$. Find ω in the leading order approximation with respect to ϵ.

6.8: Suppose the surface of discontinuity $x = 0$ separates an elastic material with $\rho = \rho_-$, $\lambda = \lambda_-$, and $\mu = \mu_-$ filling in the half-space $x < 0$ from another elastic material with $\rho = \rho_+$, $\lambda = \lambda_+$, and $\mu = \mu_+$ filling in the half-space $x > 0$.

(i) Assuming that $\boldsymbol{u} = (u, 0, 0)$ and u only depends on one spatial coordinate $x = x_1$, use the condition of continuity of $\boldsymbol{u}$ and $\boldsymbol{t}$ at $x = 0$ to derive two boundary conditions for u at $x = 0$.

(ii) Suppose there is a plane harmonic longitudinal wave propagating in the half-space $x < 0$ in the positive x-direction. This wave is defined by $u = a\cos[\omega(t - x/c_{p-})]$. Show that a solution can be sought in the form

$$u = \begin{cases} a\cos[\omega(t - x/c_{p-})] + b\cos[\omega(t + x/c_{p-})], & x < 0, \\ c\cos[\omega(t - x/c_{p+})], & x > 0, \end{cases}$$

where $c_{p\pm} = \sqrt{(\lambda_\pm + 2\mu_\pm)/\rho_\pm}$. Use the boundary conditions at $x = 0$ to calculate b/a and c/a, and find the coefficient of reflection $R = b^2/a^2$.

6.9: Suppose there is an elastic rod of radius R and length L. The rod is made of an isotropic material. There is no body force, $\boldsymbol{b} = 0$. The rod axis coincides with the z-axis of cylindrical coordinates. Its upper end is fixed at the plane $z = L$. Suppose further that there is a thin disc with radius much larger than R attached at the lower end of the rod in such a way that it is in the horizontal plane and its centre is on the z-axis. The momentum of inertia of this disc with respect to the z-axis is I. Initially there are no stresses in the rod. Then the disc is rotated by a small angle α_0, so the displacement in the rod at $z = 0$ is $\boldsymbol{u} = (0, \alpha_0 r, 0)$. The upper plane does not move, so the displacement is zero at $z = L$. After that the disc is released and allowed to move freely. As a result it starts to oscillate about the equilibrium position. Determine the frequency of this oscillation. You can neglect the moment of inertia of the rod with respect to the z-axis.

[Hint: Use the results obtained in the solution to problem **6.4**.]

6.10: Suppose a viscoelastic medium fills in the half-space $x > 0$ in Cartesian coordinates x, y, z. The motion of this medium is governed by Eq. (6.6.2). The surface bounding the medium in the x-direction oscillates with a small amplitude a, so that its position is given by $x = a \sin(\omega t)$. This oscillation drives a longitudinal wave in the viscoelastic medium. Determine the dependence of this wave amplitude on x.

6.11: Suppose a viscoelastic medium fills in the half-space $x > 0$ in Cartesian coordinates x, y, z. The motion of this medium is governed by Eq. (6.6.2). The surface bounding the medium in the x-direction oscillates with a small amplitude a in the y-direction, so that the displacement of any point in this plane is given by $v = a \sin(\omega t)$. This oscillation drives a transversal wave in the viscoelastic medium. Determine the dependence of this wave amplitude on x.

6.12: (Hunter, see Further Reading). Consider the rod described in problem **6.4**. The yield stress of the material that the rod is made of is Y.

(i) Determine the critical angle α_Y when the yield is initiated and the corresponding momentum is applied to the upper plane.

(ii) Calculate the momentum applied to the upper plane for $\alpha > \alpha_Y$.

Chapter 7
Viscous Incompressible Fluids

7.1 Main Assumptions and Governing Equations

In Chap. 5 we studied the motion of ideal fluids. In such fluids the layers can slide against each other without friction. In real fluids there is always some internal friction, which is called viscosity. In this chapter we study the motion of viscous fluids. As we mentioned in Chap. 5, we can often neglect viscosity. In this chapter we obtain an exact criterion which tells us when we can do this. But even when we can neglect the viscosity almost everywhere in the region occupied by a fluid, we cannot do this near its boundaries. As we pointed out in Sect. 4.6 of Chap. 4, the velocity of a viscous fluid at a rigid boundary must coincide with the boundary velocity. In particular, it must be zero at an immovable rigid boundary. Hence, near such a boundary there should be a thin layer where the fluid velocity varies from its value at the vicinity of the boundary to zero at the boundary.

In a viscous fluid the stress tensor is the sum of two terms. The first term describes the pressure and it is the same as in ideal fluids. The second term describes the viscosity. Hence, the stress tensor in a viscous fluid is given by

$$\mathsf{T} = -p\mathsf{I} + \mathsf{T}', \tag{7.1.1}$$

where, as before, p is the pressure, and T' is the *viscosity* tensor.

If a viscous fluid moves as a rigid body than there is no friction between its different parts. This implies that T' must depend on the velocity gradient tensor L and $\mathsf{T}' = 0$ when $\mathsf{L} = 0$. We now consider a rigid rotation of the fluid with constant angular velocity. In this case the fluid velocity is given by $\boldsymbol{v} = \boldsymbol{\Omega} \times \boldsymbol{x} = \boldsymbol{e}_i \varepsilon_{ikl} \Omega_k x_l$, where $\boldsymbol{\Omega}$ is a constant vector. Obviously, in this case there is also no friction among different parts of the fluid, meaning that $\mathsf{T}' = 0$. For this particular motion the components of the symmetric and anti-symmetric parts of L are

$$D_{ij} = \frac{1}{2}\left(\frac{\partial v_i}{\partial x_j} + \frac{\partial v_j}{\partial x_i}\right) = \frac{1}{2}(\varepsilon_{ikj} + \varepsilon_{jki})\Omega_k = 0,$$

© Springer Nature Switzerland AG 2019
M. S. Ruderman, *Fluid Dynamics and Linear Elasticity*, Springer Undergraduate Mathematics Series, https://doi.org/10.1007/978-3-030-19297-6_7

$$W_{ij} = \frac{1}{2}\left(\frac{\partial v_i}{\partial x_j} - \frac{\partial v_j}{\partial x_i}\right) = \frac{1}{2}(\varepsilon_{ikj} - \varepsilon_{jki})\Omega_k = -\varepsilon_{ijk}\Omega_k.$$

Since W_{ij} has only three non-trivial components, we can always fix them and find $\boldsymbol{\Omega}$ such that the last relation is satisfied. Then it follows that $\mathbf{T}' = 0$ when $D_{ij} = 0$, which implies that $\mathbf{T}'$ only depends on the symmetric part of $\mathbf{L}$, which is the strain-rate tensor $\mathbf{D}$. When the velocity gradients are not very large we can approximate the tensor-valued function $\mathbf{T}'(\mathbf{D})$ by the first term of its Taylor series expansion and thus obtain the linear relation between T'_{ij} and D_{ij}.

Below we only consider isotropic fluids. In Chap. 6 we obtained the general form of the linear relation between the stress tensor and the infinitesimal strain tensor $\mathbf{E}$, valid in an isotropic medium (see Eq. (6.2.12)). When deriving this relation we did not use the fact that $\mathbf{E}$ is the infinitesimal strain tensor. It could be any symmetric tensor. Hence, we can state that if $\mathbf{T}$ is a linear function of any symmetric second order tensor, then, in an isotropic medium, it is a linear superposition of this tensor and its first invariant multiplied by the unit tensor. In linear elasticity we denoted the coefficients of this linear superposition by 2μ and λ. In the case of a viscous fluid we denote them by 2η and $\zeta - \frac{2}{3}\eta$. As a result, we obtain

$$\mathbf{T}' = \left(\zeta - \frac{2}{3}\eta\right) I_1(\mathbf{D})\mathbf{I} + 2\eta\mathbf{D}, \tag{7.1.2}$$

or, in components,

$$T'_{ij} = \eta\left(\frac{\partial v_i}{\partial x_j} + \frac{\partial v_j}{\partial x_i} - \frac{2}{3}\delta_{ij}\frac{\partial v_k}{\partial x_k}\right) + \zeta\delta_{ij}\frac{\partial v_k}{\partial x_k}. \tag{7.1.3}$$

The coefficients η and ζ are called the *dynamic* and *second* viscosity, respectively. Equations (7.1.1) and (7.1.2) (or (7.1.3)) are the constitutive equations for viscous fluids. Using (7.1.1) and (7.1.3) we rewrite the momentum equation (4.4.4) as

$$\boxed{\rho\left(\frac{\partial \boldsymbol{v}}{\partial t} + (\boldsymbol{v}\cdot\nabla)\boldsymbol{v}\right) = -\nabla p + \eta\nabla^2\boldsymbol{v} + \left(\zeta + \frac{1}{3}\eta\right)\nabla(\nabla\cdot\boldsymbol{v}) + \rho\boldsymbol{b}.} \tag{7.1.4}$$

Note that we use the Eulerian description.

In what follows we study *homogeneous incompressible* viscous fluids and assume that $\rho = \text{const}$, which is an additional constitutive equation. Then it follows from the mass conservation equation (3.8.2) that

$$\nabla\cdot\boldsymbol{v} = 0. \tag{7.1.5}$$

Using this equation and taking $\boldsymbol{b} = 0$ we rewrite (7.1.4) as

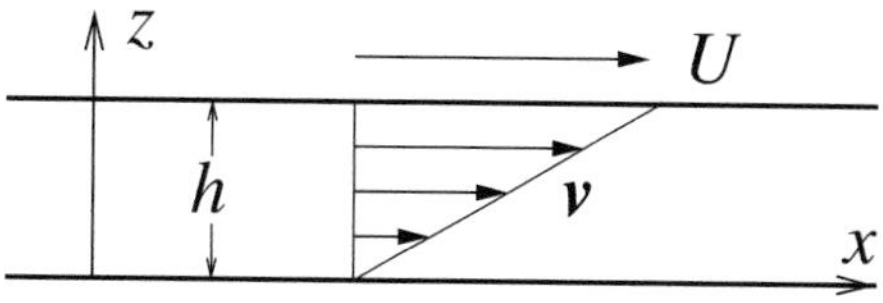

Fig. 7.1 Couette flow between two parallel planes. The lower plane is at rest, while the upper plane is moving with speed U in a direction parallel to the lower plane

$$\boxed{\rho\left(\frac{\partial \boldsymbol{v}}{\partial t} + (\boldsymbol{v}\cdot\nabla)\boldsymbol{v}\right) = -\nabla p + \eta\nabla^2\boldsymbol{v}.} \tag{7.1.6}$$

Since the second viscosity ζ is not present in the system of equations (7.1.5) and (7.1.6), the dynamic viscosity η is simply called the viscosity. Equation (7.1.6) is called the *Navier–Stokes equation* after the famous scientists, French Claude Louis Marie Henri Navier, and Anglo-Irish George Gabriel Stokes. It is a fundamental equation in fluid mechanics.

Equations (7.1.5) and (7.1.6) constitute a closed system of equations for $\boldsymbol{v}$ and p. To solve this system we also have to specify the initial and boundary conditions. In this respect recall that a viscous fluid cannot slide along rigid boundaries, so that its velocity at a rigid boundary has to be equal to the velocity of the boundary.

Example Derive the equation describing the temporal evolution of $\boldsymbol{\omega} = \frac{1}{2}\nabla\times\boldsymbol{v}$.

Solution: Using Eq. (5.3.1) we transform Eq. (7.1.6) into

$$\frac{\partial \boldsymbol{v}}{\partial t} + 2\boldsymbol{\omega}\times\boldsymbol{v} = -\nabla\left(\frac{p}{\rho} + \frac{1}{2}\|\boldsymbol{v}\|^2\right) + \frac{\eta}{\rho}\nabla^2\boldsymbol{v}.$$

Now we apply curl to this equation to obtain

$$\frac{\partial \boldsymbol{\omega}}{\partial t} + \nabla\times(\boldsymbol{\omega}\times\boldsymbol{v}) = \frac{\eta}{\rho}\nabla^2\boldsymbol{\omega}. \tag{7.1.7}$$

We have

$$\begin{aligned}\nabla\times(\boldsymbol{\omega}\times\boldsymbol{v}) &= \boldsymbol{e}_i\varepsilon_{ijk}\frac{\partial}{\partial x_j}(\varepsilon_{klm}\omega_l v_m) = \boldsymbol{e}_i(\delta_{il}\delta_{jm} - \delta_{im}\delta_{jl})\frac{\partial(\omega_l v_m)}{\partial x_j}\\ &= \boldsymbol{e}_i\frac{\partial(\omega_i v_j)}{\partial x_j} - \boldsymbol{e}_i\frac{\partial(\omega_j v_i)}{\partial x_j} = (\boldsymbol{v}\cdot\nabla)\boldsymbol{\omega} - (\boldsymbol{\omega}\cdot\nabla)\boldsymbol{v} + \boldsymbol{\omega}\nabla\cdot\boldsymbol{v} - \boldsymbol{v}\nabla\cdot\boldsymbol{\omega}.\end{aligned}$$

Substituting this expression into Eq. (7.1.7) and taking into account that $\nabla\cdot\boldsymbol{v} = \nabla\cdot\boldsymbol{\omega} = 0$ we finally arrive at

$$\frac{\partial \boldsymbol{\omega}}{\partial t} + (\boldsymbol{v}\cdot\nabla)\boldsymbol{\omega} = (\boldsymbol{\omega}\cdot\nabla)\boldsymbol{v} + \frac{\eta}{\rho}\nabla^2\boldsymbol{\omega}. \tag{7.1.8}$$

7.2 Couette Flow

Let us now consider the simplest exact solution to the system of equations (7.1.5) and (7.1.6), called *Couette flow* after the renowned French scientist Maurice Couette, who first experimentally studied a similar, but slightly more complicated flow between two coaxial rotating cylinders. If the gap between the cylinders is much smaller than the radius of the inner cylinder, then we can neglect the curvature and substitute the cylindrical surfaces by planes. After that we arrive at the flow shown in Fig. 7.1. The lower plane is at rest, while the upper one is moving with speed U in the positive x-direction. Here we use the notation $x = x_1$, $y = x_2$, and $z = x_3$. Let us look for a solution in the form $\boldsymbol{v} = u(z)\boldsymbol{e}_x$. Then $\nabla \cdot \boldsymbol{v} = 0$, so that (7.1.5) is satisfied. Since the motion of the upper plane is stationary, we assume that the fluid motion is also stationary, i.e. u is independent of t. It is easy to verify that $(\boldsymbol{v} \cdot \nabla)\boldsymbol{v} = 0$. Then the three components of (7.1.6) are:

$$\frac{\partial p}{\partial x} = \eta \frac{d^2 u}{dz^2}, \quad \frac{\partial p}{\partial y} = 0, \quad \frac{\partial p}{\partial z} = 0. \tag{7.2.1}$$

It follows from the second and third equation in (7.2.1) that $p = p(x)$. Then the first equation in (7.2.1) yields

$$p = \eta \frac{d^2 u}{dz^2} x + p_0,$$

where p_0 is a constant. We assume that p is bounded. This is only possible if

$$\frac{d^2 u}{dz^2} = 0,$$

which implies that $u = Az + B$ with A and B being constants. Then it follows that $p = p_0$. At the rigid boundaries the fluid velocity has to coincide with the velocities of these boundaries, which gives

$$u = 0 \ \text{ at } \ z = 0, \quad u = U \ \text{ at } \ z = h.$$

Using these conditions we determine A and B and, eventually, obtain

$$u = \frac{Uz}{h}. \tag{7.2.2}$$

Example Calculate the force exerted by the fluid on the upper boundary.

Solution: Since the external unit normal vector at this boundary is $-\boldsymbol{e}_z$ with components $-\delta_{j3}$, this force is

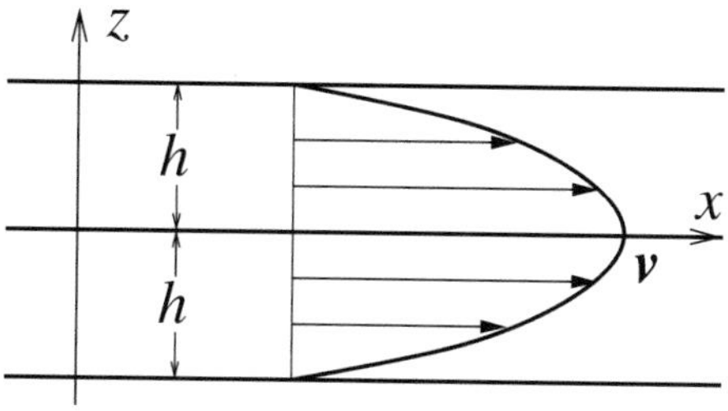

Fig. 7.2 Poiseuille flow between two parallel planes. The two planes are at rest. There is a constant pressure gradient in a direction parallel to the flow velocity

$$-\mathsf{T}\cdot\boldsymbol{e}_z = p\boldsymbol{e}_z - \eta\boldsymbol{e}_i\left(\frac{\partial v_i}{\partial x_j}+\frac{\partial v_j}{\partial x_i}\right)\delta_{j3}$$
$$= p\boldsymbol{e}_z - \eta\boldsymbol{e}_i\left(\frac{\partial v_i}{\partial z}+\frac{\partial v_3}{\partial x_i}\right) = p\boldsymbol{e}_z - \eta\frac{du}{dz}\,\boldsymbol{e}_x = p\boldsymbol{e}_z - \eta(U/h)\boldsymbol{e}_x. \quad (7.2.3)$$

The most important is the tangential component of this force $-\eta(U/h)\boldsymbol{e}_x$. It shows that, in order to keep the upper boundary moving, one needs to apply to this boundary the force $\eta U/h$ per square meter acting in the direction of motion.

7.3 Poiseuille Flow

Another very simple solution to the system of equations (7.1.5) and (7.1.6) is called *Poiseuille flow*, after the renowned French scientist Jean Léonard Marie Poiseuille, who first studied this flow. Poiseuille studied the flow in a pipe with a circular cross-section. We further simplify this problem and consider a flow between two immovable parallel planes (see Fig. 7.2). Again we assume that the flow is stationary and $\boldsymbol{v} = u(z)\boldsymbol{e}_x$. Then again $\nabla\cdot\boldsymbol{v} = 0$ and $(\boldsymbol{v}\cdot\nabla)\boldsymbol{v} = 0$, and Eq. (7.1.6) reduces to Eq. (7.2.1). However now we do not impose the condition that p is bounded. Instead we assume that the pressure gradient is constant, $\partial p/\partial x = -\Delta p = \text{const}$. Then it follows from the first equation in Eq. (7.2.1) that

$$u = -\frac{\Delta p}{2\eta}\,z^2 + Az + B, \quad (7.3.1)$$

where A and B are arbitrary constants. The velocity must satisfy the boundary conditions at the rigid boundaries,

$$u = 0 \ \text{ at } \ z = \pm h.$$

Using these conditions we determine A and B and eventually arrive at

$$u = \frac{\Delta p}{2\eta}\left(h^2 - z^2\right). \quad (7.3.2)$$

Let us calculate the water flux through a plane perpendicular to the flow velocity. This flux per unit length in the y-direction is

$$Q = \rho \int_{-h}^{h} u\,dz = \frac{2\rho h^3 \Delta p}{3\eta}. \tag{7.3.3}$$

Example Calculate the force acting on a unit area of the lower plane.

Solution: The unit vector at the lower plane is $\boldsymbol{e}_z$. Then the calculation is the same as the calculation of the force acting on the upper plane in the Couette flow, which is given by Eq. (7.2.3). Hence, we skip the intermediate steps and obtain

$$\mathsf{T}\cdot\boldsymbol{e}_z = -p\boldsymbol{e}_z + \eta\boldsymbol{e}_x \left.\frac{du}{dz}\right|_{z=-h} = -p\boldsymbol{e}_z + h\boldsymbol{e}_x \Delta p. \tag{7.3.4}$$

Again the most important is the tangential component of this force equal to $h\Delta p$. The same tangential force acts on the upper plane. Hence, to keep the planes at rest one needs to apply to each plane the force equal to $h\Delta p$ per unit area in a direction opposite to the velocity.

7.4 The Reynolds Number and Stokes' Approximation

Usually, when we consider the motion of a fluid we can introduce the characteristic spatial scale L and the characteristic speed V. For example, if we consider the motion of a sphere of radius R with velocity U in a fluid, then $L = R$ and $V = U$. Let us consider a stationary motion ($\partial/\partial t = 0$) and introduce the dimensionless variables

$$\boldsymbol{x}^* = \boldsymbol{x}/L, \quad \boldsymbol{v}^* = \boldsymbol{v}/V, \quad p^* = p/\rho V^2.$$

Then we rewrite (7.1.6) as

$$(\boldsymbol{v}^* \cdot \nabla^*)\boldsymbol{v}^* = -\nabla^* p^* + \mathrm{Re}^{-1}\nabla^{*2}\boldsymbol{v}^*, \tag{7.4.1}$$

where $\nabla^* = (\partial/\partial x_1^*,\ \partial/\partial x_2^*,\ \partial/\partial x_3^*)$, and $\mathrm{Re} = \rho LV/\eta$ is called the *Reynolds number*, after the renowned English scientist Osborne Reynolds. We see that the Navier–Stokes equation for stationary motions written in the dimensionless variables contains only one dimensionless parameter Re. Hence, the solution to Eq. (7.4.1) must have the form $\boldsymbol{v}^* = \boldsymbol{f}(\boldsymbol{x}^*, \mathrm{Re})$ implying that $\boldsymbol{v} = V\boldsymbol{f}(\boldsymbol{x}/L, \mathrm{Re})$. Let us assume, for example, that the solution to the Navier–Stokes equation describing the flow about a sphere of radius R in an unbounded fluid moving with speed U at large distances from the sphere is $\boldsymbol{v} = \boldsymbol{F}(\boldsymbol{x})$. Let us now consider the same problem, but with a sphere of radius $R' = 2R$. To have the same Reynolds number we need to decrease the flow speed by a factor of 2 and take it to be $U' = U/2$. Then the solution describing the flow near this sphere with flow speed U' far from the sphere is given by $\boldsymbol{v} = \frac{1}{2}\boldsymbol{F}(\boldsymbol{x}/2)$.

When $Re \gg 1$ we can neglect the last term on the right-hand side of (7.4.1) and arrive at the stationary Euler equation, which we write in the dimensional variables as

$$\rho(\boldsymbol{v}\cdot\nabla)\boldsymbol{v} = -\nabla p. \tag{7.4.2}$$

On the other hand, when $Re \ll 1$, we can neglect the left-hand side of (7.4.1) in comparison with the second term on the right-hand side. We cannot neglect $\nabla^* p^*$ because p^* can be large. Now, returning to the original variables, we obtain *Stokes' approximation* to the Navier–Stokes equation (7.1.6):

$$\nabla p = \eta\nabla^2\boldsymbol{v}. \tag{7.4.3}$$

This equation is much simpler than (7.1.6), and many interesting problems have been solved with the use of the Stokes' approximation.

In particular, this approximation was used to calculate the resistance force acting on a sphere moving with a constant velocity in a viscous fluid. Let us solve this problem. We assume that the sphere is moving with constant velocity $\boldsymbol{U}$. We introduce the reference frame moving together with the sphere. In this reference frame the fluid motion is stationary and we can use Eq. (7.4.3). We also introduce spherical coordinates r, θ, ϕ with the polar axis in the direction of the sphere's velocity. Then far from the sphere the fluid velocity is $-\boldsymbol{U}$.

We make a viable assumption that the solution is axisymmetric and independent of ϕ, and the ϕ-component of the velocity is zero. Then, using Eq. (2.13.17), we reduce Eq. (7.1.5) to

$$\frac{\partial(r^2 v_r)}{\partial r} + \frac{r}{\sin\theta}\frac{\partial(v_\theta\sin\theta)}{\partial\theta} = 0. \tag{7.4.4}$$

This equation implies that the velocity components can be expressed in terms of the flux function ψ,

$$v_r = \frac{1}{r^2\sin\theta}\frac{\partial\psi}{\partial\theta}, \quad v_\theta = -\frac{1}{r\sin\theta}\frac{\partial\psi}{\partial r}. \tag{7.4.5}$$

It follows from the condition $\boldsymbol{v} \to -\boldsymbol{U} = -U(\boldsymbol{e}_r\cos\theta - \boldsymbol{e}_\theta\sin\theta)$ as $r \to \infty$ that

$$\frac{\psi}{r^2} \to -\frac{1}{2}U\sin^2\theta \quad \text{as} \quad r \to \infty, \tag{7.4.6}$$

where we have taken into account that ψ is defined up to an arbitrary additive constant. At the surface of the sphere the velocity must be zero. This condition gives

$$\psi = \psi_0, \quad \frac{\partial\psi}{\partial r} = 0 \quad \text{at} \quad r = R, \tag{7.4.7}$$

where ψ_0 is a constant. Using Eq. (2.13.18) yields

$$\nabla \times \boldsymbol{v} = -\frac{\boldsymbol{e}_\phi}{r}\left[\frac{1}{\sin\theta}\frac{\partial^2\psi}{\partial r^2} + \frac{1}{r^2}\frac{\partial}{\partial\theta}\left(\frac{1}{\sin\theta}\frac{\partial\psi}{\partial\theta}\right)\right]. \tag{7.4.8}$$

It follows from Eqs. (6.3.12) and (7.1.5) that $\nabla^2\boldsymbol{v} = -\nabla\times\nabla\times\boldsymbol{v}$. Using this relation and Eqs. (2.13.11), (2.13.18), and (7.4.8) we obtain from Eq. (7.4.3)

$$\begin{aligned}\frac{\partial p}{\partial r} &= \frac{\eta}{r^2\sin\theta}\frac{\partial}{\partial\theta}\left[\frac{\partial^2\psi}{\partial r^2} + \frac{\sin\theta}{r^2}\frac{\partial}{\partial\theta}\left(\frac{1}{\sin\theta}\frac{\partial\psi}{\partial\theta}\right)\right],\\ \frac{\partial p}{\partial\theta} &= -\frac{\eta}{\sin\theta}\frac{\partial}{\partial r}\left[\frac{\partial^2\psi}{\partial r^2} + \frac{\sin\theta}{r^2}\frac{\partial}{\partial\theta}\left(\frac{1}{\sin\theta}\frac{\partial\psi}{\partial\theta}\right)\right].\end{aligned} \tag{7.4.9}$$

Using cross-differentiation to eliminate p from these equations yields the equation for ψ,

$$\left(\frac{\partial^2}{\partial r^2} + \frac{\sin\theta}{r^2}\frac{\partial}{\partial\theta}\frac{1}{\sin\theta}\frac{\partial}{\partial\theta}\right)^2\psi = 0. \tag{7.4.10}$$

We look for the solution to this equation in the form $\psi = X(r)\sin^2\theta$. Substituting this expression into Eq. (7.4.10) we obtain

$$\frac{d^4X}{dr^4} - \frac{4}{r^2}\frac{d^2X}{dr^2} + \frac{8}{r^3}\frac{dX}{dr} - \frac{8X}{r^4} = 0. \tag{7.4.11}$$

Since $\psi = 0$ for $\theta = 0$, it follows that $\psi_0 = 0$ in the boundary condition (7.4.7). Then the boundary conditions (7.4.6) and (7.4.7) imply that

$$X = \frac{dX}{dr} = 0 \quad \text{at } r = R, \qquad \frac{X}{r^2} \to -\frac{U}{2} \quad \text{as } r \to \infty. \tag{7.4.12}$$

We look for a solution to Eq. (7.4.11) in the form $X = r^m$. Substituting this expression into Eq. (7.4.11) we obtain

$$(m-1)(m-2)(m^2-3m-4) = 0. \tag{7.4.13}$$

The solutions to this equation are $m = -1,\ 1,\ 2,\ 4$. Hence, the general solution to Eq. (7.4.11) is

$$X = \frac{C_1}{r} + C_2 r + C_3 r^2 + C_4 r^4, \tag{7.4.14}$$

where C_1, C_2, C_3, and C_4 are constant. It follows from the boundary condition at $r \to \infty$ that $C_4 = 0$ and $C_3 = -U/2$. Using these results and Eq. (7.4.14) we obtain from the boundary conditions at $r = R$

$$\frac{C_1}{R} + C_2 R = \frac{1}{2}UR^2, \qquad \frac{C_1}{R^2} - C_2 = -UR. \tag{7.4.15}$$

We obtain from these equations $C_1 = -\frac{1}{4}UR^3$ and $C_2 = \frac{3}{4}UR$. Then the expression for the flux function is

$$\psi = -\frac{U}{4}\left(\frac{R^3}{r} - 3Rr + 2r^2\right)\sin^2\theta. \tag{7.4.16}$$

Using this result we obtain from Eqs. (7.4.5) and (7.4.9) the expressions for the velocity components and pressure,

$$v_r = -\frac{U}{2}\left(\frac{R^3}{r^3} - \frac{3R}{r} + 2\right)\cos\theta, \quad v_\theta = -\frac{U}{4}\left(\frac{R^3}{r^3} + \frac{3R}{r} - 4\right)\sin\theta, \tag{7.4.17}$$

$$p = \frac{3\eta RU\cos\theta}{2r^2} + p_0, \tag{7.4.18}$$

where p_0 is a constant. It follows from Eqs. (3.5.4), (3.5.6) and (7.1.2) that in an incompressible fluid $\mathbf{T}' = \eta\left[\nabla \boldsymbol{v} + (\nabla \boldsymbol{v})^T\right]$. Then, taking into account that the normal vector at the surface of the sphere is $\boldsymbol{e}_r$ and using Eqs. (2.13.15) and (2.13.16) we obtain that the surface traction at the surface of the sphere is given by

$$\boldsymbol{t} = -p\boldsymbol{e}_r + \eta\left[\nabla \boldsymbol{v} + (\nabla \boldsymbol{v})^T\right]\cdot \boldsymbol{e}_r = \left(-p + 2\eta\frac{\partial v_r}{\partial r}\right)\boldsymbol{e}_r + \eta\left(\frac{\partial v_\theta}{\partial r} + \frac{1}{r}\frac{\partial v_r}{\partial \theta} - \frac{v_\theta}{r}\right)\boldsymbol{e}_\theta, \tag{7.4.19}$$

where all the quantities are calculated at $r = R$. Substituting Eqs. (7.4.17) and (7.4.18) into this expression we obtain that the surface traction at the surface of the sphere is given by

$$\boldsymbol{t} = \frac{3\eta U}{2R}(-\boldsymbol{e}_r\cos\theta + \boldsymbol{e}_\theta\sin\theta) - p_0\boldsymbol{e}_r. \tag{7.4.20}$$

Introducing the auxiliary Cartesian coordinates with the origin at the centre of the sphere and the z-axis in the direction of the sphere's velocity we obtain that the Cartesian components of the surface traction at the surface of the sphere are

$$\begin{aligned} t_x &= -p_0\sin\theta\cos\phi, \\ t_y &= -p_0\sin\theta\sin\phi, \\ t_z &= -\frac{3\eta U}{2R} - p_0\cos\theta. \end{aligned} \tag{7.4.21}$$

The force acting on the sphere is equal to the integral of $\boldsymbol{t}$ over its surface. Since this integral involves an integration with respect to ϕ from 0 to 2π, it is obvious that the x and y-component of this force are equal to zero. Hence, only the z-component of the force is non-zero. It is given by

$$F = -R^2 \int_0^{2\pi} d\phi \int_0^{\pi} \left(\frac{3\eta U}{2R} + p_0 \cos\theta \right) \sin\theta \, d\theta = -6\pi\eta R U. \tag{7.4.22}$$

The minus sign indicates that this force is acting in the direction opposite to the direction of the sphere's velocity.

Example Estimate the speed of the stationary motion of a spherical dust particle falling in the air under the action of gravity.

Solution: Since the motion is stationary, the particle velocity U is constant. This means that two forces acting on the particle, gravity and resistance, are in equilibrium. Let the particle radius and mass be R and M. The gravity force acting on the particle is gM. Using this result and Eq. (7.4.22) we obtain

$$6\pi\eta R U = gM,$$

which gives

$$U = \frac{gM}{6\pi\eta R}. \tag{7.4.23}$$

Substituting $g \approx 9.8$ m s^{-2} and the typical dynamic viscosity for the air $\eta \approx 1.72 \times 10^{-5}$ kg m^{-1}s^{-1} into Eq. (7.4.23) we obtain

$$U \approx 3.02 \times 10^4 M/R \text{ m s}^{-1}, \tag{7.4.24}$$

where R is measured in meters and M in kilograms.

To obtain (7.4.24) we have used Stokes' approximation, which is only valid if Re $\ll 1$. Let us introduce the particle density ρ_p. Then $M = \frac{4}{3}\pi\rho_p R^3$ and, using (7.4.24), we obtain

$$\text{Re} = \frac{\rho R U}{\eta} = \frac{2\rho\rho_p g R^3}{9\eta^2},$$

where ρ is the density of the air. To satisfy the condition $Re \ll 1$ we then have to take

$$R \ll \left(\frac{9\eta^2}{2\rho\rho_p g} \right)^{1/3}.$$

Now, using the typical values $\rho_p \approx 2 \times 10^3$ kg m^{-3} and $\rho \approx 1.3$ kg m^{-3} we obtain $R \ll 4.4 \times 10^{-5}$ m $= 44\ \mu$ (micron). Hence, (7.4.23) is applicable to dust particles with radius of the order of ten microns or smaller.

For a particle with $R = 10\ \mu$ we obtain from (7.4.24) $U \approx 0.025$ m s^{-1}.

7.5 The Boundary Layer

As we have already pointed out in the previous section, when $\text{Re} \gg 1$ we can neglect viscosity and describe the fluid motion by the Euler equation. When the motion is stationary this equation reduces to Eq. (7.4.2). However this equation cannot be used to describe the motion near rigid boundaries. The only boundary condition that can be imposed on the motion of an ideal fluid at a rigid boundary is that the normal velocity component is zero. We cannot impose any condition on the tangential velocity component. On the other hand, the velocity of a viscous fluid at a rigid boundary must coincide with the boundary velocity. In particular, it must be zero at an immovable boundary. Hence, we can expect that the tangential velocity component changes from a finite value to zero in a narrow layer near this boundary. This layer is called the *boundary layer*. The motion in this layer is characterised by large gradients in the direction orthogonal to the boundary. As a result, the viscous term in the Navier–Stokes equation cannot be neglected in the boundary layer even when $\text{Re} \gg 1$.

As we have already stated, in a boundary layer we can only expect the presence of large gradients in the direction orthogonal to a rigid boundary. The characteristic scale of the velocity variation in the directions parallel to the boundary is still the same as it is far from the boundary, which is L. This observation enables us to substantially simplify the Navier–Stokes equation. The theory of viscous fluid motion in boundary layers was first developed by the renowned German scientist Ludwig Prandtl. Below we derive the equations describing fluid motion in boundary layers. For simplicity we assume that the rigid boundary is flat and the motion is two-dimensional, that is, it is independent of x_3 in Cartesian coordinates $x_1,\ x_2,\ x_3$, and $v_3 = 0$. It is convenient to use the notation $x = x_1$, $y = x_2$, $u = v_1$, and $v = v_2$. Then Eqs. (7.1.5) and (7.1.6) reduce to

$$\frac{\partial u}{\partial x} + \frac{\partial v}{\partial y} = 0, \tag{7.5.1}$$

$$u\frac{\partial u}{\partial x} + v\frac{\partial u}{\partial y} = -\frac{1}{\rho}\frac{\partial p}{\partial x} + \nu\left(\frac{\partial^2 u}{\partial x^2} + \frac{\partial^2 u}{\partial y^2}\right), \tag{7.5.2}$$

$$u\frac{\partial v}{\partial x} + v\frac{\partial v}{\partial y} = -\frac{1}{\rho}\frac{\partial p}{\partial y} + \nu\left(\frac{\partial^2 v}{\partial x^2} + \frac{\partial^2 v}{\partial y^2}\right), \tag{7.5.3}$$

where $\nu = \eta/\rho$ is the *kinematic viscosity*. We again introduce the dimensionless variables, however slightly differently than in the previous section. In the boundary layer the second term on the right-hand side of Eq. (7.4.1) must be of the order of unity. Obviously this is only possible when the characteristic scale of the velocity variation in the direction orthogonal to the rigid surface, which is the y-direction, is $\text{Re}^{-1/2}L$. Then, comparing terms in Eq. (7.5.1) we conclude that the ratio of v and u must be of the order of $\text{Re}^{-1/2}$. In accordance with this analysis we introduce the dimensionless variables as

$$x^* = \frac{x}{L}, \quad y^* = \frac{\mathrm{Re}^{1/2} y}{L}, \quad u^* = \frac{u}{V}, \quad v^* = \frac{\mathrm{Re}^{1/2} v}{V}, \quad p^* = \frac{p}{\rho V^2}. \tag{7.5.4}$$

Then we rewrite Eqs. (7.5.2) and (7.5.3) in the dimensionless form as

$$u^* \frac{\partial u^*}{\partial x^*} + v^* \frac{\partial u^*}{\partial y^*} = -\frac{\partial p^*}{\partial x^*} + \mathrm{Re}^{-1} \frac{\partial^2 u^*}{\partial x^{*2}} + \frac{\partial^2 u^*}{\partial y^{*2}}, \tag{7.5.5}$$

$$\mathrm{Re}^{-1} \left(u^* \frac{\partial v^*}{\partial x^*} + v^* \frac{\partial v^*}{\partial y^*} \right) = -\frac{\partial p^*}{\partial y^*} + \mathrm{Re}^{-2} \frac{\partial^2 v^*}{\partial x^{*2}} + \mathrm{Re}^{-1} \frac{\partial^2 v^*}{\partial y^{*2}}. \tag{7.5.6}$$

Now we neglect small terms of the order of Re^{-1} and Re^{-2} in Eqs. (7.5.5) and (7.5.6). This yields

$$u^* \frac{\partial u^*}{\partial x^*} + v^* \frac{\partial u^*}{\partial y^*} = -\frac{\partial p^*}{\partial x^*} + \frac{\partial^2 u^*}{\partial y^{*2}}, \tag{7.5.7}$$

$$\frac{\partial p^*}{\partial y^*} = 0. \tag{7.5.8}$$

Then we return to the original dimensional variables to obtain

$$u \frac{\partial u}{\partial x} + v \frac{\partial u}{\partial y} = -\frac{1}{\rho} \frac{\partial p}{\partial x} + \nu \frac{\partial^2 u}{\partial y^2}, \tag{7.5.9}$$

$$\frac{\partial p}{\partial y} = 0. \tag{7.5.10}$$

The last equation is especially important. It states that the pressure does not change across the boundary layer. The full solution describing the flow is obtained by matching two solutions, one describing the flow of an ideal fluid far from the rigid boundary, and the second describing the flow in the boundary layer near the rigid boundary. The matching is made in the overlap region defined by the condition $\mathrm{Re}^{-1/2} L \ll y \ll L$. In this region both solutions are valid. The solution describing the ideal flow is practically independent of y in the overlap region, implying that we can take in this region $v = 0$ and $u = U(x)$, where $U(x)$ is the velocity of the ideal flow at the rigid boundary. In an ideal flow Bernoulli's integral (5.3.3) is valid. Then we obtain

$$p = -\frac{\rho}{2} U^2 + p_0, \tag{7.5.11}$$

where p_0 is an arbitrary constant. Substituting this expression into Eqs. (7.5.9) we arrive at

$$u \frac{\partial u}{\partial x} + v \frac{\partial u}{\partial y} - \nu \frac{\partial^2 u}{\partial y^2} = U \frac{dU}{dx}. \tag{7.5.12}$$

Equations (7.5.11) and (7.5.12) describe the motion of the fluid in the boundary layer. They need to be supplemented with boundary conditions. We have the boundary condition at the rigid surface,

$$u = v = 0 \quad \text{at} \quad y = 0. \tag{7.5.13}$$

In addition the velocity in the boundary layer must tend to the velocity of the ideal fluid far from the boundary. This conditions is written as

$$u \to U(x) \quad \text{as} \quad y \to \infty. \tag{7.5.14}$$

One important observation is that Eqs. (7.5.7) and (7.5.8) do not contain the Reynolds number. Obviously, Eqs. (7.5.1) written in the dimensionless variables does not contain the Reynolds number either. Finally, the boundary conditions also do not contain it. This implies that the solution written in the dimensionless variables is independent of Re. Hence, when we change Re, the flow is subject to a similarity transformation where x and u do not change, while y and v vary inversely proportional to $\mathrm{Re}^{1/2}$.

Example We consider the *Blasius problem*. There is a semi-infinite thin plate and a viscous fluid is moving with constant velocity U parallel to the plate at both its sides. Describe the motion in the boundary layer near the plate.

Solution: We introduce Cartesian coordinates $x,\ y$ with the x-axis coinciding with the plate. The edge of the plate is at the coordinate origin. The flow is symmetric with respect to the plate implying that we can only consider the flow at one side of the plate, that is, in the region $y \geq 0$.

The characteristic speed in this problem is $V = U$. It follows from the analysis in terms of dimensionless variables that u/U must be a function of x/L and $\mathrm{Re}^{1/2}y/L = y(U/\nu L)^{1/2}$. However in this problem there is no characteristic length. This implies that u/U must only depend on such a combination of x/L and $\mathrm{Re}^{1/2}y/L$ that does not contain L. This combination is

$$\xi = y\sqrt{\frac{U}{\nu x}}. \tag{7.5.15}$$

It follows from Eq. (7.5.1) that the velocity components can be expressed in terms of the flux function ψ as

$$u = \frac{\partial \psi}{\partial y}, \qquad v = -\frac{\partial \psi}{\partial x}. \tag{7.5.16}$$

Taking into account that u is a function of ξ we obtain from the first of these equations that

$$\psi = \int u(\xi)\, dy = \sqrt{\frac{\nu x}{U}} \int u(\xi)\, d\xi = \sqrt{\nu x U}\, f(\xi). \tag{7.5.17}$$

Substituting this expression into Eq. (7.5.16) we obtain

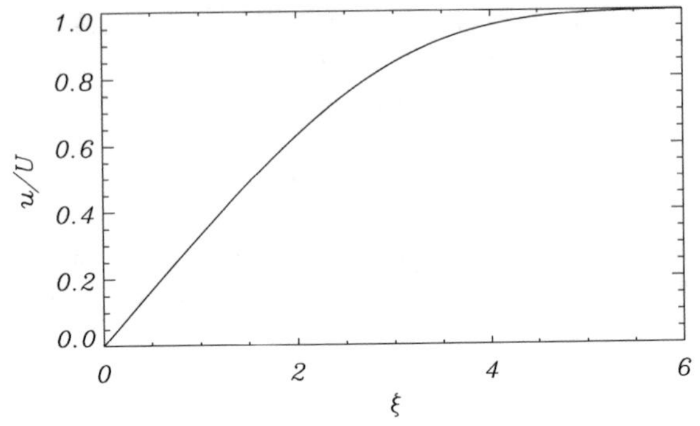

Fig. 7.3 Dependence of u/U on ξ in the boundary layer

$$u = Uf'(\xi), \quad v = \frac{1}{2}\sqrt{\frac{\nu U}{x}}(\xi f' - f), \tag{7.5.18}$$

where the prime indicates the derivative with respect to ξ. Substituting these expression for u and v into Eq. (7.5.12) yields the equation for f,

$$ff'' + 2f''' = 0. \tag{7.5.19}$$

We note that $U = \text{const}$, implying that the right-hand side of Eq. (7.5.12) is zero. It follows from the condition $u = v = 0$ at $y = 0$ that

$$f(0) = f'(0) = 0. \tag{7.5.20}$$

The condition (7.5.14) reduces to

$$f' \to 1 \quad \text{as} \quad \xi \to \infty. \tag{7.5.21}$$

Equation (7.5.19) with the boundary conditions (7.5.20) and (7.5.21) was solved numerically. In Fig. 7.3 the dependence of $u/U = f'$ on ξ is shown. We see that $u \approx U$ for $\xi \geq 4$, which can be rewritten as $y \geq 4\sqrt{\nu x/U}$. We can consider $4\sqrt{\nu x/U}$ as the characteristic thickness of the boundary layer. We see that this thickness increases with the distance from the plate edge.

Let us calculate the tangential component of the force acting on the plate. This force per unit area of the plate surface is equal to the x-component of the surface traction t_x. The normal unit vector to the plate is $\boldsymbol{e}_y = \boldsymbol{e}_2$ with components (0, 1, 0). Using Eqs. (7.1.1), (7.1.3), (7.5.1), (7.5.18), and (7.5.20) we obtain for the surface traction

$$t_x = \eta\left(\frac{\partial u}{\partial y} + \frac{\partial v}{\partial x}\right) = U\sqrt{\frac{\rho\eta U}{x}}\, f''(0). \tag{7.5.22}$$

Using the numerically obtained value $f''(0) \approx 0.332$ we transform this expression into

$$t_x \approx 0.332U\sqrt{\frac{\rho\eta U}{x}}. \tag{7.5.23}$$

This expression is singular at $x = 0$. However, the singularity is integrable, and the force acting on the rectangular area on the plate that has unit length in the y-direction and is bounded by the lines $x = 0$ and $x = l$ in the x-direction is finite (see Problem **7.7**).

Problems

7.1: Suppose there are two coaxial cylinders with radii R_1 and R_2, where $R_1 < R_2$. Both cylinders rotate, the internal with angular speed Ω_1, and the external with angular speed Ω_2. The gap between the cylinders is filled with a viscous liquid with dynamic viscosity η. The flow between the cylinders is called the *Couette cylindrical flow*. Determine the velocity of this flow assuming that it is in the azimuthal direction and both the velocity and pressure only depend on the distance from the cylinder axis. Calculate the tangential component of the surface traction at each cylinder.

7.2: Suppose there are two coaxial cylinders with radii R_1 and R_2, where $R_1 < R_2$. The external cylinder is at rest and the internal cylinder moves with constant speed V along its axis. The gap between the cylinders is filled with a viscous liquid with dynamic viscosity η. Determine the flow between the cylinders. Calculate the tangential forces acting on each cylinder per unit length of the cylinder axis.

7.3: *Cylindrical Poiseuille flow* is the flow of a viscous liquid in a cylinder with velocity parallel to the cylinder's axis. Suppose the cylinder's radius is R and the constant pressure gradient along its axis is Δp. Determine the velocity in the cylinder. Calculate the tangential force acting on the cylinder's surface per unit length along the cylinder's axis.

7.4: Solve the same problem, however for the Poiseuille flow in the gap between two coaxial cylinders with radii R_1 and R_2, where $R_1 < R_2$.

7.5: Suppose there are two spheres with radii R_1 and R_2, where $R_1 < R_2$, and with the same centre. The internal sphere is at rest, while the external rotates with a constant angular velocity of magnitude Ω. The gap between the spheres is filled with a viscous liquid with dynamic viscosity η. Use Stokes' approximation to determine the velocity of the fluid. Calculate the momentum of the force that must be applied to the external sphere to keep it rotating with a constant angular velocity. For what values of parameters is Stokes' approximation applicable?

7.6: A falling ball viscometer is a device used to measure viscosity. It consists of a long vertical tube filled with a viscous fluid. There is a small ball moving downward with constant speed under the action of gravity. You are given that the radius of the ball is $R = 5$ mm, the density of its material is $\rho_b = 2 \times 10^3$ kg m^{-3}, the density of the fluid is $\rho_f = 10^3$ kg m^{-3}, and the ball speed is $U = 0.1$ m s^{-1}. Use Stokes' approximation to determine the dynamic viscosity of the fluid. Calculate the Reynolds number and thus conclude whether Stokes' approximation is applicable.

7.7: Suppose there is a flow of a viscous fluid with dynamic viscosity η and density ρ near a rigid plate of length l. At large distance from the plate the flow velocity is parallel to the plate and its magnitude is constant and equal to U. You are given that $\mathrm{Re} = lU/\nu \gg 1$, where $\nu = \eta/\rho$. Calculate the tangential force acting on the plate per unit length in the direction perpendicular to the fluid velocity.

Chapter 8
Ideal Barotropic Fluids

8.1 A Glimpse of the Kinetic Theory of Gases

In this chapter we study the motion of compressible fluids. We will interchangeably call them either fluids of gases. In the Introduction we briefly discussed why we can often consider various media as continua when studying their motions. When we study the motion of dense and nearly incompressible fluids, like water or oil, usually there is no question that we can consider them as continua. The situation with the description of motions of gases is more complex. To understand when we can consider a gas as a continuum, we need to take a very brief excursion into the kinetic theory of gases. In this excursion we often do not give rigorous definitions and rely on intuition. For example, we will often use the words "large number". Usually we mean a number of the order of a billion or larger.

The basis for the kinetic theory of gases was laid out by the Swiss scientist Daniel Bernoulli, who published a book on the subject, *Hydrodynamica*, in 1738. However, a rigorous mathematical basis for this theory was only developed much later, in the 19th century, through the efforts of other researchers, Scots John James Waterston and James Clerk Maxwell, Germans August Krönig and Rudolf Clausius, and Austrian Ludwig Boltzmann.

In our brief introduction we consider a domain $\mathcal{V}$ filled with a gas that consists of a large number N of atoms and/or molecules. They are in a permanent chaotic motion. In what follows we call both atoms and molecules particles. Each particle is characterised by its spatial position $\boldsymbol{x}$ and velocity $\boldsymbol{v}$. Hence, it is convenient to describe a particle's motion as the motion of a point in six-dimensional phase space, where the first three coordinates determine the particle spatial position, and the next three coordinates determine its velocity. We now consider a small volume dX inside $\mathcal{V}$ that contains the point $\boldsymbol{x}$, and a small volume dV in the space of velocities that contains the point $\boldsymbol{v}$. We assume that, although the volumes are small, the number of particles that have spatial positions inside dX and velocities inside dV is very large. Since dX and dV are small, the number of particles with spatial positions inside dX and velocities inside dV is proportional to $dX\,dV$. The coefficient of proportionality

© Springer Nature Switzerland AG 2019

M. S. Ruderman, *Fluid Dynamics and Linear Elasticity*, Springer Undergraduate Mathematics Series, https://doi.org/10.1007/978-3-030-19297-6_8

f is called the *distribution* function; f is a function of $\boldsymbol{x}$ and $\boldsymbol{v}$, and it also depends on time t. Hence, it is a function of seven variables, $f(t, \boldsymbol{x}, \boldsymbol{v})$. The number of particles that are in the volume dX and have velocities inside dV is $f\,dX\,dV$. The condition that dX and dV are small means that the particles are distributed homogeneously in the volume $dX\,dV$, that is, if we, for example, split both volumes in half, that is, take the volumes $\frac{1}{2}dX$ and $\frac{1}{2}dV$, then the number of particles that are in the volume $\frac{1}{2}dX$ and have velocities inside $\frac{1}{2}dV$ will be one quarter of the number of particles in $dX\,dV$. This is equivalent to the condition that the size of dX is much smaller than the characteristic scale of variation of $f(t, \boldsymbol{x}, \boldsymbol{v})$ with respect to $\boldsymbol{x}$, and the size of dV is much smaller than the characteristic scale of variation of $f(t, \boldsymbol{x}, \boldsymbol{v})$ with respect to $\boldsymbol{v}$. The function $f(t, \boldsymbol{x}, \boldsymbol{v})$ satisfies the so-called *Boltzmann equation*, named after the Austrian physicist Ludwig Eduard Boltzmann. Since it contains seven independent variables, it is extremely difficult to solve this equation even using contemporary supercomputers.

The gas density is given by

$$\rho(t, \boldsymbol{x}) = m \int f(t, \boldsymbol{x}, \boldsymbol{v})\, dV, \tag{8.1.1}$$

where m is the mass of a particle (we assume that all particles are the same) and the integral is taken over the whole velocity space V. Similarly we define the velocity (often called the bulk velocity) of the elementary volume dX,

$$\bar{\boldsymbol{v}} = \frac{m}{\rho} \int \boldsymbol{v} f(t, \boldsymbol{x}, \boldsymbol{v})\, dV. \tag{8.1.2}$$

Quantities like ρ and $\bar{\boldsymbol{v}}$ are called *macroscopic* in the kinetic theory of gases. The condition that the size of dX is much smaller than the characteristic scale of variation of $f(t, \boldsymbol{x}, \boldsymbol{v})$ with respect to $\boldsymbol{x}$ implies that the macroscopic quantities do not vary inside dX.

When the gas inside the volume $\mathcal{V}$ is isolated from any external impact then, after some time, it becomes homogeneous, meaning that its density ρ will be constant. The distribution function will only depend on $\boldsymbol{v}$. It will be

$$f(\boldsymbol{v}) = f_{\mathrm{M}}(\boldsymbol{v}) \equiv \frac{N}{X}\left(\frac{m}{2\pi kT}\right)^{3/2} \exp\left(-\frac{mv^2}{2kT}\right), \tag{8.1.3}$$

where $v = \|\boldsymbol{v}\|$, m is the mass of a particle, X is the volume of the domain $\mathcal{V}$, N is the number of particles in this domain, $k \approx 1.38 \times 10^{-23}$ N m K^{-1} is the Boltzmann constant, and T is the temperature. This distribution function is called the *Maxwell distribution* after the great scientist James Clerk Maxwell. Substituting $f = f_{\mathrm{M}}$ into Eqs. (8.1.1) and (8.1.2) we obtain $\rho = mN/X$ and $\bar{\boldsymbol{v}} = 0$. To clarify the physical sense of the temperature we calculate the kinetic energy of particles in the domain $\mathcal{V}$. It is equal to

$$\mathcal{E} = \frac{mX}{2} \int v^2 f_{\rm M}(\boldsymbol{v})\, d\boldsymbol{v} = \frac{mX}{2} 4\pi \int_0^\infty v^4 f_{\rm M}(\boldsymbol{v})\, dv = \frac{3}{2} NkT. \tag{8.1.4}$$

In fact, the description of particle motion in six-dimensional phase space is only possible when we consider the particles as rigid non-rotating smooth balls colliding with each other. The condition that the balls are smooth implies that they remain non-rotating after collisions. In particular, Eq. (8.1.4) is only valid for gases consisting of particles that can be considered as such balls. In this case each particle has three degrees of freedom and Eq. (8.1.4) shows that the energy per one degree of freedom and per one particle is $\frac{1}{2}kT$.

Very often molecules in gases have a more complex shape. For example, air molecules are usually modelled as two equal balls connected by a bar, so they look like a kind of dumbbell. In principle, it has six degrees of freedom, the first three related to the motion of its centre of mass, and the next three related to rotation about three mutually orthogonal axis of inertia. However, its moment of inertia with respect to the axis connecting the centres of balls is much smaller than the momentum of inertia with respect to any axis going through the centre of the bar and orthogonal to it. Hence, the rotation about the axis connecting the centres of the balls can be neglected and such molecules are assumed to have five degrees of freedom. In a gas consisting of such molecules we obtain $\mathcal{E} = \frac{5}{2}NkT$, and the distribution function is a function of nine variables. In what follows we will only consider the simplest case of gases that can be considered as a collection of rigid non-rotating smooth balls.

When a gas is moving under the action of external forces or being driven by an initial pulse, it is, in general, inhomogeneous and its distribution function is different from the Maxwell distribution. We introduce the characteristic spatial scale L and time $\widetilde{T}$ of a problem under consideration. We also introduce the mean free path of particles, that is, the mean distance that a particle travels between two collisions, $l_{\rm col}$, and the mean collisional time $t_{\rm col}$. The ratio ${\rm Kn} = l_{\rm col}/L$ is called the *Knudsen number* after the Danish physicist Martin Knudsen who introduced it. We now assume that ${\rm Kn} \ll 1$. Then we can choose such dX that its linear size is much smaller than L, but much greater than $l_{\rm col}$. The first condition implies that dX is almost homogeneous, that is, all macroscopic quantities do not change inside dX. The second condition means that if we assume that dX is a liquid volume with each of its points moving with local mean velocity $\bar{\boldsymbol{v}}$, then only a very small number of particles close to its boundary can leave or enter dX during the mean collision time. In the limit $l_{\rm col} \to 0$ the volume dX always contains the same particles. Moreover, due to collisions the distribution function in dX always remains very close to the Maxwell distribution, meaning that $f(t, \boldsymbol{x}, \boldsymbol{v}) = f_{\rm M}(t, \boldsymbol{x}, \boldsymbol{v}) + \delta f(t, \boldsymbol{x}, \boldsymbol{v})$, where $|\delta f(t, \boldsymbol{x}, \boldsymbol{v})| \ll f_{\rm M}(t, \boldsymbol{x}, \boldsymbol{v})$. But now the Maxwell distribution is in the reference frame moving with the mean velocity of dX, so it has the form

$$f_{\rm M}(\boldsymbol{v}) = \frac{\rho(t, \boldsymbol{x})}{m} \left(\frac{m}{2\pi kT(t, \boldsymbol{x})} \right)^{3/2} \exp\left(-\frac{m[\boldsymbol{v} - \bar{\boldsymbol{v}}(t, \boldsymbol{x})]^2}{2kT(t, \boldsymbol{x})} \right). \tag{8.1.5}$$

This expression defines the *local Maxwell distribution*. It is straightforward to verify that $\rho(t, \boldsymbol{x})$ and $\bar{\boldsymbol{v}}(t, \boldsymbol{x})$ are the gas density and bulk velocity defined by Eqs. (8.1.1) and (8.1.2), respectively.

In the kinetic theory of gases the Boltzmann equation for the distribution function is solved using the regular perturbation method with Kn as a small parameter. In the first-order approximation the equations of mass, momentum, and energy conservation are obtained. In particular, the stress tensor in the momentum equation is given by

$$T_{ij} = -m \int (v_i - \bar{v}_i)(v_j - \bar{v}_j) f_{\mathrm{M}}(\boldsymbol{v})\, d\boldsymbol{v} = -\frac{k}{m}\rho T \delta_{ij}. \tag{8.1.6}$$

On the other hand, we know that when the stress tensor is proportional to the unit tensor, the proportionality coefficient is the pressure taken with the minus sign, that it $T_{ij} = -p\delta_{ij}$. Hence, we obtain the relation

$$\boxed{p = \frac{k}{m}\rho T.} \tag{8.1.7}$$

This relation is called the *Clapeyron law*, after the French scientist Benoît Paul Émile Clapeyron. It is also called the *ideal gas law*, or *perfect gas law*.

In the next order approximation $\delta f(t, \boldsymbol{x}, \boldsymbol{v})$ is calculated and the expressions for the viscosity and heat flux related to thermal conduction are obtained. But we stop our brief introduction to the kinetic theory of gases here. The main conclusion is that we only can consider a gas as a continuum and describe its motion by hydrodynamic equations when Kn $\ll 1$ and $t_{\mathrm{col}} \ll \widetilde{T}$. Very often it is assumed that the second condition follows from the first one. Although, as a rule, this is true, sometimes this is not the case. Hence, we should consider these two conditions as independent.

8.2 Introduction to Thermodynamics

In Chaps. 3 and 4 we derived the mass conservation and momentum equations for a continuum. Then, to obtain a closed set of equations describing the motion of an incompressible fluid it was enough to assume that the stress tensor is proportional to the unit tensor. However, to describe the motion of compressible gases we need to obtain equations relating internal characteristics of a gas, like pressure and temperature, with other quantities. In principle, when Kn $\ll 1$ we can obtain the full hydrodynamic description of gas motion using the kinetic theory of gases. However, there is a much simpler way to obtain these relations. They are given by a branch of science called *thermodynamics*.

Thermodynamics is a branch of physics which deals with the energy and work of a system. Its development began when the idea that all physical substances consist of atoms and molecules had only started to emerge and was not generally accepted. Hence, in particular, it considers all gases as continua and only deals with their macroscopic characteristics. Thermodynamics originated in the seventeenth century, mainly in the works of the German scientist Otto von Guericke, and English scientists Robert Boyle and Robert Hooke. However, it attained its present shape much later, at the end of the eighteenth and beginning of the nineteenth century due to efforts of such researchers as French Sandi Carnot, Scots William Rankine, William Thomson (Lord Kelvin), and James Clerk Maxwell, American Josiah Willard Gibbs, Austrian Ludwig Boltzmann, Germans Max Planck, and Rudolf Clausius, and Dutch Johannes van der Waals.

Thermodynamics only deals with the large scale response of a system which we can observe and measure in experiments. The basis of thermodynamics is formed by three fundamental laws. The *first law* of thermodynamics is the relation between the internal energy of a system U, the heat Q supplied to a system, and the work W done by external forces on a system:

$$U = Q + W. \tag{8.2.1}$$

Before it became clear that all material bodies consist of molecules and atoms scientists believed that a hot object contained a concentrated amount of a hypothetical massless "fluid" that they named "caloric", and this fluid is the agent that transfers the heat. Although this fluid does not exist, the theory of caloric fluid managed to explain many observed phenomena. In particular, heat transfer was correctly described by this theory.

To formulate the *second law* of thermodynamics we first need to define the macroscopic variable called *temperature*. To introduce the temperature we need to formulate three principles:

1. The temperature of an object can affect some physical properties of the object, such as the length of a solid, or the gas pressure in a closed vessel, or the electrical resistance of a wire.

2. Two objects are in *thermodynamic equilibrium* when they have the same temperature.

3. If two objects of different temperatures are brought into contact with one another, they will eventually establish a thermodynamic equilibrium.

With these three thermodynamic principles, we can construct a device for measuring temperature, a *thermometer*, which assigns a number to the temperature of an object. When the thermometer is brought into contact with another object, it quickly establishes a thermodynamic equilibrium. By measuring the thermodynamic effect on some physical property of the thermometer at some fixed conditions, like the boiling point and freezing point of water, we can establish a scale for assigning temperature values. The number assigned to the temperature depends on what we pick for the reference condition. Below we use the temperature scale which has been named after Lord Kelvin. In this scale the temperature is always positive. The zero temperature in this scale corresponds to -273.15 degree Celsius. This temperature is also called absolute zero. At this temperature molecules and atoms stop moving. The step of one degree is the same as in the Celsius scale, that is, it is the difference between the freezing and boiling points of water divided by 100.

Let us consider a cyclic process where a gas state slowly changes and at the end returns to its initial state. Then the *Clausius theorem* (also called the *Clausius inequality*) (1854) states that

$$\oint \frac{\delta Q}{T} \le 0. \tag{8.2.2}$$

It is postulated that there are reversible processes where the Clausius inequality becomes an equality, that is, in Eq. (8.2.2) there is an equality sign, and any two states of a system can be connected by a reversible process. This implies that an integral along a path $\mathcal{L}$ relating two states, $\int_{\mathcal{L}} \delta Q/T$, is independent of $\mathcal{L}$. Then we can introduce a function of the system state S such that

$$dS = \frac{\delta Q}{T}. \tag{8.2.3}$$

This function is called the *entropy*. We consider two states of a system and a path $\mathcal{L}$ connecting these two states. Then we return from the final state to the initial state by a reversible path. The integral along this path is equal to the difference between the entropy in the final and initial states with the minus sign, $-\Delta S$. The two paths, $\mathcal{L}$ and its reverse, constitute a cyclic process for which the inequality in Eq. (8.2.2) is valid. As a result we obtain

$$\int_{\mathcal{L}} \frac{\delta Q}{T} \le \Delta S. \tag{8.2.4}$$

If a system is isolated then there is no heat supplied to the system, that is, $\delta Q = 0$. Then it follows from Eq. (8.2.4) that $\Delta S \ge 0$. On the basis of this result we can formulate the *second law* of thermodynamics as the statement that the entropy of an isolated system cannot decrease.

The *third law* of thermodynamics was formulated by the German scientist Walther Hermann Nernst. It states that the entropy of a system at absolute zero is exactly equal to zero. “Absolute zero” means that the temperature is equal to zero in the Kelvin temperature scale.

8.3 The Energy Equation

We take the scalar product of the momentum Eq. (4.4.3) with $\boldsymbol{v}$ to obtain the mechanical energy equation

$$\frac{1}{2}\frac{D\|\boldsymbol{v}\|^2}{Dt} = \frac{1}{\rho}\boldsymbol{v}\cdot(\nabla\cdot\mathbf{T}) + \boldsymbol{v}\cdot\boldsymbol{b}. \tag{8.3.1}$$

This equation states that the rate of change of the kinetic energy of a unit mass of fluid is equal to the rate at which work is done on the fluid by the internal and body forces.

Let us now derive the equation describing the variation of the total energy, kinetic and internal. Below we use U to denote the internal energy per unit mass, and S to denote the entropy per unit mass. We consider a material volume V, that is, the volume that consists of the same particles of the fluid at any time. The surface of this volume is Σ. The rate of change of the kinetic plus internal energy of V must be equal to the rate at which work is done on the fluid by internal and body forces plus the rate at which the heat is added to the fluid. The heat can be split into two parts. The first part is generated inside the fluid at a rate ϵ per unit mass (e.g. by chemical reactions). The second part is due to the heat flux $\boldsymbol{q}$ across the surface (e.g. due to thermal conduction). As a result we obtain

$$\frac{d}{dt}\int_V\left(\frac{\|\boldsymbol{v}\|^2}{2}+U\right)\rho\,dV=\int_\Sigma(\boldsymbol{v}\cdot\mathbf{T})\cdot d\boldsymbol{\Sigma}+\int_V(\boldsymbol{v}\cdot\boldsymbol{b})\rho\,dV+\int_V\epsilon\rho\,dV-\int_\Sigma\boldsymbol{q}\cdot d\boldsymbol{\Sigma}. \tag{8.3.2}$$

In this equation $\boldsymbol{q}$ is the heat flux through a unit surface directed outward from V, $d\boldsymbol{\Sigma}=\boldsymbol{n}\,d\Sigma$, $\boldsymbol{n}$ is the external unit normal to the surface Σ, and we have used the fact that the force acting on a unit area of surface Σ is $\mathbf{T}\cdot\boldsymbol{n}$. Using Gauss' divergence theorem, Eq. (2.1.4), and the Lemma from Sect. 3.8 we obtain the equation for total energy variation in the differential form,

$$\rho\frac{D}{Dt}\left(\frac{\|\boldsymbol{v}\|^2}{2}+U\right)=\nabla\cdot(\boldsymbol{v}\cdot\mathbf{T})+\rho\boldsymbol{v}\cdot\boldsymbol{b}+\rho\epsilon-\nabla\cdot\boldsymbol{q}. \tag{8.3.3}$$

We can also derive the equation for the internal energy alone by dividing this equation by ρ and then subtracting Eq. (8.3.1). This gives

$$\frac{DU}{Dt}=\frac{1}{\rho}\mathbf{D}:\mathbf{T}+\epsilon-\frac{1}{\rho}\nabla\cdot\boldsymbol{q}, \tag{8.3.4}$$

where $\mathbf{D}$ is the strain-rate tensor defined by Eq. (3.5.6). We recall that the colon indicates the double contraction of tensors. When deriving this equation we took into account that $\mathbf{T}$ is a symmetric tensor, which enabled us to substitute $\mathbf{D}:\mathbf{T}$ for $\mathbf{T}:\nabla\boldsymbol{v}$.

Below we consider the stress tensor given by Eq. (7.1.1) with $\mathbf{T}'$ defined by Eq. (7.1.3) and assume that there are no sources of internal heat, which implies that $\epsilon=0$. Then we transform Eq. (8.3.4) into

$$\boxed{\rho\frac{DU}{Dt}=\frac{p}{\rho}\frac{D\rho}{Dt}+\mathbf{D}:\mathbf{T}'-\nabla\cdot\boldsymbol{q},} \tag{8.3.5}$$

where we used Eq. (3.8.3) to substitute $(1/\rho)D\rho/Dt$ for $-\nabla\cdot\boldsymbol{v}$.

Consider now a volume of gas that very slowly (quasi-statically) evolves in time so that its internal energy changes by dU. Since the evolution is very slow the velocity is very small and thus we can take $\mathbf{T}'=0$. The last term on the right-hand side of Eq. (8.3.5) describes the heat flux, meaning that $-(1/\rho)(\nabla\cdot\boldsymbol{q})\,dt=\delta Q$. Then, taking into account that $\delta Q = T\,dS$, we obtain from Eq. (8.3.5)

$$dU = \frac{p}{\rho^2}\,d\rho + T\,dS. \tag{8.3.6}$$

The first term on the right-hand side of this equation describes the work of pressure on the volume, so $dW = (p/\rho^2)d\rho$. Hence, in fact Eq. (8.3.6) is simply another form of the first law of thermodynamics. It follows from Eq. (8.3.6) that

$$T = \left(\frac{\partial U}{\partial S}\right)_\rho, \quad p = \rho^2\left(\frac{\partial U}{\partial \rho}\right)_S, \tag{8.3.7}$$

where the subscripts ρ and S indicate that the partial derivatives are calculated at constant ρ and constant S, respectively.

We define the specific heat at constant pressure C_p as the amount of heat required to increase the temperature of a volume with unit mass by one degree without changing the pressure. Hence $\delta Q = C_p\,dT$ and, using $\delta Q = T\,dS$, we obtain

$$C_p = T\left(\frac{\partial S}{\partial T}\right)_p. \tag{8.3.8}$$

Similarly, we define the specific heat at constant volume C_v as the amount of heat required to increase the temperature of a volume with unit mass by one degree without changing the density, and obtain

$$C_v = T\left(\frac{\partial S}{\partial T}\right)_\rho. \tag{8.3.9}$$

Below we consider a *perfect* gas. This is a gas for which the Clapeyron law Eq. (8.1.7) is satisfied. In the macroscopic theory of gas motion it is usually written in a slightly different form,

$$\boxed{p = \frac{\widetilde{R}}{\tilde{\mu}}\rho T,} \tag{8.3.10}$$

where $\widetilde{R} = k/m_p \approx 8.25 \times 10^3\ \mathrm{m^2\,s^{-2}\,K^{-1}}$ is the universal gas constant and $\tilde{\mu} = m/m_p$ is the ratio of the mass of a molecule or an atom of the gas to the proton mass m_p. For example, $\tilde{\mu} \approx 29$ for dry air. The second condition that a perfect gas must satisfy is that C_p and C_v are constant. Real gases are not perfect. But in a large range of pressure, density, and temperature variation the Clapeyron law is satisfied with high accuracy and C_p and C_v are almost constant. Hence, the model of a perfect gas provides a very good approximate description of real gases.

The state of a perfect gas is completely defined if its pressure, density, and temperature are given. It is enough to fix any two of them. Then the third one is defined by the Clapeyron law. Using Eq. (8.3.10) we rewrite Eqs. (8.3.8) and (8.3.9) as

$$\left(\frac{\partial S}{\partial \rho}\right)_p = -\frac{C_p}{\rho}, \quad \left(\frac{\partial S}{\partial p}\right)_\rho = \frac{C_v}{p}. \tag{8.3.11}$$

It follows from the first of these equations that $S = -C_p \ln \rho + f(p)$, where $f(p)$ is an arbitrary function. Substituting this result into the second equation we obtain $f(p) = C_v \ln p + \text{const}$. Taking the constant in this expression equal to zero we obtain

$$\boxed{S = C_v \ln \frac{p}{\rho^\gamma}, \quad \gamma = \frac{C_p}{C_v}.} \tag{8.3.12}$$

The quantity γ is called the *adiabatic exponent*. The origin of this term will be explained below. We note that the expression for S given by Eq. (8.3.12) does not satisfy the third law of thermodynamics. Indeed, $S \to -\infty$ as $T \to 0$. Obviously, adding a constant to the expression for S does not help. The reason for this problem is that the model of a perfect gas is not applicable at very low temperatures. At such temperatures quantum effects become important and $C_v \to 0$ as $T \to 0$.

Now we can obtain the expression for the internal energy of a perfect gas. Using Eqs. (8.3.10) and (8.3.12) to express p and T in terms of ρ and S we transform Eq. (8.3.7) into

$$\left(\frac{\partial U}{\partial S}\right)_\rho = \frac{\tilde{\mu}\rho^{\gamma-1}}{\widetilde{R}} e^{S/C_v}, \quad \left(\frac{\partial U}{\partial \rho}\right)_S = \rho^{\gamma-2} e^{S/C_v}. \tag{8.3.13}$$

These two equations for one function U are only compatible if the mixed second derivatives of U with respect to S and ρ calculated using the first and second equation coincide. A simple calculation shows that this condition reduces to

$$\boxed{C_p - C_v = \frac{\widetilde{R}}{\tilde{\mu}}.} \tag{8.3.14}$$

Integrating the first equation in (8.3.13) we obtain

$$U = \frac{\tilde{\mu} C_v \rho^{\gamma-1}}{\widetilde{R}} e^{S/C_v} + f(\rho), \tag{8.3.15}$$

where $f(\rho)$ is an arbitrary function. The internal energy is related to the chaotic motion of gas particles. We have seen in Sect. 8.1 that the temperature is also a measure of the chaotic motion energy. When $T \to 0$ the chaotic motion stops. Hence, it is natural to impose the condition that $U \to 0$ as $T \to 0$. Substituting Eq. (8.3.15) into the second equation in (8.3.13) yields the equation for $f(\rho)$. Using Eq. (8.3.14) and the condition that $U \to 0$ as $T \to 0$ we obtain $f(\rho) = 0$. Then, with the aid of Eqs. (8.3.10), (8.3.12), and (8.3.14) we eventually arrive at

$$\boxed{U = \frac{p}{(\gamma - 1)\rho} = C_v T.} \tag{8.3.16}$$

Let us consider a volume of the gas with unit mass and assume that the temperature is constant in this volume. The number of particles in this volume is $N = 1/m$, where m is the mass of a particle constituting the gas. In accordance with the analysis in Sect. 8.1 the internal energy of the gas in this volume is $U = \frac{n}{2}NkT$, where n is the degree of freedom of one particle. Using the relation $\widetilde{R}/\tilde{\mu} = k/m$ we rewrite this expression as $U = \frac{n}{2}\widetilde{R}T/\tilde{\mu}$. Comparing this expression with Eq. (8.3.16) yields $2C_v = n\widetilde{R}/\tilde{\mu}$. Using Eqs. (8.3.12) and (8.3.14) we obtain $2\gamma = n(\gamma - 1)$, which gives $\gamma = 1 + 2/n$. Since $n \geq 3$ it follows from this relation that $1 < \gamma \leq 5/3$.

We now consider the motion of an inviscid ($\mathbf{T}' = 0$) gas in the absence of external forces ($\boldsymbol{b} = 0$) and heat conduction ($\boldsymbol{q} = 0$). Then, using Eq. (8.3.16) we reduce the energy Eq. (8.3.5) to

$$\frac{D}{Dt}\left(\frac{p}{\rho^\gamma}\right) = 0. \tag{8.3.17}$$

It follows from this equation and Eq. (8.3.12) that the entropy of a Lagrangian particle is conserved (recall that a Lagrangian particle is not a physical particle constituting a gas, that is, an atom or molecule, but an infinitesimal volume of a gas considered as a continuum). Such a gas motion is called *adiabatic*. The exponent γ is called adiabatic because it is involved in the equation describing adiabatic motions.

If the entropy is homogeneous at the initial time, i.e. it is independent of the spatial variables, then it follows from Eq. (8.3.17) that it remains homogeneous and independent of time. In this case Eq. (8.3.17) reduces to a very simple form,

$$\boxed{p = p_0\left(\frac{\rho}{\rho_0}\right)^\gamma,} \tag{8.3.18}$$

where the constants p_0 and ρ_0 are some reference pressure and density. Such gas flows are called *isentropic*.

Equations (8.3.17) and (8.3.18) provide a sufficiently good approximation for gases when the Reynolds number is large and we can neglect viscosity, and the motion is sufficiently fast so that there is not enough time for two neighbouring Lagrangian particles to exchange energy through thermal conduction. In Chap. 5 we considered water as an incompressible fluid. However, it is only an approximation. Although water compressibility is very small, it is still not zero. It is its compressibility that allows sound to propagate through water. When studying this propagation we can also assume that the pressure is a function of the density, however it is not given by Eq. (8.3.18). Fluid motions where the pressure is a function of the density are called *barotropic*. An isentropic motion of a fluid is a particular case of a barotropic motion. Below in this section we only consider barotropic motions of fluids. These motions are described by the mass conservation Eq. (3.8.2) or Eq. (3.8.3), the momentum Eq. (5.1.2) or Eq. (5.1.3), and the equation $p = p(\rho)$. In the particular case of isentropic motion the function $p(\rho)$ is given by Eq. (8.3.18). The relation $p = p(\rho)$ together with $\mathbf{T} = -p\,\mathbf{I}$ are the constitutive equations for ideal compressible barotropic fluids.

8.4 Sound Waves

It is difficult to overestimate the importance of sound waves in our life. Our spoken communication is based on sound waves. Sound waves of very high frequency (so-called ultrasound) are widely used in medicine for diagnostics of various diseases. Sea navigation uses sound waves by employing the technique known as sonar. There are many other examples of the use of sound waves. Below we study the propagation of sound waves in air. We also briefly explain how the results that we obtain can be translated to other media, like water.

We consider the motion of an ideal barotropic fluid (i.e. without viscosity and thermal conditions) and neglect the body force ($\boldsymbol{b} = 0$). We assume that initially it is homogeneous and at rest, so that $\boldsymbol{v} = 0$ and $\rho = \rho_0$. Then it is slightly perturbed, so that $\boldsymbol{v} \neq 0$ but small, and $\rho = \rho_0 + \delta\rho$ with $|\delta\rho| \ll \rho_0$. These assumptions enable us to neglect all terms except the linear ones with respect to $\boldsymbol{v}$ and $\delta\rho$ in Eqs. (3.8.2) and (5.1.3). Hence, we neglect the term $(\boldsymbol{v} \cdot \nabla)\boldsymbol{v}$ on the left-hand side of (5.1.3), and write

$$\rho\boldsymbol{v} \approx \rho_0\boldsymbol{v}, \quad \frac{1}{\rho}\nabla p = \frac{a_s^2}{\rho}\nabla\rho \approx \frac{a_s^2(\rho_0)}{\rho_0}\nabla\delta\rho, \tag{8.4.1}$$

where we introduced the notation

$$\boxed{a_s^2 = \frac{dp}{d\rho}.} \tag{8.4.2}$$

The quantity a_s is called the *sound speed*. We will see below that sound waves propagate with speed $a_s(\rho_0)$. When the fluid motion is isentropic we obtain using Eq. (8.3.18)

$$a_s^2 = \frac{\gamma p_0}{\rho_0}\left(\frac{\rho}{\rho_0}\right)^{\gamma-1}. \tag{8.4.3}$$

In particular,

$$a_s^2(\rho_0) = \frac{\gamma p_0}{\rho_0}. \tag{8.4.4}$$

Now, using Eq. (8.4.1) we reduce the system of Eqs. (3.8.2) and (5.1.3) to

$$\frac{\partial(\delta\rho)}{\partial t} + \rho_0\nabla\cdot\boldsymbol{v} = 0, \tag{8.4.5}$$

$$\rho_0\frac{\partial\boldsymbol{v}}{\partial t} = -a_s^2\nabla\delta\rho, \tag{8.4.6}$$

where we wrote a_s instead of $a_s(\rho_0)$. Assuming that a_s is constant together with neglecting $(\boldsymbol{v} \cdot \nabla)\boldsymbol{v}$ linearises the momentum equation.

Let us differentiate Eq. (8.4.5) with respect to time and take the divergence of Eq. (8.4.6):

$$\frac{\partial^2 \delta\rho}{\partial t^2} + \rho_0 \frac{\partial(\nabla \cdot \boldsymbol{v})}{\partial t} = 0,$$

$$\rho_0 \frac{\partial(\nabla \cdot \boldsymbol{v})}{\partial t} = -a_s^2 \nabla^2 \delta\rho.$$

Now we eliminate $\partial(\nabla \cdot \boldsymbol{v})/\partial t$ from these two equations to obtain

$$\frac{\partial^2 \delta\rho}{\partial t^2} - a_s^2 \nabla^2 \delta\rho = 0. \tag{8.4.7}$$

We have obtained the well-known wave equation. It describes the propagation of sound waves with sound speed a_s.

We now assume that $\boldsymbol{v}$ and $\delta\rho$ only depend on one spatial variable $x_1 = x$. Then Eq. (8.4.7) reduces to

$$\frac{\partial^2 \delta\rho}{\partial t^2} - a_s^2 \frac{\partial^2 \delta\rho}{\partial x^2} = 0. \tag{8.4.8}$$

This equation is similar to Eqs. (6.4.1) and (6.4.4) describing one-dimensional longitudinal and transverse waves propagating in an isotropic elastic material. The general solution to Eq. (8.4.8) in an unbounded domain is given by the d'Alembert formula:

$$\delta\rho = f(t - x/a_s) + g(t + x/a_s), \tag{8.4.9}$$

where f and g are arbitrary functions.

The pressure can also be written as a sum of the unperturbed value and perturbation: $p = p_0 + \delta p$. The pressure and density perturbations are related by

$$\delta p = \left.\frac{dp}{d\rho}\right|_{\rho=\rho_0} \delta\rho = a_s^2\, \delta\rho. \tag{8.4.10}$$

It follows from Eq. (8.4.6) that, for one-dimensional motion, only one component of the velocity, $v_1 = v$, is non-zero. Then, substituting Eq. (8.4.9) into Eq. (8.4.6), we obtain

$$\rho_0 \frac{\partial v}{\partial t} = a_s\big[f'(t - x/a_s) - g'(t + x/a_s)\big] = a_s \frac{\partial}{\partial t}[f(t - x/a_s) - g(t + x/a_s)].$$

Integrating this expression with respect to time yields

$$v = \frac{a_s}{\rho_0}[f(t - x/a_s) - g(t + x/a_s)]. \tag{8.4.11}$$

Strictly speaking, we have to add an arbitrary function of x to the right-hand side of this expression. However, we assume that, when there is no density perturbation ($f = g = 0$), the velocity is also zero. This condition eliminates an arbitrary function on the right-hand side of Eq. (8.4.11). Using Eq. (8.4.10) yields

$$\delta p = a_s^2[f(t - x/a_s) + g(t + x/a_s)]. \tag{8.4.12}$$

Sound waves are studied by the branch of science called *acoustics*. Harmonic waves play a special role in wave theory in general, and in acoustics in particular. In these waves $f(t)$ and $g(t)$ are linear combinations of $\sin(\omega t)$ and $\cos(\omega t)$. The period of a harmonic wave is $2\pi/\omega$. The number of periods per one second, $\omega/2\pi$, is called the *wave frequency* and measured in Hz (Hertz after the German scientist Heinrich Hertz). Hence, the period of a wave of frequency 1 Hz is 1 s. Note that ω is also often called the "frequency." To distinguish between ω and $\omega/2\pi$, ω is sometimes called the *cyclic frequency*. The sounds that we hear are superpositions of a huge number of harmonic waves with various frequencies propagating in various directions.

The best human ears can hear harmonic sounds with frequencies between 20 Hz and 20 kHz (1 kHz $= 10^3$ Hz). Waves with frequencies below 20 Hz are called *infrasound* waves, while waves with frequencies above 20 Hz are called *ultrasound* waves. The latter are widely used in medical examinations.

Under normal conditions $\gamma \approx 1.4$ for air. Then, substituting $p_0 = p_a = 10^5$ N m^{-2} and $\rho_0 = \rho_a = 1.3$ kg m^{-3} into Eq. (8.4.4) we find $a_s \approx 330$ m s$^{-1} \approx 1200$ km/hour. In Chaps. 5 and 7 we considered water as an example of an incompressible fluid. However, as we pointed out, this is an approximation. In fact, water is slightly compressible, and it is its compressibility that enables sound waves to propagate in water. The relation between the pressure and density in water is not described by Eq. (8.4.4), however, it is a good approximation to consider water as a barotropic fluid. The speed of sound in water is about 1480 m s^{-1}. Finally, we note that the longitudinal or P-waves in solids studied in Chap. 6 can also be considered as sound waves because their propagation is related to the density variation.

Example Suppose a sphere of radius R is immersed into an ideal immovable compressible fluid with the unperturbed density ρ_0. The sphere radius harmonically oscillates with frequency ω, so that it is equal to $R[1 + \epsilon \sin(\omega t)]$, where $\epsilon \ll 1$. This radial oscillation drives a sound wave in the fluid. Find the spatial and temporal density and velocity variations outside the sphere.

Solution: Since the problem is spherically symmetric it is natural to assume that the velocity and density only depend on r in the spherical coordinates with the origin at the centre of the sphere. In addition, we assume that the velocity has the form $\boldsymbol{v} = v(r)\boldsymbol{e}_r$. Then, using Eqs. (2.13.11) and (2.13.17) we reduce Eqs. (8.4.6) and (8.4.7) to

$$\rho_0 \frac{\partial v}{\partial t} = -a_s^2 \frac{\partial(\delta\rho)}{\partial r}, \tag{8.4.13}$$

$$\frac{\partial^2 \delta\rho}{\partial t^2} - \frac{a_s^2}{r^2}\frac{\partial}{\partial r} r^2 \frac{\partial(\delta\rho)}{\partial r} = 0. \tag{8.4.14}$$

At the surface of the sphere the radial velocity must be equal to the velocity of the surface, which is equal to the time derivative of the sphere's radius. In the linear theory we can write this condition at the unperturbed sphere surface. As a result we obtain

$$v = \epsilon\omega R\cos(\omega t) \quad \text{at} \quad r = R. \tag{8.4.15}$$

Making the variable substitution $\delta\rho = \chi/r$ we reduce Eq. (8.4.14) to

$$\frac{\partial^2\chi}{\partial t^2} - a_s^2\frac{\partial^2\chi}{\partial r^2} = 0. \tag{8.4.16}$$

We impose the condition that there are only waves outgoing from the sphere. This implies that we should take a solution to Eq. (8.4.16) given by $\chi = f(t - r/a_s)$. It follows from Eqs. (8.4.13) and (8.4.15) that χ satisfies the boundary condition

$$a_s^2\left(\frac{\partial\chi}{\partial r} - \frac{\chi}{R}\right) = \epsilon\rho_0\omega^2R^2\sin(\omega t) \quad \text{at} \quad r = R.$$

Substituting $\chi = f(t - r/a_s)$ into this boundary condition yields

$$a_s\frac{df}{ds} + \frac{a_s^2 f}{R} = -\epsilon\rho_0\omega^2R^2\sin(\omega(s + R/a_s)) \quad \text{at} \quad s = t - R/a_s, \tag{8.4.17}$$

where $s = t - r/a_s$. The density must be bounded as $r \to \infty$, implying that $f(s)/s$ must be bounded as $s \to -\infty$. The solution to Eq. (8.4.17) satisfying this condition is

$$f(s) = \frac{\epsilon\rho_0\omega^2R^3}{a_s(a_s^2 + \omega^2R^2)}[\omega R\cos(\omega(s + R/a_s)) - a_s\sin(\omega(s + R/a_s))].$$

Then we obtain

$$\delta\rho = \frac{\epsilon\rho_0\omega^2R^3}{a_s r(a_s^2 + \omega^2R^2)}\left[\omega R\cos\left(\omega\left(t - \frac{r-R}{a_s}\right)\right) - a_s\sin\left(\omega\left(s - \frac{r-R}{a_s}\right)\right)\right].$$

Substituting this expression into Eq. (8.4.13) and integrating the obtained equation with respect to t yields

$$v = \frac{\epsilon\omega R^3}{r^2(a_s^2 + \omega^2R^2)}\left[(a_s^2 + \omega^2Rr)\cos\left(\omega\left(t - \frac{r-R}{a_s}\right)\right)\right.$$
$$\left. - a_s\omega(r - R)\sin\left(\omega\left(s - \frac{r-R}{a_s}\right)\right)\right].$$

Strictly speaking, we have to add an additive function of r to this equation. However, using Eq. (8.4.5) we can prove that the mean value of v with respect to time over the period is zero. It follows from this condition that this function of r must be zero.

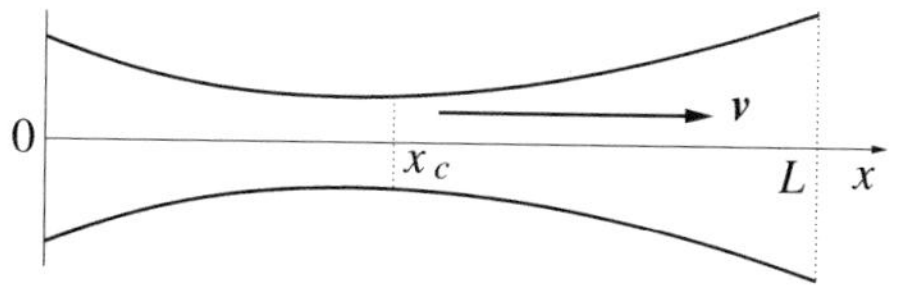

Fig. 8.1 Sketch of a de Laval nozzle

8.5 A One-Dimensional Stationary Motion of Gas. The de Laval Nozzle

In this section we consider a one-dimensional stationary motion of gas. Our main aim is to describe the motion of gas in a device called the *de Laval nozzle.* A de Laval nozzle (or *convergent-divergent nozzle*, *CD nozzle* or *con-di nozzle*) is a tube with a variable cross-section with an area that first decreases and then, after reaching a minimum, starts to increase. This device is used to accelerate a gas passing through it to a supersonic speed by converting the internal energy of the flow into kinetic energy. Because of this, the nozzle is used in some types of steam turbines and rocket engine nozzles. It is also used in supersonic jet engines. A slightly modified theory of the de Laval nozzle is used in astrophysics to describe solar and stellar winds.

We consider an isentropic motion of an ideal gas in a straight channel with a variable cross-section. We assume that the motion is stationary, that is, all variables are independent of time. We also assume that the characteristic scale of the cross-section variation is much larger than its size. For example, if the channel has a circular cross-section of variable radius, then the maximum radius must be much smaller than the characteristic scale of radius variation. A typical channel cross-section by a plane containing the channel axis is shown in Fig. 8.1. Below we use Cartesian coordinates with the x_1 axis coinciding with the channel axis.

Since the motion is stationary, Eq. (3.8.2) reduces to

$$\frac{\partial(\rho v_1)}{\partial x_1} + \frac{\partial(\rho v_2)}{\partial x_2} + \frac{\partial(\rho v_3)}{\partial x_3} = 0. \tag{8.5.1}$$

We now integrate this equation over a cross-section defined by the equation $x_1 =$ const. Let the cross-section be Σ and the contour C be its boundary. Using Green's theorem we obtain

$$\frac{\partial}{\partial x_1} \int_\Sigma \rho v_1 \, d\Sigma - \oint_C \rho(v_3 \, dx_2 - v_2 \, dx_3) = 0. \tag{8.5.2}$$

Since the wall of the channel is almost parallel to the x_1-axis, the same is true for the velocity. This implies that v_2 and v_3 are very small and we can neglect them. We also assume that we can neglect the variation of v_1 and ρ in the cross-section. Then we can rewrite Eq. (8.5.2) in the approximate form

$$\frac{d}{dx}(\rho v A) = 0,$$

where $A(x)$ is the cross-section area and we introduced the notation $x = x_1$ and $v = v_1$. It follows from this equation that

$$\rho v A = \rho_0 v_0 A_0, \tag{8.5.3}$$

where the subscript '0' indicates that a quantity is calculated at $x = 0$. The first component of the momentum Eq. (5.1.3) reads

$$v_1 \frac{\partial v_1}{\partial x_1} + v_2 \frac{\partial v_1}{\partial x_2} + v_3 \frac{\partial v_1}{\partial x_3} = -\frac{1}{\rho}\frac{\partial p}{\partial x_1}.$$

Since we can neglect the dependence of ρ on x_2 and x_3 it follows from Eq. (8.3.18) that the same is true for p. Since v_2 and v_3 are small and v_1 only weakly depends on x_2 an x_3, we can neglect the second and third terms on the left-hand side of this equation. Then it reduces to

$$v\frac{dv}{dx} = -\frac{1}{\rho}\frac{dp}{dx}.$$

Taking into account that $a_s^2 = dp/d\rho$ we obtain from this equation

$$v\,dv + \frac{a_s^2}{\rho}\,d\rho = 0. \tag{8.5.4}$$

Differentiating Eq. (8.5.3) and dividing the result by $\rho v A$ yields

$$\frac{dv}{v} + \frac{d\rho}{\rho} + \frac{dA}{A} = 0. \tag{8.5.5}$$

Eliminating $d\rho/\rho$ from Eqs. (8.5.4) and (8.5.5) we obtain the *Hugoniot equation* (named after the French scientist Pierre Henri Hugoniot)

$$\boxed{\left(M^2 - 1\right)\frac{dv}{v} = \frac{dA}{A}, \quad M = \frac{v}{a_s},} \tag{8.5.6}$$

where M is the *Mach number*, named after the Austrian physicist and philosopher Ernst Mach. A flow with $M < 1$ is called *subsonic*, while it is called *supersonic* when $M > 1$.

Using Eq. (8.5.6) we can draw a very important conclusion. First we consider the case $M < 1$. Then it follows that the flow velocity increases when $A(x)$ decreases ($dA < 0$), and it decreases when $A(x)$ increases ($dA > 0$). This is in accordance with our intuition: when the cross-section area decreases we need to increase the velocity to keep the mass flux constant. However, when $M > 1$ the velocity behaves contrary to our intuition. It increases when $A(x)$ increases and decreases when $A(x)$ decreases.

The reason for this is that, in a supersonic flow, the density increases very fast when the velocity decreases, and it decreases very fast when the velocity increases. As a result, the increase/decrease of the density overcomes the decrease/increase of the velocity.

Integrating Eq. (8.5.4) and using Eq. (8.4.3) we obtain

$$\frac{v^2}{2}+\frac{a_s^2}{\gamma-1}=\frac{v_0^2}{2}+\frac{a_0^2}{\gamma-1}, \tag{8.5.7}$$

where $a_0=a_s(\rho_0)$. Again using Eq. (8.4.3) we obtain from this equation

$$\frac{\rho}{\rho_0}=\left[\frac{2+(\gamma-1)M_0^2}{2+(\gamma-1)M^2}\right]^{\frac{1}{\gamma-1}}. \tag{8.5.8}$$

Then, using Eqs. (8.3.18) and (8.5.6)–(8.5.8) yields

$$\frac{p}{p_0}=\left[\frac{2+(\gamma-1)M_0^2}{2+(\gamma-1)M^2}\right]^{\frac{\gamma}{\gamma-1}}, \tag{8.5.9}$$

$$\frac{v}{v_0}=\frac{M}{M_0}\left[\frac{2+(\gamma-1)M_0^2}{2+(\gamma-1)M^2}\right]^{1/2}. \tag{8.5.10}$$

Substituting Eqs. (8.5.8) and (8.5.10) into Eq. (8.5.3) we obtain

$$\frac{A}{A_0}=\frac{M_0}{M}\left[\frac{2+(\gamma-1)M^2}{2+(\gamma-1)M_0^2}\right]^{\frac{\gamma+1}{2(\gamma-1)}}. \tag{8.5.11}$$

When M_0 is given Eqs. (8.5.8)–(8.5.11) implicitly define M, ρ/ρ_0, p/p_0, and v/v_0 as functions of A/A_0.

Although we can use Eq. (8.5.11) to study the dependence of M on A, it is more convenient to use the relation between these two quantities written in the differential form. Using Eq. (8.5.10) we obtain from Eq. (8.5.6)

$$\frac{dA}{A}=\frac{2(M^2-1)}{2+(\gamma-1)M^2}\frac{dM}{M}. \tag{8.5.12}$$

Then, using the relation $dA=A'(x)\,dx$ and introducing the variable τ we obtain from this equation the autonomous system of two first-order differential equations

$$\frac{dx}{d\tau}=\frac{M^2-1}{2+(\gamma-1)M^2},\quad \frac{dM}{d\tau}=\frac{MA'(x)}{2A(x)}. \tag{8.5.13}$$

In these equations $A(x)$ is considered as a given function. We assume that $A(x)$ varies as shown in Fig. 8.1. It decreases in the interval $(0, x_c)$ and then increases in the interval (x_c, L). It follows that the system of Eq. (8.5.13) has exactly one singular point, $x = x_c$ and $M = 1$. The Jacobian matrix of the system (8.5.13) calculated at this point is

$$\begin{pmatrix} 0 & \dfrac{2}{\gamma+1} \\ \dfrac{A''(x_c)}{2A(x_c)} & 0 \end{pmatrix}.$$

The eigenvalues of this matrix are

$$\lambda_{1,2} = \pm\sqrt{\frac{A''(x_c)}{(\gamma+1)A(x_c)}}.$$

Since $A(x)$ takes its minimum at x_c, it follows that $A''(x_c) > 0$. Then λ_1 and λ_2 are real and have different signs, which implies that $(x_c, 1)$ is a saddle. A typical phase portrait of the system (8.5.13) is shown in Fig. 8.2. We see that there are two curves, $M = l_1(x)$ and $M = l_2(x)$, that pass through the critical point. The curve l_1 corresponds to the transition from subsonic to supersonic flow, while the curve l_2 to the transition from supersonic to subsonic flow. We introduce the notation $l_1(0) = M_-$ and $l_2(0) = M_+$. The two curves divide the phase plane into four domains. The curves in the lower domain correspond to flows that are everywhere subsonic. In these flows M increases in the interval $(0, x_c)$ and then decreases in the interval (x_c, L). The curves in the upper domain correspond to flows that are everywhere supersonic. In these flows M decreases in the interval $(0, x_c)$ and then increases in the interval (x_c, L). Finally, the integral curves in the left and right domains have no physical sense because any of these curves gives either no value of M for a given x, or two different values of M for a given x.

From the inspection of Fig. 8.2 we can conclude the following. If $M_0 \in (0, M_-)$, then the flow is everywhere subsonic. If $M_0 = M_-$ then the flow is transitional. It is subsonic in $(0, x_c)$ and supersonic in (x_c, L). Flows with $M_0 \in (M_-, M_+)$ do not exist. If $M = M_+$ then the flow is also transitional, but now it is supersonic in $(0, x_c)$ and subsonic in (x_c, L). Finally, if $M \in (M_+, \infty)$ then the flow is everywhere supersonic.

We see that, for a given function $A(x)$, there are infinitely many solutions describing flows that are everywhere either subsonic or supersonic, but exactly one solution describing the subsonic-supersonic transition, and exactly one solution describing the supersonic-subsonic transition. However, the stability analysis shows that the supersonic-subsonic solution is unstable, meaning that a continuous transition from a supersonic to a subsonic flow is impossible.

In accordance with Eq. (8.5.12) M is a monotonically increasing function of A when $M > 1$. Substituting $M = 1$, $M_0 = M_-$, and $A = A_c = A(x_c)$ into Eq. (8.5.11) we obtain the equation determining M_-,

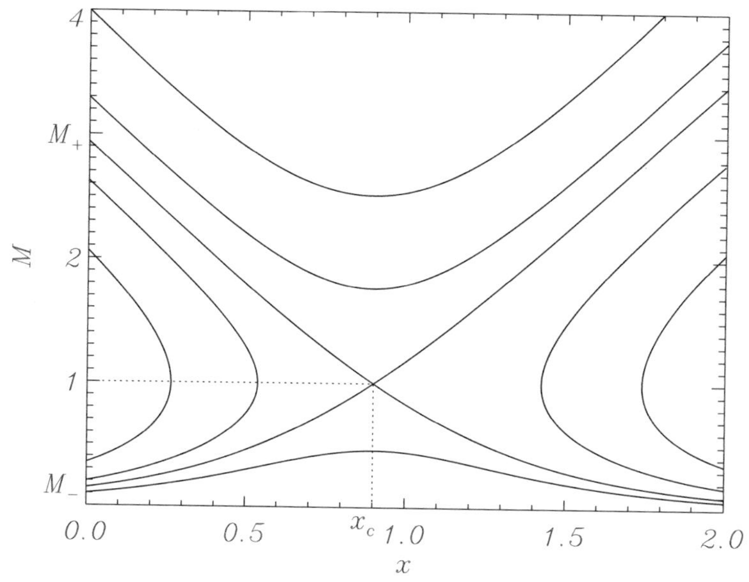

Fig. 8.2 Typical phase portrait of the system (8.5.13)

$$\frac{A_c}{A_0} = M_- \left[\frac{\gamma + 1}{2 + (\gamma - 1)M_-^2}\right]^{\frac{\gamma+1}{2(\gamma-1)}}. \qquad (8.5.14)$$

To obtain the dependences of ρ, p, and v on M in the transitional flow we have to substitute $M_0 = M_-$ into Eqs. (8.5.8)–(8.5.10).

Let us assume that $A_0 \gg A_c$. Then it follows from Eq. (8.5.14) that $M_- \ll 1$. Since $M_- \ll 1$ it follows that the flow at the nozzle entrance is strongly subsonic and the gas there is almost at rest. Substituting M_- for M_0 in Eq. (8.5.10) and taking into account that $M_- \ll 1$ we reduce it to

$$v \approx a_{s0} M \left[\frac{2}{2 + (\gamma - 1)M^2}\right]^{1/2}. \qquad (8.5.15)$$

It is easy to show that the right-hand side of this equation is a monotonically increasing function of M. Taking $A \to \infty$ we obtain from Eq. (8.5.11) that $M \to \infty$. Then it follows from Eq. (8.5.15) that $v \to v_{max} = a_{s0}\sqrt{2/(\gamma - 1)}$. For example, for air $\gamma \approx 1.4$. If we take $a_{s0} = 330$ m s^{-1} as a typical sound speed when p_0 is the atmospheric pressure and the temperature is moderate (say, between 0° and 20° C), then $v_{max} \approx 740$ m s^{-1}. Note that $\rho \to 0$ and $p \to 0$ as $M \to \infty$, so to reach its maximum value at the nozzle exit it must flow out in a vacuum.

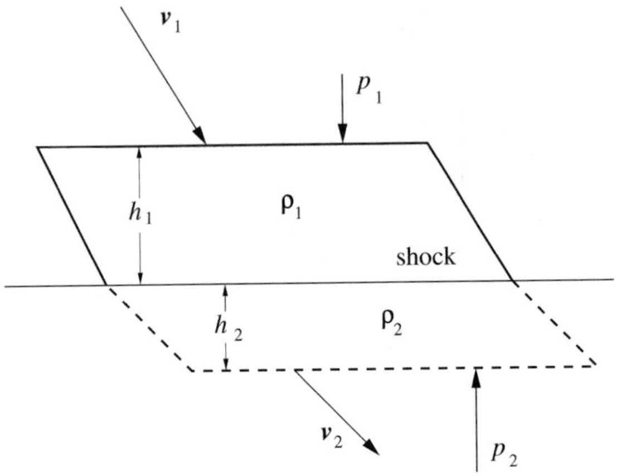

Fig. 8.3 Illustration of the derivation of the Rankine–Hugoniot relations. The thick solid lines show a liquid volume at time t, while the thick dashed lines show this volume at time $t + \Delta t$

8.6 Shock Waves

A *shock wave*, or simply shock, is characterised by an abrupt, nearly discontinuous change in pressure, temperature and density of a gas. Shocks appear, for example, when an airplane is moving at a supersonic speed in the air. When it flies relatively low and passes us we hear a clap. This is a shock that hits our ears. The same happens when we hear a roll of thunder. The effects of shocks are not always so innocent. Shocks caused by explosions can convert big building into dust, throw heavy vehicles into the air like carton boxes, and cause heavy contusions or even death.

Although in reality a shock is an almost abrupt but still continuous change of all quantities in a gas, the transition layer where this change occurs is so thin (its thickness is of the order of the mean free path of atoms or molecules), that it can be considered as a discontinuity. The mass conservation, momentum, and energy equations written in differential form are not valid for discontinuous motions of a continuum. However, the same equations written in an integral form are valid even for such flows. We use them to derive the relations between the variables at the two sides of a shock.

For simplicity, we assume that the shock surface is a plane. However the relations that we will derive are also valid when the shock surface is curved because a small part of this surface can be approximated by a tangential plane. We study the flow with the discontinuity in a reference frame where the discontinuity is at rest. We consider a liquid volume, that is, a volume always consisting of the same particles. At the initial time the volume is at one side of the discontinuity and has the form of an oblique prism with a unit square at the base. One of the two bases of the prism is on the discontinuity and the prism height is h_1 (see Fig. 8.3). The liquid volume is assumed to be so small that we can neglect the variation of any quantity inside it. The density, pressure and velocity of the liquid in the prism are ρ_1, p_1, and $\boldsymbol{v}_1$. The prism edges are parallel to $\boldsymbol{v}_1$. Since the prism base has unit size, it follows that its volume is equal to its height h_1. After the time interval $\Delta t = h_1/v_{n1}$, where v_{n1} is the velocity component normal to the discontinuity, all the liquid that was inside the prism at the initial time crosses the discontinuity and forms the new prism with the same base and edges parallel to $\boldsymbol{v}_2$. This prism's height is $h_2 = v_{2n}\,\Delta t = v_{2n}h_1/v_{n1}$ and its volume is h_2. Since the liquid mass is the same in the two prisms, we obtain $\rho_1 h_1 = \rho_2 h_2$, which reduces to

$$\boxed{\rho_1 v_{n1} = \rho_2 v_{n2}.} \tag{8.6.1}$$

Now we use the momentum equation. The variation of the momentum is equal to the impulse of force. The forces acting on the side surfaces of the prisms are in balance, which implies that the momentum component tangential to the discontinuity is conserved. Then it follows that

$$\boxed{\boldsymbol{v}_{\tau 1} = \boldsymbol{v}_{\tau 2},} \tag{8.6.2}$$

where $\boldsymbol{v}_\tau$ is the velocity component tangential to the discontinuity. In particular, this equation implies that the vectors $\boldsymbol{n}$, $\boldsymbol{v}_1$, and $\boldsymbol{v}_2$ are *coplanar*, where $\boldsymbol{n}$ is the unit vector normal to the surfaces of discontinuity. The variation of the momentum component normal to the discontinuity is $\rho_2 h_2 v_{n2} - \rho_1 h_1 v_{n1}$. The impulse of the force acting on the fluid volume is $(p_1 - p_2)\Delta t = (p_1 - p_2)h_1/v_{n1}$. As a result we obtain

$$\rho_2 h_2 v_{n2} - \rho_1 h_1 v_{n1} = \frac{(p_1 - p_2)h_1}{v_{n1}},$$

which, with the aid of the relation $h_2 = h_1 v_{n2}/v_{n1}$, reduces to

$$\boxed{\rho_1 v_{n1}^2 + p_1 = \rho_2 v_{n2}^2 + p_2.} \tag{8.6.3}$$

Finally, we proceed to the energy equation. The work done by the pressure forces applied to the prism bases is $p_1 h_1 - p_2 h_2$. The work done by the pressure forces applied to the side surfaces of the prisms is proportional to h_1^2. Hence, taking h_1 sufficiently small enables us to neglect this work. The volume energy variation is equal to the work done by external forces, which gives

$$h_2\rho_2\left(U_2 + \frac{\|\boldsymbol{v}_2\|^2}{2}\right) - h_1\rho_1\left(U_1 + \frac{\|\boldsymbol{v}_1\|^2}{2}\right) = p_1 h_1 - p_2 h_2.$$

Using the relation $h_2 = h_1 v_{n2}/v_{n1}$ and Eqs. (8.6.1) and (8.6.2) we reduce this equation to

$$\boxed{\rho_1 v_{n1}\left(U_1 + \frac{v_{n1}^2}{2}\right) + p_1 v_{n1} = \rho_2 v_{n2}\left(U_2 + \frac{v_{n2}^2}{2}\right) + p_2 v_{n2}.} \tag{8.6.4}$$

Equations (8.6.1)–(8.6.4) are called the *Rankine–Hugoniot relations* after two scientists, English William John Macquorn Rankine and French Pierre Henri Hugoniot, who first formulated them. Eliminating ρ_2 and p_2 from Eqs. (8.6.1), (8.6.3), and (8.6.4), and using the expression for the internal energy U given by Eqs. (8.3.16) we obtain the equation for v_{n2},

$$(\gamma+1)\rho_1 v_{n1} v_{n2}^2 - 2\gamma(\rho_1 v_{n1}^2 + p_1)v_{n2} + 2\gamma p_1 v_{n1} + (\gamma-1)\rho_1 v_{n1}^3 = 0. \quad (8.6.5)$$

One solution to this equation is $v_{n2} = v_{n1}$. This solution corresponds to the case when there is no discontinuity. The second solution is

$$\frac{v_{n2}}{v_{n1}} = 1 - \frac{2(M_1^2-1)}{(\gamma+1)M_1^2}, \quad (8.6.6)$$

where the Mach number was calculated using the normal component of the velocity, $M_1 = v_{n1}/a_{s1}$. With the aid of this result we obtain from Eqs. (8.3.10), (8.6.1) and (8.6.3)

$$\frac{\rho_2}{\rho_1} = 1 + \frac{2(M_1^2-1)}{2+(\gamma-1)M_1^2}, \quad \frac{p_2}{p_1} = 1 + \frac{2\gamma(M_1^2-1)}{\gamma+1}, \quad (8.6.7)$$

$$\frac{T_2}{T_1} = 1 + \frac{2(\gamma-1)(\gamma M_1^2+1)(M_1^2-1)}{(\gamma+1)^2 M_1^2}. \quad (8.6.8)$$

The condition $p_2 > 0$ imposes the restriction on M_1,

$$M_1^2 > \frac{\gamma-1}{2\gamma}. \quad (8.6.9)$$

Using Eqs. (8.6.6) and (8.6.7) we obtain

$$M_2^2 = 1 - \frac{(\gamma+1)(M_1^2-1)}{2\gamma M_1^2 - (\gamma-1)}. \quad (8.6.10)$$

It follows from Eqs. (8.6.6), (8.6.7), and (8.6.10) that

$$v_{n2} > v_{n1}, \quad \rho_2 < \rho_1, \quad p_2 < p_1, \quad M_2 > 1 \quad (8.6.11)$$

when $M_1 < 1$, and

$$v_{n2} < v_{n1}, \quad \rho_2 > \rho_1, \quad p_2 > p_1, \quad M_2 < 1 \quad (8.6.12)$$

when $M_1 > 1$. Since the fluid density drops when a fluid passes through a shock with $M_1 < 1$, such shocks are called *rarefaction* shocks. On the other hand, the fluid density increases when a fluid passes through a shock with $M_1 > 1$, so such shocks are called *compression* shocks. It is worth giving the limiting values of the velocity, density, pressure, and second Mach number as $M_1 \to \infty$:

$$\frac{v_{n2}}{v_{n1}} \to \frac{\gamma - 1}{\gamma + 1}, \quad \frac{\rho_2}{\rho_1} \to \frac{\gamma + 1}{\gamma - 1}, \quad \frac{p_2}{p_1} \to \infty, \quad M_2^2 \to \frac{\gamma - 1}{2\gamma}. \tag{8.6.13}$$

Lemma 8.1 *Rarefaction shocks do not exist.*

Proof If p and ρ in the flow vary continuously then, in accordance with Eq. (8.3.18), p_2 and ρ_2 would be related by the equation

$$p_2 = p_1 \left(\frac{\rho_2}{\rho_1}\right)^{\gamma}. \tag{8.6.14}$$

The curve in the (ρ_2, p_2)-plane determined by this equation is called the *adiabatic curve*. Eliminating M_1 from Eq. (8.6.7) we obtain

$$p_2 = p_1 \frac{(\gamma + 1)\rho_2 - (\gamma - 1)\rho_1}{(\gamma + 1)\rho_1 - (\gamma - 1)\rho_2}. \tag{8.6.15}$$

The curve in the (ρ_2, p_2)-plane determined by this equation is called the *shock adiabatic curve*. It is obvious that both the adiabatic curve and the shock adiabatic curve contain the point (ρ_1, p_1).

A small fluid volume crossing a shock is an isolated system (it is also called thermally insulated). In accordance with the second law of thermodynamics its entropy cannot decrease. It follows from Eq. (8.3.12) that the entropy does not change along the adiabatic curve. Let us calculate the entropy jump across the shock. Using Eq. (8.3.12) we obtain

$$S_2 - S_1 = C_v \ln \left[\frac{p_2}{p_1}\left(\frac{\rho_1}{\rho_2}\right)^{\gamma}\right] \equiv C_v \ln F. \tag{8.6.16}$$

Using Eq. (8.6.7) we obtain

$$F = F(M_1^2) = \frac{2\gamma M_1^2 - (\gamma - 1)}{\gamma + 1}\left[\frac{2 + (\gamma - 1)M_1^2}{(\gamma + 1)M_1^2}\right]^{\gamma}. \tag{8.6.17}$$

Differentiating this expression yields

$$F'(M_1^2) = \frac{2\gamma(\gamma - 1)(M_1^2 - 1)^2}{(\gamma + 1)^2 M_1^4}\left[\frac{2 + (\gamma - 1)M_1^2}{(\gamma + 1)M_1^2}\right]^{\gamma - 1} > 0. \tag{8.6.18}$$

Since $F(1) = 1$ it follows that $S_2 > S_1$ ($S_2 < S_1$) when $M_1 > 1$ ($M_1 < 1$), meaning that the entropy increases in compression shocks and decreases in rarefaction shocks.

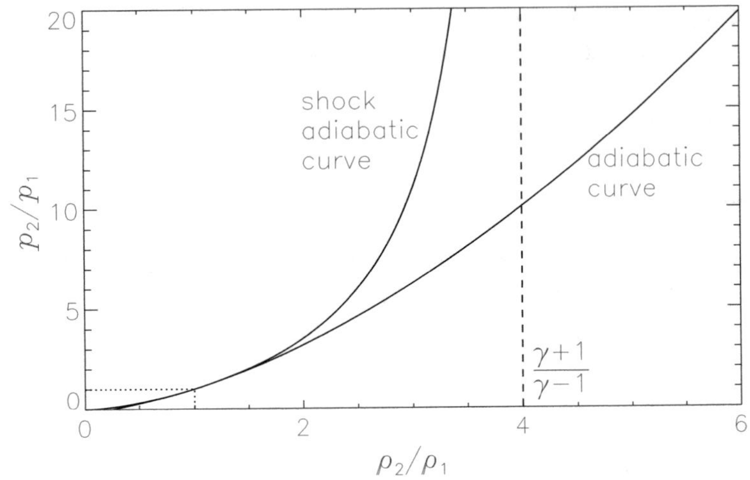

Fig. 8.4 Adiabatic curve and shock adiabatic curve

Hence, we conclude that the existence of rarefaction shocks contradicts the second law of thermodynamics. The lemma is proved. □

It is expedient to consider the graphs of the adiabatic curve and shock adiabatic curve. We have $S_2 = S_1$ at any point (ρ_2, p_2) on the adiabatic curve. Since S_2 is a monotonically increasing function of p_2 at fixed ρ_2, it follows that $S_2 > S_1$ in the domain above the adiabatic curve, and $S_2 < S_1$ in the domain below the adiabatic curve in the ρ_2, p_2-plane. Since $S_2 < S_1$ in rarefaction shocks ($\rho_2 < \rho_1$) it follows that the part of the shock adiabatic curve corresponding to rarefaction shocks is below the adiabatic curve. In compressional shocks ($\rho_2 > \rho_1$) $S_2 > S_1$, which implies that the part of the shock adiabatic curve corresponding to compressional shocks is above the adiabatic curve. The two curves are shown in Fig. 8.4. It is not very clear from this figure that the shock adiabatic curve is below the adiabatic curve when $\rho_2/\rho_1 < 1$. To make this clearer we zoom in on a part of this figure (see Fig. 8.5).

Since $M_1 > 1$, it follows from Eqs. (8.3.16), (8.6.6), and (8.6.8) that, when a gas passes through a shock, its velocity decreases and the internal energy increases. This implies that the gas kinetic energy is converted into the internal energy inside the shock, which is in reality not a discontinuity but a narrow transitional layer where viscosity and thermal conduction operate.

Often we need to study propagation of shocks through an immovable gas (see Fig. 8.6). In that case we arrive at the picture studied in this section if we introduce a new reference frame moving together with the shock. The gas before the shock is immovable in the reference frame x, while in the reference frame x' the shock is immovable. The shock propagates through the immovable gas with velocity V. The tangential velocity component is zero in both reference frames. The density and pressure are the same in both reference frames. The normal velocity component is zero before the shock and it is v_{n2} after the shock in the reference frame x. Then it is

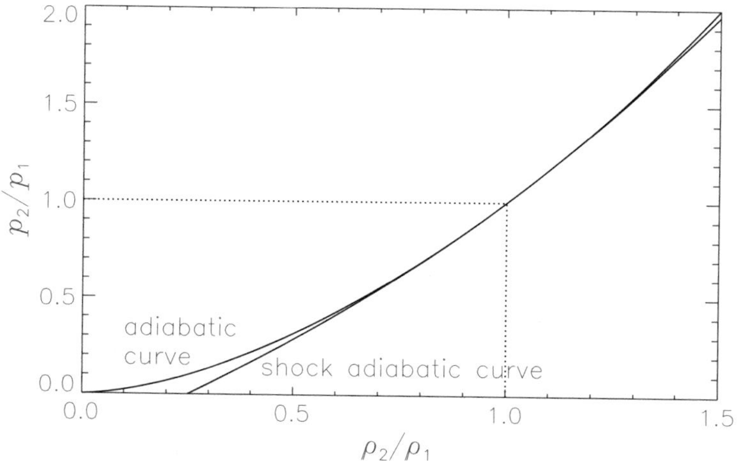

Fig. 8.5 A magnified part of Fig. 8.4

Fig. 8.6 Introducing the reference frame where a shock propagates through a fluid at rest

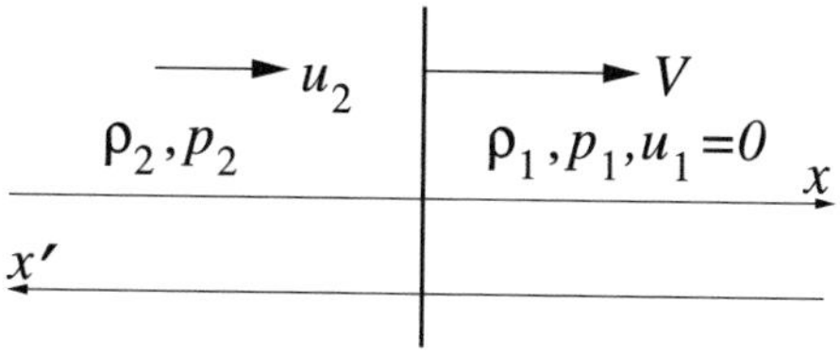

equal to $v'_{n1} = V$ before the shock and $v'_{n2} = V - v_{n2}$ in the reference frame x'. Note that in this reference frame the velocity is in the positive x'-direction. In the reference frame x' we have the same flow as was studied before. Consequently u'_{n2}/u'_{n1}, ρ_2/ρ_1, and p_2/p_1 are given by Eqs. (8.6.6) and (8.6.7). Then we immediately obtain

$$\frac{V - v_{n2}}{V} = \frac{2 + (\gamma - 1)M_1^2}{(\gamma + 1)M_1^2},$$

where $M_1 = V/a_{s1}$, and $a_{s1}^2 = \gamma p_1/\rho_1$. This equation reduces to

$$v_{n2} = \frac{2V}{(\gamma + 1)}\left(1 - \frac{1}{M_1^2}\right). \tag{8.6.19}$$

We see that, when propagating, the shock drags the fluid behind it.

Example (*Piston problem*) In a straight channel a piston starts to move instantaneously with constant speed U. Initially the gas in the channel is at rest (see Fig. 8.7). Describe the gas motion in the channel.

Solution: Since no signal can propagate instantaneously, it follows that the fluid far from the piston is at rest. We studied the motion of a gas in a channel in Sect. 8.5.

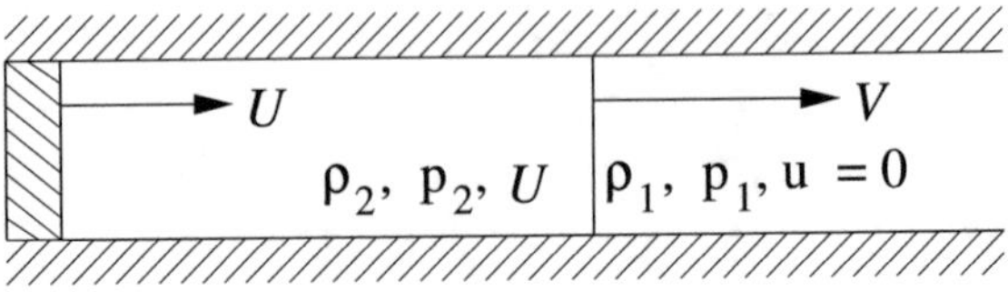

Fig. 8.7 Illustration of the piston problem

It follows from Eq. (8.5.6) that $v =$ const when the channel cross-section is constant ($A =$ const). As a result, we obtain that the gas is at rest before the piston ($v_1 = 0$) where the flow parameters are continuous. Then it follow from Eqs. (8.5.8) and (8.5.9) that the density and pressure do not change in the region where they are continuous. On the other hand, the flow velocity at the piston must be equal to the piston speed U. The only way for the speed to change from zero to U is through a shock. Hence, we conclude that there must be a shock moving ahead of the piston. Since behind the shock the flow velocity must be equal to the piston speed, $u_2 = U$, it follows from Eq. (8.6.19) that

$$U = \frac{2V}{\gamma + 1}\left(1 - \frac{1}{M_1^2}\right), \tag{8.6.20}$$

where V is the shock velocity. Using the relation $M_1 = V/a_{s1}$ we reduce this equation to

$$2V^2 - (\gamma + 1)UV - 2a_{s1}^2 = 0. \tag{8.6.21}$$

Since $V > 0$, we choose the positive root of this quadratic equation:

$$V = \frac{1}{4}\left[(\gamma + 1)U + \sqrt{(\gamma + 1)^2U^2 + 16a_{s1}^2}\right]. \tag{8.6.22}$$

Substituting $M_1 = V/a_{s1}$ and using Eqs. (8.6.7) and (8.6.21) we obtain

$$\rho_2 = \frac{\rho_1 V}{V - U}, \quad p_2 = p_1\frac{2V + (\gamma - 1)U}{2V - (\gamma + 1)U}. \tag{8.6.23}$$

8.7 Simple Waves

In this section we study special solutions to the one-dimensional equations of compressible fluids. We assume that the density, pressure and velocity only depend on one spatial variable x_1, and use below the notation $x = x_1$. We also assume that the velocity has only one component $v = v_1$. In this case the mass and momentum conservation equations reduce to

$$\frac{\partial \rho}{\partial t} + \frac{\partial(\rho v)}{\partial x} = 0, \tag{8.7.1}$$

$$\frac{\partial v}{\partial t} + v\frac{\partial v}{\partial x} = -\frac{1}{\rho}\frac{\partial p}{\partial x}. \tag{8.7.2}$$

The flow is assumed to be isentropic, so p and ρ are related by Eq. (8.3.18). In what follows we use the velocity and sound speed as dependent variables. To simplify the notation we drop the subscript "s" of the sound speed and denote it as a. Using Eqs. (8.3.18) and (8.4.3) we express ρ and p in terms of a,

$$\rho = \rho_0\left(\frac{a}{a_0}\right)^{\frac{2}{\gamma-1}}, \quad p = p_0\left(\frac{a}{a_0}\right)^{\frac{2\gamma}{\gamma-1}}. \tag{8.7.3}$$

Recall that $a_0^2 = \gamma p_0/\rho_0$. Using Eq. (8.7.3) we transform Eqs. (8.7.1) and (8.7.2) into

$$\frac{\partial a}{\partial t} + v\frac{\partial a}{\partial x} + \frac{\gamma-1}{2}a\frac{\partial v}{\partial x} = 0, \tag{8.7.4}$$

$$\frac{\partial v}{\partial t} + v\frac{\partial v}{\partial x} + \frac{2a}{\gamma-1}\frac{\partial a}{\partial x} = 0. \tag{8.7.5}$$

Now we introduce the *Riemann invariants*,

$$R_+ = v + \frac{2a}{\gamma-1}, \qquad R_- = v - \frac{2a}{\gamma-1}. \tag{8.7.6}$$

Using the Riemann invariants we transform Eqs. (8.7.4) and (8.7.5) into

$$\frac{\partial R_+}{\partial t} - \frac{\partial R_-}{\partial t} + \frac{R_+ + R_-}{2}\left(\frac{\partial R_+}{\partial x} - \frac{\partial R_-}{\partial x}\right) + \frac{\gamma-1}{4}(R_+ - R_-)\left(\frac{\partial R_+}{\partial x} + \frac{\partial R_-}{\partial x}\right) = 0, \tag{8.7.7}$$

$$\frac{\partial R_+}{\partial t} + \frac{\partial R_-}{\partial t} + \frac{R_+ + R_-}{2}\left(\frac{\partial R_+}{\partial x} + \frac{\partial R_-}{\partial x}\right) + \frac{\gamma-1}{4}(R_+ - R_-)\left(\frac{\partial R_+}{\partial x} - \frac{\partial R_-}{\partial x}\right) = 0. \tag{8.7.8}$$

Adding and subtracting these equations we obtain

$$\frac{\partial R_+}{\partial t} + \frac{\gamma+1}{4}R_+\frac{\partial R_+}{\partial x} + \frac{3-\gamma}{4}R_-\frac{\partial R_+}{\partial x} = 0, \tag{8.7.9}$$

$$\frac{\partial R_-}{\partial t} + \frac{\gamma+1}{4}R_-\frac{\partial R_-}{\partial x} + \frac{3-\gamma}{4}R_+\frac{\partial R_-}{\partial x} = 0. \tag{8.7.10}$$

A solution to this system of equations where either $R_+ = \text{const}$ or $R_- = \text{const}$ is called a *simple wave*. Let us take $R_- = -2a_0/(\gamma-1)$. Then

$$a = a_0 + \frac{\gamma-1}{2}v, \qquad R_+ = 2v + \frac{2a_0}{\gamma-1}. \tag{8.7.11}$$

Equation (8.7.10) is satisfied identically, while Eq. (8.7.9) reduces to

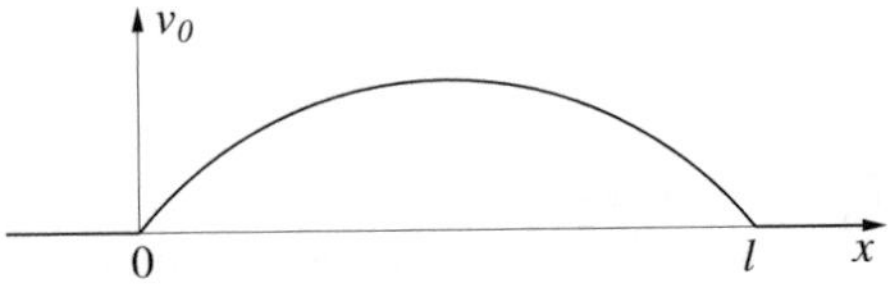

Fig. 8.8 The initial condition defined by Eq. (8.7.15)

$$\boxed{\frac{\partial v}{\partial t} + \left(a_0 + \frac{\gamma + 1}{2} v\right) \frac{\partial v}{\partial x} = 0.} \tag{8.7.12}$$

Consider the line $x = x(t)$ in the tx-plane determined by the equation

$$\boxed{\frac{dx}{dt} = a_0 + \frac{\gamma + 1}{2} v(t, x).} \tag{8.7.13}$$

This line is called a *characteristic* of Eq. (8.7.12). In accordance with the theorem on the existence of solutions to differential equations we can find a characteristic passing through any point (t_0, x_0) in the domain on the tx-plane where $v(t, x)$ is defined as a single-valued function with continuous first derivatives. Using Eqs. (8.7.12) and (8.7.13) we obtain that on a characteristic

$$\frac{dv}{dt} = \frac{\partial v}{\partial t} + \frac{\partial v}{\partial x}\frac{dx}{dt} = \frac{\partial v}{\partial t} + \left(a_0 + \frac{\gamma + 1}{2} v\right) \frac{\partial v}{\partial x} = 0, \tag{8.7.14}$$

implying that v does not vary along a characteristic. Then it follows from Eq. (8.7.13) that the characteristics are straight lines.

Using characteristics we can find the solution to Eq. (8.7.12) when the initial condition is given, that is, when v is given at $t = 0$. To show how to do this we consider an example. Let at $t = 0$

$$v = v_0(x) = U \begin{cases} 0, & x < 0, \\ \sin(\pi x/\ell), & 0 < x < \ell, \\ 0, & x > \ell. \end{cases} \tag{8.7.15}$$

The function $v_0(x)$ is shown in Fig. 8.8.

Consider a characteristic passing through the point $(0, x_0)$. On this characteristic $v = v_0(x_0)$. Hence its equation is

$$x = x_0 + t\left(a_0 + \frac{\gamma + 1}{2} v_0(x_0)\right). \tag{8.7.16}$$

The characteristic determined by Eq. (8.7.16) is shown in Fig. 8.9. We see that $v(t, x) = v(0, x_0)$ where the points (t, x) and $(0, x_0)$ are on the same characteristic. The equations of the two characteristics shown by dashed lines are $x = a_0 t$ and $x = \ell + a_0 t$. If a characteristic is below the lower of the two characteristics, or above

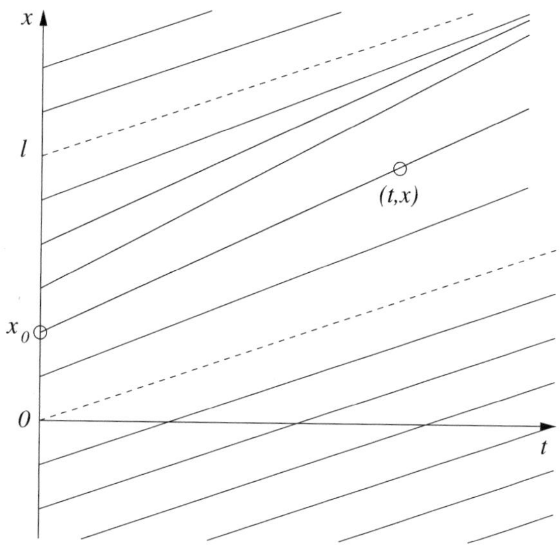

Fig. 8.9 Characteristics of Eq. (8.7.12)

the upper one, then it crosses the x-axis either below 0 or above ℓ. Consequently, $v = 0$ on such a characteristic. This implies that the fluid is at rest in regions $x < a_0 t$ and $x > \ell + a_0 t$.

To calculate $v(t, x)$ we need to find the coordinate of the intersection point with the x-axis, $x_0(t, x)$, of a characteristic passing through the point (t, x); $x_0(t, x)$ is defined by Eq. (8.7.16). If there is only one characteristic passing through each point (t, x), then we will find the unique single-valued solution to Eq. (8.7.12). However if two or more characteristics intersect at (t, x), then we will find two different values of v at the same point, which is unphysical.

First of all we note that in the regions $x < a_0 t$ and $x > \ell + a_0 t$ the characteristics are parallel to the corresponding boundary characteristics shown by the dashed lines in Fig. 8.9. Hence, they cannot intersect. However we can see in Fig. 8.9 that the characteristics are not parallel in the region $a_0 t < x < \ell + a_0 t$, so that they can intersect. Consider two characteristics that start at x_1 and x_2 on the x-axis, $0 < x_1 < x_2 < \ell$. Their equations are

$$x = x_1 + t\left(a_0 + \frac{\gamma + 1}{2} U \sin \frac{\pi x_1}{\ell}\right), \quad x = x_2 + t\left(a_0 + \frac{\gamma + 1}{2} U \sin \frac{\pi x_2}{\ell}\right).$$

It is easy to show that they intersect at the moment of time t_{int} defined by

$$t_{\text{int}} = \frac{-2(x_2 - x_1)}{U(\gamma + 1)[\sin(\pi x_2/\ell) - \sin(\pi x_1/\ell)]}.$$

If $\sin(\pi x_1/\ell) < \sin(\pi x_2/\ell)$, then $t_{\text{int}} < 0$, and we disregard this intersection since we only consider $t > 0$. However $t_{\text{int}} > 0$ when $\sin(\pi x_1/\ell) > \sin(\pi x_2/\ell)$. In that case we cannot find a single-valued solution to Eq. (8.7.12) for $t \geq t_{\text{int}}$.

In fact, we cannot find a single-valued solution to Eq. (8.7.12) for $t \geq t_c$, where t_c is the moment of time when the intersection of two characteristics occurs for the first time. Obviously $t_c = \min\limits_{x_1,x_2} t_{\text{int}}$. We calculate this minimum in two steps. First we find $\min\limits_{x_1} t_{\text{int}}$ keeping x_2 fixed. Then we obtain

$$t_c = \min_{x_2}\left(\min_{x_1} t_{\text{int}}\right).$$

To calculate $\min\limits_{x_1} t_{\text{int}}$ we rewrite the expression for t_{int} as

$$t_{\text{int}} = \frac{2\ell}{\pi U(\gamma+1)} \frac{\pi(x_2 - x_1)/2\ell}{\sin[\pi(x_2 - x_1)/2\ell]} \frac{-1}{\cos[\pi(x_1 + x_2)/2\ell]}.$$

The condition that $t_{\text{int}} > 0$ is equivalent to $x_1 + x_2 > l$. It is well known that the function $\sin x/x$ monotonically decreases when $0 < x < \pi$, so that the second multiplier in this expression is a monotonically decreasing function of x_1; $-\cos[\pi(x_1 + x_2)/2\ell]$ is a monotonically increasing function of x_1 when $x_1 + x_2 > \ell$. Hence, t_{int} is a monotonically decreasing function of x_1, and

$$\min_{x_1} t_{\text{int}} = \lim_{x_1 \to x_2} t_{\text{int}} = \frac{-2\ell}{\pi U(\gamma+1)\cos(\pi x_2/\ell)}.$$

Now

$$t_c = \min_{x_2} \frac{-2\ell}{\pi U(\gamma+1)\cos(\pi x_2/\ell)} = \frac{2\ell}{\pi U(\gamma+1)}, \tag{8.7.17}$$

and this minimum is taken at $x_2 = l$. Hence, the intersection time t_{int} takes this minimum value at $x_1 \to x_2 = \ell$.

A single-valued solution to Eq. (8.7.12) only exists for $t < t_c$. The Eq. (8.7.17) for t_c can be also obtained in a different way. To obtain the solution to Eq. (8.7.12) we need to find a unique solution to Eq. (8.7.16) considered as an equation for x_0. This is only possible when the right-hand side of this equation is a monotonic function of x_0. Now

$$\frac{d}{dx_0}\left[x_0 + t\left(a_0 + \frac{\gamma+1}{2}\,v_0(x_0)\right)\right] = 1 + \frac{t}{t_c}\cos\frac{\pi x_0}{\ell}.$$

This expression is positive for any x_0 when $t < t_c$, so that the right-hand side of Eq. (8.7.16) is a monotonically increasing function and we can solve this equation with respect to x_0 for any x. However, for $t > t_c$ the derivative changes sign at $x_0 = (\ell/\pi)[\pi - \arccos(t_c/t)]$, the right-hand side of Eq. (8.7.16) is a non-monotonic function of x_0, and we cannot find a unique solution to Eq. (8.7.16) at least for some values of x.

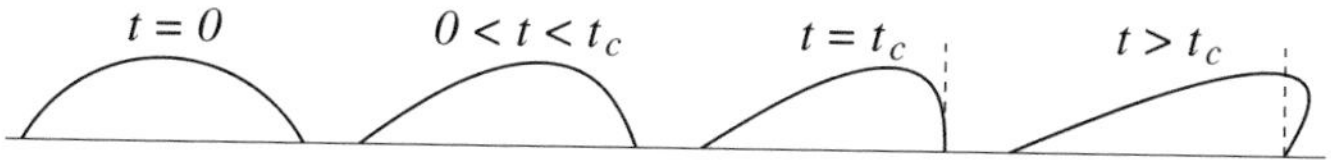

Fig. 8.10 Evolution of the solution to Eq. (8.7.12) with the initial condition (8.7.15)

It is instructive to calculate the derivative of $v(t, x)$ with respect to x at $t = t_c$:

$$\left.\frac{\partial v}{\partial x}\right|_{t=t_c} = \frac{\partial v_0(x_0(t_c, x))}{\partial x} = \frac{dv_0}{dx_0}\frac{\partial x_0(t_c, x)}{\partial x} = \frac{U\pi}{\ell}\cos\frac{\pi x_0}{\ell}\frac{\partial x_0(t_c, x)}{\partial x}.$$

Differentiating Eq. (8.7.16) we obtain

$$1 = \frac{\partial x_0}{\partial x}\left(1 + \frac{(\gamma+1)t}{2}\frac{dv_0}{dx_0}\right) = \frac{\partial x_0}{\partial x}\left(1 + \frac{\pi U t(\gamma+1)}{2\ell}\cos\frac{\pi x_0}{\ell}\right).$$

Then

$$\left.\frac{\partial x_0}{\partial x}\right|_{t=t_c} = \frac{1}{1+\cos(\pi x_0/\ell)},$$

and

$$\left.\frac{\partial v}{\partial x}\right|_{t=t_c} = \frac{U\pi}{\ell}\frac{\cos(\pi x_0/\ell)}{1+\cos(\pi x_0/\ell)}.$$

We see that

$$\left.\frac{\partial v}{\partial x}\right|_{t=t_c} \to -\infty \quad \text{as} \quad x_0 \to \ell.$$

The phenomenon of the formation of an infinite gradient in a solution is called a *gradient catastrophe*.

The evolution of the solution to Eq. (8.7.12) with the initial condition (8.7.15) is shown in Fig. 8.10. If we formally extend the solution beyond $t > t_c$, then we obtain a multivalued solution, which is unphysical. What happens to the wave beyond t_c? At t close to t_c there is a very large gradient near $x = \ell$. As a result, near this point dissipation related to viscosity and thermal conduction becomes important. This dissipation prevents the formation of the infinite gradient at $x = x_0$. The dissipation only operates in a very narrow layer. If we use an approximation of an ideal fluid, then this narrow layer is substituted by a discontinuity, which is a shock. Hence, at $t = t_c$ a shock starts to form near $x = \ell$. First this shock intensity grows. However, simultaneously, the energy related to the simple wave dissipates and, eventually, the wave amplitude tends to zero as $t \to \infty$. We will consider an example of wave damping due to shock formation in the next section.

8.8 Nonlinear Acoustics

In Sect. 8 we studied sound waves. As we have already stated, the branch of science that studies acoustic waves is called *acoustics*. When studying acoustic waves we assumed that their amplitudes are small and linearised the hydrodynamic equations. However, sometimes the amplitudes of sound waves are so large that the linear description is no longer valid and we need to take nonlinear terms in the hydrodynamic equations into account. The branch of acoustics that studies nonlinear acoustic waves is called *nonlinear acoustics*. In our analysis we closely follow the discussion given by Rudenko and Soluyan (see Further Reading).

Simple waves play a very important role in nonlinear acoustics. However, the approach to studying simple waves in nonlinear acoustics is slightly different from that used in the previous section. Usually it is assumed that a sound wave is driven at some surface and then propagates from this surface. Hence, instead of studying temporal wave evolution like in the previous section, one needs to study the wave's spatial evolution. When one-dimensional problems are considered it is assumed that a sound wave is driven at $x = 0$ and then propagates in the region $x > 0$. Hence, the boundary condition $v = v_0(t)$ at $x = 0$ is given, and this condition is valid for any time t, both negative and positive. The equation describing a simple wave is solved for $x > 0$.

Usually it is assumed in nonlinear acoustics that, although a linear description of sound waves is not valid, the wave amplitude is still sufficiently small. This enables us to keep the quadratic terms with respect to the wave amplitude in the hydrodynamic equations and neglect higher order terms proportional to the wave amplitude cubed and to higher powers of the amplitude. In particular, using Eq. (8.7.12) we obtain

$$v\frac{\partial v}{\partial x} = -\frac{2v}{2a_0 + (\gamma + 1)v}\frac{\partial v}{\partial t} \approx -\frac{v}{a_0}\frac{\partial v}{\partial t}.$$

With the aid of this result we transform Eq. (8.7.12) into

$$\boxed{\frac{\partial v}{\partial x} + \frac{1}{a_0}\left(1 - \frac{\gamma + 1}{2a_0}v\right)\frac{\partial v}{\partial t} = 0.} \tag{8.8.1}$$

To solve this equation we use a method slightly different from that used in the previous section. The characteristic equation for Eq. (8.8.1) is

$$\frac{dt}{dx} = \frac{1}{a_0} - \frac{\gamma + 1}{2a_0^2}v(t, x). \tag{8.8.2}$$

We assume that we have the boundary condition $v = \Phi(t)$ at $x = 0$. Since $v = \text{const}$ on a characteristic, it follows from Eq. (8.8.2) that

$$t = \Phi^{-1}(v) + \frac{x}{a_0}\left(1 - \frac{\gamma + 1}{2a_0}v\right), \tag{8.8.3}$$

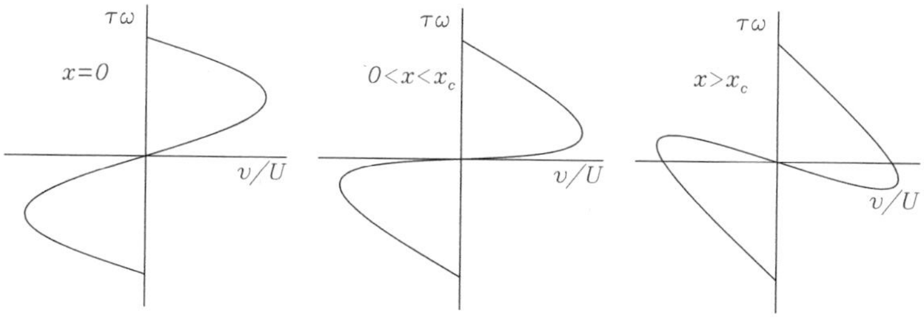

Fig. 8.11 Evolution of the dependence of $\omega\tau$ on v determined by Eq. (8.8.5) with an increase in x

where $\Phi^{-1}(v)$ is the function inverse to $\Phi(t)$.

Example A harmonic sound wave is driven at $x = 0$, where the velocity is defined by the boundary condition

$$v = U\sin(\omega t) \quad \text{at} \quad x = 0. \tag{8.8.4}$$

Find the solution to Eq. (8.8.1) valid for $x > 0$ and thus describe the spatial evolution of the wave.

Solution: Using the boundary condition (8.8.4) we transform Eq. (8.8.3) into

$$\omega\tau = -\frac{\gamma+1}{2a_0^2}\omega x v + \arcsin\frac{v}{U}, \quad \tau = t - \frac{x}{a_0}. \tag{8.8.5}$$

The dependence of $\omega\tau$ on v determined by Eq. (8.8.5) for various values of x is shown in Fig. 8.11. We see that the inverse function determines a single-valued smooth function $v(\omega\tau)$ when $x < x_c$, where x_c is given by

$$x_c = \frac{2a_0}{(\gamma+1)\omega M}, \quad M = \frac{U}{a_0}. \tag{8.8.6}$$

The assumption that the wave amplitude is small implies that $M \ll 1$. When $x = x_c$ we obtain $\partial\tau/\partial v = 0$ at $v = 0$, which implies that $\partial v/\partial\tau = \infty$ at $\tau = 0$. When $x > x_c$ the dependence of v on $\omega\tau$ becomes multivalued, which is unphysical. Hence, we can only obtain a smooth solution to Eq. (8.8.1) satisfying the boundary condition (8.8.4) when $x < x_c$. As we have already explained in the previous section, a shock appears in the wave profile at $x = x_c$. This shock then exists for $x > x_c$.

We now determine the position and intensity of the shock. We use the mass conservation Eq. (8.6.1) at the shock. It is written in the reference frame where the shock is at rest. To obtain the mass conservation in the reference frame where the shock propagates we need to substitute $v_1 - v_{\rm sh}$ and $v_2 - v_{\rm sh}$ for v_1 and v_2, respectively, in Eq. (8.6.1), where $v_{\rm sh}$ is the shock velocity. Then we obtain from Eq. (8.6.1)

$$v_{\rm sh} = \frac{\rho_2 v_2 - \rho_1 v_1}{\rho_2 - \rho_1}. \tag{8.8.7}$$

The motion before and after the shock is described by the simple wave solution implying, in particular, that the sound speed and velocity are related by Eq. (8.7.11), and the density is expressed in terms of the sound speed by Eq. (8.7.3). Using these two equations we obtain

$$\frac{\rho_{1,2}}{\rho_0} = \left(1 + \frac{\gamma-1}{2}\frac{v_{1,2}}{a_0}\right)^{\frac{2}{\gamma-1}} = 1 + \frac{v_{1,2}}{a_0} + \frac{3-\gamma}{4}\frac{v_{1,2}^2}{a_0^2} + \mathcal{O}(M^3). \tag{8.8.8}$$

Substituting this result into Eq. (8.8.7) yields

$$v_{\rm sh} = a_0 + \frac{\gamma+1}{4}(v_1 + v_2) + \mathcal{O}(M^2). \tag{8.8.9}$$

Let the shock arrive at spatial position x at time $t_{\rm sh}$. We introduce $\tau_{\rm sh} = t_{\rm sh} - x/a_0$. Since $dt_{\rm sh}/dx = 1/v_{\rm sh}$, we obtain with the aid of Eq. (8.8.9)

$$\frac{d\tau_{\rm sh}}{dx} = \frac{1}{v_{\rm sh}} - \frac{1}{a_0} = \frac{a_0 - v_{\rm sh}}{a_0^2} + \mathcal{O}(M^2) = -\frac{\gamma+1}{4a_0^2}(v_1 + v_2) + \mathcal{O}(M^2). \tag{8.8.10}$$

Consider the shaded area in Fig. 8.12. To calculate the size of this area (taking into account the sign) we introduce a parametrisation of the curve $\tau(v)$. We choose its length s as a parameter, so that $v = v(s)$ and $\tau = \tau(s)$. This curve intersects the line $\tau = \tau_{\rm sh}$ at $s = s_1$ and $s = s_2$, so $v_1 = v(s_1)$, $v_2 = v(s_2)$, and $\tau(s_1) = \tau(s_2) = \tau_{\rm sh}$. Then we obtain $S(x) = \int_{s_1}^{s_2} [\tau(s) - \tau_{\rm sh}](\partial v/\partial s)\, ds$ (taking into account the sign). We note that $v(s)$ and $\tau(s)$ also depend on x as well as s_1 and s_2. The function $v(s)$ and $\tau(s)$ are related by Eq. (8.8.5). Then, using Eq. (8.8.10) and taking into account that $\tau(s_1) = \tau(s_2) = \tau_{\rm sh}$ we obtain up to terms of order M

$$\frac{dS}{dx} = \int_{s_1}^{s_2} \left(\frac{\partial \tau}{\partial x} - \frac{d\tau_{\rm sh}}{dx}\right)\frac{\partial v}{\partial s} ds + \int_{s_1}^{s_2} [\tau(s) - \tau_{\rm sh}]\frac{\partial^2 v}{\partial s \partial x} ds. \tag{8.8.11}$$

Differentiating Eq. (8.8.5) with respect to s and x yields

$$\frac{\partial \tau}{\partial s} = \left(-\frac{\gamma+1}{2a_0^2}x + \frac{1}{\omega\sqrt{U^2 - v^2}}\right)\frac{\partial v}{\partial s}, \tag{8.8.12}$$

$$\frac{\partial \tau}{\partial x} = -\frac{\gamma+1}{2a_0^2}\left(v + x\frac{\partial v}{\partial x}\right) + \frac{1}{\omega\sqrt{U^2 - v^2}}\frac{\partial v}{\partial x}. \tag{8.8.13}$$

Using integration by parts and Eq. (8.8.12) we obtain

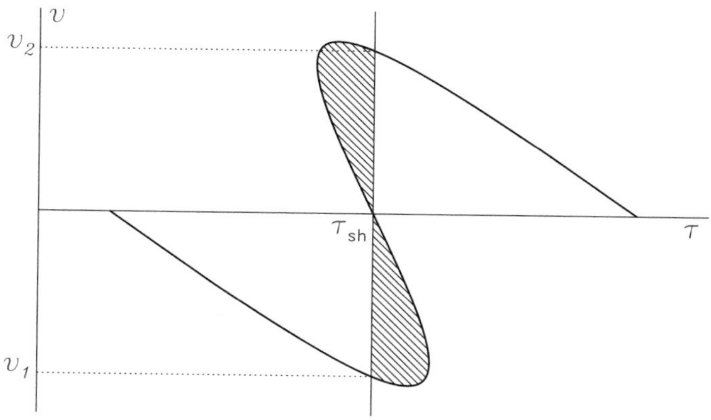

Fig. 8.12 Illustration of the determination of the spatial shock position. The two shaded areas have the same sizes when $\tau_{sh} = 0$

$$\int_{s_1}^{s_2} [\tau(s) - \tau_{sh}] \frac{\partial^2 v}{\partial s \partial x} ds = \int_{s_1}^{s_2} \left(\frac{\gamma+1}{2a_0^2} x - \frac{1}{\omega\sqrt{U^2 - v^2}} \right) \frac{\partial v}{\partial s} \frac{\partial v}{\partial x} ds. \quad (8.8.14)$$

Substituting Eqs. (8.8.10), (8.8.12), and (8.8.14) into Eq. (8.8.11) yields

$$\frac{dS}{dx} = -\int_{s_1}^{s_2} \left(\frac{\gamma+1}{2a_0^2} v + \frac{d\tau_{sh}}{dx} \right) \frac{\partial v}{\partial s} ds = \frac{\gamma+1}{4a_0^2} \left[(v_1 + v_2)v - v^2 \right] \Big|_{s_1}^{s_2} = 0. \quad (8.8.15)$$

Since $S(x_c) = 0$, it follows from this result that $S(x) \equiv 0$. It is straightforward to see that the derivation can be repeated for any form of the second term on the right-hand side of Eq. (8.8.5) that behaves similarly to $\arcsin(v/U)$. Hence, this result holds for a wide range of boundary conditions at $x = 0$.

Using Eq. (8.8.5) we obtain

$$S(x) = -\int_{s_1}^{s_2} \left(\frac{\gamma+1}{2a_0^2} xv - \frac{1}{\omega} \arcsin \frac{v}{U} \right) \frac{\partial v}{\partial s} ds - \tau_{sh}(v_2 - v_1). \quad (8.8.16)$$

Calculating the integral in this equation yields

$$\int_{s_1}^{s_2} \left(\frac{\gamma+1}{2a_0^2} xv - \frac{1}{\omega} \arcsin \frac{v}{U} \right) \frac{\partial v}{\partial s} ds = \frac{\gamma+1}{4a_0^2} x(v_2^2 - v_1^2)$$
$$+ \frac{1}{\omega} \left(\sqrt{U^2 - v_1^2} - \sqrt{U^2 - v_2^2} - v_2 \arcsin \frac{v_2}{U} + v_1 \arcsin \frac{v_1}{U} \right). \quad (8.8.17)$$

We now consider S as a function of two variables, x and τ_{sh}. It is obvious from Fig. 8.12 that, when τ_{sh} increases, the absolute value of the area to the left of the vertical line $\tau = \tau_{sh}$ increases, while the area to the right of the vertical line $\tau = \tau_{sh}$ decreases. Since the area to the left of the vertical line $\tau = \tau_{sh}$ is negative, it follows

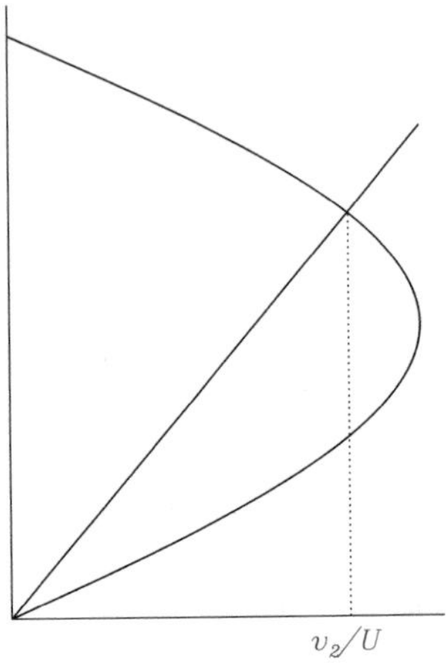

Fig. 8.13 Graphical investigation of Eq. (8.8.18). The straight line is the graph of its left-hand side, and the curve is the graph of its right-hand side

that S is a monotonically decreasing function of $\tau = \tau_{sh}$. Hence, there is no more than one value of τ_{sh} such that $S = 0$.

Since the function on the right-hand side of Eq. (8.8.5) is odd, we conclude that $v_1 = -v_2$ when $\tau_{sh} = 0$. Then it follows from Eqs. (8.8.16) and (8.8.17) that $S = 0$. Hence, finally, we obtain that $\tau_{sh} = 0$ and $v_1 = -v_2$.

We see that the velocity jump at the shock is equal to $2v_2$. It follows from Eq. (8.8.5) that the dependence of v_2 on x is determined by the equation

$$\frac{xv_2}{x_c U} = \arcsin\frac{v_2}{U}. \tag{8.8.18}$$

A graphical investigation of this equation is shown in Fig. 8.13. As x increases from x_c to infinity the angle between the straight line and the horizontal axis increases from 45°, when this line is tangent to the graph of the right-hand side of Eq. (8.8.18), to 90°. The velocity v_2 increases from zero to U at $x = \pi x_c/2$ and then decreases to zero as x increases further. The dependence of v on x for a few values of t is shown in Fig. 8.14. The dependence of v on x for other values of t are similar. Note the tooth-like shape of the graph $v(x)$ for large values of x.

Problems

8.1: Derive the equation describing the evolution of entropy.

8.2: Suppose the half-space $x < 0$ in Cartesian coordinates $x,\ y,\ z$ is filled with an ideal gas. It is bounded by a rigid wall. In the unperturbed state the gas is at rest, and its density and pressure are equal to ρ_0 and p_0, respectively. There is a harmonic sound wave of cyclic frequency ω launched at $x \to -\infty$ and propagating in the positive x-direction. The amplitude of the density perturbation in this wave is $\epsilon\rho_0$, where $\epsilon \ll 1$. Calculate the pressure imposed by the sound wave on the rigid wall.

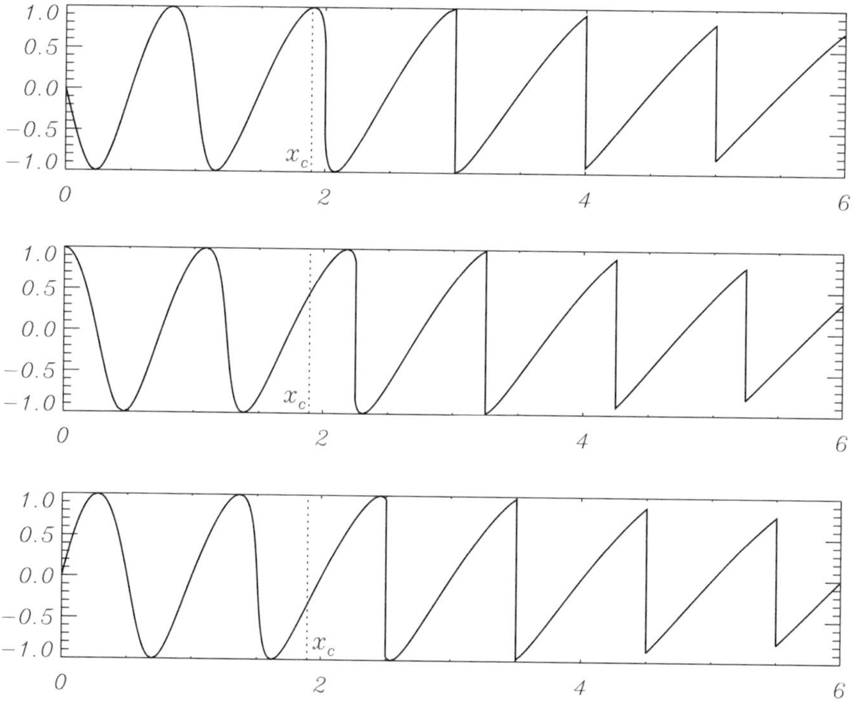

Fig. 8.14 Dependence of v on x in a sound wave driven at $x = 0$ and propagating to the region $x > 0$; v and x are given in units of U and the wavelength $\lambda = 2\pi a_0/\omega$, respectively. The top, middle, and bottom panels correspond to $\omega t = 2\pi n$, $\omega t = \pi/2 + 2\pi n$, and $\omega t = \pi + 2\pi n$, where n is any integer

8.3: Suppose the two half-spaces $x < 0$ and $x > 0$ in Cartesian coordinates $x,\ y,\ z$ are filled with ideal gases with the same pressure p_0, and densities ρ_1 and ρ_2, respectively. There is a harmonic sound wave of cyclic frequency ω launched at $x \to -\infty$ and propagating in the positive x-direction. The amplitude of the density perturbation in this wave is $\epsilon\rho_1$, where $\epsilon \ll 1$. The wave is partially reflected from the surface $x = 0$ separating the two gases and partially transmitted through this surface. Calculate the amplitudes of the density perturbation in the reflected and transmitted waves.

8.4: Suppose the area of the cross-section of a de Laval nozzle at its entrance is A_0. There is a supersonic gas flow in the nozzle. At the nozzle entrance the Mach number is 2, and the gas density and pressure are ρ_0 and p_0, respectively. The gas pressure at the exit is $p_1 = \frac{1}{2}p_0$. The gas adiabatic index is $\gamma = 1.4$. Determine the nozzle cross-section, Mach number, and density at the nozzle exit.

8.5: The shock intensity is determined by the parameter $M_1 - 1$, where M_1 is the Mach number of the flow before the shock in the reference frame where the shock is immovable. A shock is considered as weak if $M_1 - 1 \ll 1$. Show that in a weak shock the entropy jump across the shock is proportional to $(M_1 - 1)^3$.

8.6: Suppose the half-space $x < 0$ in Cartesian coordinates $x,\ y,\ z$ is filled with an ideal gas. It is bounded by a rigid wall. There is a shock propagating in the positive x-direction. The shock Mach number is $M_1 = 3$. The gas density and pressure before the shock are ρ_1 and p_1, respectively. The shock hits the wall and is reflected from it. The wall will be destroyed if the pressure imposed on the wall after the shock reflection exceeds $40p_1$. Will the wall be destroyed? You can take $\gamma = 1.4$.

8.7: In a straight channel a piston starts to move with speed $U(t)$, where $U(0) = 0$, and $U(t) = U_0 \tanh(t/T)$ with constant U_0 and T. Initially the gas in the channel is at rest. You are also given that $U_0 < a_0$, where a_0 is the speed of sound far from the piston. Determine the moment of time when the gradient catastrophe occurs.

8.8: Consider a simple wave connecting two regions with constant flow parameters. This simple wave is called a *self-similar rarefaction* wave if at the initial time all characteristics corresponding to the wave intersect at one point. Hence, in the tx-plane the characteristics consist of three sets. One set is a collection of straight lines intersecting at one point on the x-axis. It is sandwiched by two sets of parallel straight lines. The flow density, pressure and velocity before the wave are ρ_0, p_0, and $v_0 = 0$, respectively, while the velocity after the wave is $v_1 < 0$. Determine the temporal and spatial dependence of the flow parameters in the wave, and the constant density and pressure of the flow after the wave.

8.9: Suppose there is a membrane separating two half-spaces filled with gases. Both gases are at rest. The density and pressure of the left gas are ρ_1 and p_1, while the density and pressure of the right gas are ρ_2 and p_2, where $p_2/p_1 = 1 + \epsilon$ and $0 < \epsilon \ll 1$. The adiabatic exponent is the same for both gases and equal to γ. Describe the gas motion. Give your answers in the leading order approximation with respect to ϵ.

8.10: (Rudenko and Soluyan, see Further Reading). A nonlinear sound wave is driven at $x = 0$ by a periodic driver defined by $v = U \sin(\omega t)$. It propagates in the region $x > 0$. Since v is a periodic function of t, $v(t)$ can be expanded in a Fourier series with respect to t. The set of squares of Fourier coefficients is called the *spectrum* of the sound wave. Investigate the dependence of the spectrum on x for $x \le x_c$, where x_c is the distance where the shock starts to appear.

Chapter 9
Non-ideal Compressible Fluids

9.1 Governing Equations

In the previous chapter we studied the motion of an ideal compressible fluid. We neglected viscosity in the momentum equation and used the entropy conservation instead of the full energy equation. In this chapter we study how the account of viscosity and entropy variation affects fluid motion. We again use the Eulerian description of fluid motion. We completely describe the motion of a fluid if we determine the temporal and spatial dependence of the density, pressure, and velocity. For this we need to solve the system consisting of the mass conservation, momentum, and energy equation. The mass conservation Eq. (3.8.2) is quite universal and is not affected by viscosity or any processes changing the entropy. The momentum equation for viscous compressible fluids was derived in Chap. 7. It is Eq. (7.1.4). The energy equation is Eq. (8.3.5) with the energy density given by Eq. (8.3.16). In what follows we assume that there is no external force, $\boldsymbol{b} = 0$. Usually the heat flux is antiparallel to the temperature gradient and proportional to this gradient. Hence,

$$\boxed{\boldsymbol{q} = -\kappa \nabla T,} \qquad (9.1.1)$$

where κ is the coefficient of heat conduction. This coefficient depends on T, however this dependence is fairly weak. Below we assume that $\kappa = \text{const}$. In this chapter we only use Cartesian coordinates. Then, using Eqs. (3.5.6) and (7.1.3) we obtain

$$\mathsf{D} : \mathsf{T}' = \eta \left(\frac{\partial v_i}{\partial x_j} \frac{\partial v_j}{\partial x_i} + \frac{\partial v_i}{\partial x_j} \frac{\partial v_i}{\partial x_j} \right) + \left(\zeta - \frac{2\eta}{3} \right) (\nabla \cdot \boldsymbol{v})^2. \qquad (9.1.2)$$

Using Eqs. (3.8.3), (8.3.10), (8.3.14), and (8.3.16) yields

$$U = C_v T, \quad \frac{p}{\rho^2} \frac{D\rho}{Dt} = -C_v(\gamma - 1) T \, \nabla \cdot \boldsymbol{v}. \qquad (9.1.3)$$

© Springer Nature Switzerland AG 2019

M. S. Ruderman, *Fluid Dynamics and Linear Elasticity*, Springer Undergraduate Mathematics Series, https://doi.org/10.1007/978-3-030-19297-6_9

With the aid of Eqs. (9.1.1)–(9.1.3) we transform Eq. (8.3.5) into

$$\boxed{\begin{aligned}\rho C_v\left(\frac{DT}{Dt}+(\gamma-1)T\,\nabla\cdot\boldsymbol{v}\right)&=\eta\left(\frac{\partial v_i}{\partial x_j}\frac{\partial v_j}{\partial x_i}+\frac{\partial v_i}{\partial x_j}\frac{\partial v_i}{\partial x_j}\right)\\&\quad+\left(\zeta-\frac{2\eta}{3}\right)(\nabla\cdot\boldsymbol{v})^2+\kappa\nabla^2T.\end{aligned}} \tag{9.1.4}$$

The constitutive equations for the model of a compressible viscous fluid with thermal conduction are given by Eqs. (7.1.1), (7.1.2), (8.3.10), and (9.1.1).

Below we assume that η and ζ are of the same order. Then we obtain that the first and second terms on the right-hand side of Eq. (9.1.4) are of the same order, and their ratio to its left-hand side is of the order of

$$\frac{\eta\nabla\cdot\boldsymbol{v}}{\rho C_v T}\sim\frac{\eta V}{\rho C_v T L},$$

where L and V are the characteristic spatial scale and velocity of the problem. On the other hand, in accordance with Eqs. (8.3.16) and (8.4.4) $C_v T\sim p/\rho\sim a_s^2$, where a_s is the speed of sound. Assuming that $V\sim a_s$ we eventually obtain

$$\frac{\eta\nabla\cdot\boldsymbol{v}}{\rho C_v T}\sim \mathrm{Re}^{-1}.$$

Hence, when Re $\ll 1$ the terms related to viscosity can be neglected not only in the momentum Eq. (7.1.4), but also in the energy Eq. (9.1.4).

Let us now estimate the ratio of the last term on the right-hand side of Eq. (9.1.4) to its left-hand side. We obtain $\kappa\nabla^2T\sim\kappa T/L^2$ and $\rho C_v T\,\nabla\cdot\boldsymbol{v}\sim\rho C_p T V/L$, where we used the fact that $C_v\sim C_p$. Hence, the ratio of the last term on the right-hand side of Eq. (9.1.4) to its left-hand side is of the order of

$$\boxed{\frac{\kappa}{\rho C_p L V}=\mathrm{Pe}^{-1},} \tag{9.1.5}$$

where Pe is the *Peclet number*, named after the French physicist Jean Claude Eugène Péclet. When Pe $\gg 1$ we can neglect the last term describing the heat conduction in the energy Eq. (9.1.4).

One more dimensionless number widely used in hydrodynamics is the *Prandtl number* Pr, named after Ludwig Prandtl. It is defined as

$$\boxed{\mathrm{Pr}=\frac{\mathrm{Pe}}{\mathrm{Re}}=\frac{C_p\eta}{\kappa}.} \tag{9.1.6}$$

This number determines the relative importance of two dissipative processes: viscosity and thermal conduction. When $\mathrm{Pr} \ll 1$ thermal conduction dominates viscosity, while viscosity dominates thermal conduction when $\mathrm{Pr} \gg 1$. Pr is close to unity for air and many other gases.

To obtain a closed system of equations describing the motion of a viscous thermal conducting compressible fluid we must supplement the mass conservation Eq. (3.8.2), the momentum Eq. (7.1.4), and the energy Eq. (9.1.4) with the Clapeyron law given by Eq. (8.3.10).

9.2 Damping of Sound Waves

In Sect. 8.4 we studied the propagation of sound waves. Since in that section we neglected all dissipative processes, we found that sound waves propagate without changing their shape and amplitude. In this section we study the effect of viscosity and thermal conduction on sound waves. We consider the same unperturbed state as in Sect. 8.4, namely we assume that the fluid is at rest ($\boldsymbol{v} = 0$) and its density, pressure, and temperature are constant and equal to ρ_0, p_0, and T_0, respectively. We assume that there is no external force ($\boldsymbol{b} = 0$). We also denote the unperturbed speed of sound by a_s. Now, similar to Sect. 8.4, we linearise the governing equations. We denote the perturbations of the density, pressure, and temperature by $\delta\rho$, δp, and δT. The linearised mass conservation equation is the same as in Sect. 8.4, but, for convenience, we give it here,

$$\frac{\partial(\delta\rho)}{\partial t} + \rho_0 \nabla \cdot \boldsymbol{v} = 0. \tag{9.2.1}$$

Then we linearise the momentum Eq. (7.1.4). As a result we obtain

$$\rho_0 \frac{\partial \boldsymbol{v}}{\partial t} = -\nabla \delta p + \eta \nabla^2 \boldsymbol{v} + \left(\zeta + \tfrac{1}{3}\eta\right) \nabla(\nabla \cdot \boldsymbol{v}). \tag{9.2.2}$$

Linearising the energy Eq. (9.1.4) yields

$$\rho_0 C_v \left(\frac{\partial(\delta T)}{\partial t} + (\gamma - 1) T_0 \nabla \cdot \boldsymbol{v} \right) = \kappa \nabla^2 \delta T. \tag{9.2.3}$$

We note that in this equation there is no term related to the energy dissipation caused by viscosity. Finally, we linearise the Clapeyron Eq. (8.3.10) to obtain

$$\delta p = \frac{\widetilde{R}}{\tilde{\mu}} (\rho_0 \, \delta T + T_0 \, \delta\rho). \tag{9.2.4}$$

Below we only consider one-dimensional solutions to the system of Eqs. (9.2.1)–(9.2.4), and assume that all variables only depend on $x = x_1$ in Cartesian coordinates $x_1,\ x_2,\ x_3$ and $\boldsymbol{v} = (v, 0, 0)$. Then Eqs. (9.2.1)–(9.2.3) reduce to

$$\frac{\partial(\delta\rho)}{\partial t} + \rho_0 \frac{\partial v}{\partial x} = 0, \tag{9.2.5}$$

$$\rho_0 \frac{\partial v}{\partial t} = -\frac{\partial(\delta p)}{\partial x} + \left(\zeta + \tfrac{4}{3}\eta\right)\frac{\partial^2 v}{\partial x^2}, \tag{9.2.6}$$

$$\rho_0 C_v \left(\frac{\partial(\delta T)}{\partial t} + (\gamma - 1)T_0 \frac{\partial v}{\partial x}\right) = \kappa \frac{\partial^2 \delta T}{\partial x^2}. \tag{9.2.7}$$

Now we look for solutions to the system of Eqs. (9.2.4)–(9.2.7) in the form

$$\begin{aligned} v = \Re\{\hat{v}\exp(ikx - i\omega t)\},\ \delta\rho = \Re\{\hat{\rho}\exp(ikx - i\omega t)\}, \\ \delta p = \Re\{\hat{p}\exp(ikx - i\omega t)\},\ \delta T = \Re\{\hat{T}\exp(ikx - i\omega t)\}, \end{aligned} \tag{9.2.8}$$

where $\Re$ indicates the real part of a complex quantity, and $\hat{v},\ \hat{\rho},\ \hat{p},\ \hat{T},\ k$, and ω are (in general complex) constants. Substituting these expressions into the set of Eqs. (9.2.4)–(9.2.7) we obtain

$$\hat{p} = \frac{\widetilde{R}}{\tilde{\mu}}(\rho_0\hat{T} + T_0\hat{\rho}), \tag{9.2.9}$$

$$\omega\hat{\rho} - \rho_0 k\hat{v} = 0, \tag{9.2.10}$$

$$\rho_0\omega\hat{v} = k\hat{p} - i\left(\zeta + \tfrac{4}{3}\eta\right)k^2\hat{v}, \tag{9.2.11}$$

$$\rho_0 C_v\left[\omega\hat{T} - (\gamma - 1)T_0 k\hat{v}\right] = -i\kappa k^2\hat{T}. \tag{9.2.12}$$

Equations (9.2.9)–(9.2.12) constitute a system of linear homogeneous algebraic equations for the variables $\hat{v},\ \hat{\rho},\ \hat{p}$, and $\hat{T}$. This system has non-trivial solutions only if its determinant is equal to zero. This condition is written as

$$\omega^3 - a_s^2\omega k^2 = \frac{i\kappa a_s^2 k^4}{\gamma\rho_0 C_v} - \frac{i\omega^2 k^2}{\rho_0}\left(\zeta + \tfrac{4}{3}\eta + \frac{\kappa}{C_v}\right) + \left(\zeta + \tfrac{4}{3}\eta\right)\frac{\kappa\omega k^4}{\rho_0^2 C_v}. \tag{9.2.13}$$

When deriving this equation we used the relations

$$\frac{\widetilde{R}}{\tilde{\mu}} = C_v(\gamma - 1), \quad a_s^2 = \frac{\widetilde{R}}{\tilde{\mu}}\gamma T_0, \tag{9.2.14}$$

which follow from Eqs. (8.3.10), (8.3.14), (8.4.4), and the relation $\gamma = C_p/C_v$. Let us estimate the terms in Eqs. (9.2.13). We can consider k^{-1} as the characteristic spatial scale of the problem and a_s as the characteristic speed. Since ω^{-1} is the characteristic time of the problem, ω/k is of the order of a_s. Then, in this particular case, the Reynolds and Peclet numbers can be defined as $\mathrm{Re} = \rho_0 a_s/k\eta$ and $\mathrm{Pe} = \gamma\rho_0 a_s C_v/k\kappa$, where we used the relation $C_p = \gamma C_v$. Using these definitions we easily obtain that the ratio of the first, second, and third terms of the right-hand side of Eq. (9.2.13) to the second term of the left-hand side of this equation are of the order of Pe^{-1}, $\mathrm{Re}^{-1} + \mathrm{Pe}^{-1}$, and $\mathrm{Re}^{-1}\mathrm{Pe}^{-1}$, respectively, where we used the assumption that ζ and η are of the same order. Below we assume that $\mathrm{Re} \gg 1$ and $\mathrm{Pe} \gg 1$, which implies that the right-hand side of Eq. (9.2.13) is much smaller than the left-hand side. In addition, this implies that the third term on the right-hand side is much smaller than the first and second ones, and thus can be neglected. The fact that the right-hand side of Eq. (9.2.13) is much smaller than the left-hand side enables us to use the regular perturbation method to solve Eq. (9.2.13).

First we assume that a harmonic perturbation with wavenumber k is launched at the initial moment of time in the whole space. This implies that k is a given real quantity. In the first-order approximation we neglect the right-hand side and take the solution corresponding to the sound wave propagating in the positive x-direction, which is $\omega = a_s k$. In the second-order approximation we look for the solution in the form $\omega = a_s k + \delta\omega$, where $|\delta\omega| \ll a_s k$. To calculate $\delta\omega$ we substitute $\omega = a_s k$ into the right-hand side of Eq. (9.2.13). Then we obtain

$$\delta\omega = -\frac{ik^2}{2\rho_0}\left(\zeta + \tfrac{4}{3}\eta + \frac{\kappa(\gamma - 1)}{\gamma C_v}\right) \equiv -i\chi_t k^2. \tag{9.2.15}$$

Since the solution to the set of Eqs. (9.2.9)–(9.2.12) is defined up to multiplication by an arbitrary complex constant, we can always choose one of the four quantities, $\hat{v}$, $\hat{\rho}$, $\hat{p}$, and $\hat{T}$, to be real. We choose $\hat{\rho}$ to be real. Then, using Eq. (9.2.15) we obtain

$$\delta\rho = \hat{\rho}\, e^{-\chi_t k^2 t}\cos[k(x - a_s t)]. \tag{9.2.16}$$

We do not give the expressions for v, δp, and δT. They can be easily found using Eqs. (9.2.8)–(9.2.11).

We see that viscosity and thermal conduction cause the damping of sound waves. The damping time t_{damp} is defined as the time it takes for the wave amplitude to decrease by a factor of e. Hence,

$$t_{\mathrm{damp}} = \frac{1}{\chi_t k^2} \sim \frac{\mathrm{Re}}{a_s k} \sim \frac{\mathrm{Pe}}{a_s k}, \tag{9.2.17}$$

where we used the fact that, for the majority of gases, Re and Pe are of the same order. We note that the damping time is proportional to the wavelength, $2\pi/k$, squared. Hence, the smaller the wavelength is, the faster the wave damps.

Now we consider another problem. We assume that the sound wave is launched at $x = 0$ and propagates in the half-space $x > 0$. In that case ω is a given real quantity and we need to solve Eq. (9.2.13) to determine k. Again we use the regular perturbation method and obtain in the first-order approximation $k = \omega/a_s$. Then we look for the solution to Eq. (9.2.13) in the form $k = \omega/a_s + \delta k$, where $|\delta k| \ll \omega/a_s$. To calculate δk we substitute $k = \omega/a_s$ into the right-hand side of Eq. (9.2.13). Then we obtain

$$\delta k = \frac{i\omega^2}{2\rho_0 a_s^3}\left(\zeta + \tfrac{4}{3}\eta + \frac{\kappa(\gamma - 1)}{\gamma C_v}\right) \equiv i\chi_s\omega^2. \tag{9.2.18}$$

Again taking $\hat{\rho}$ to be real yields

$$\delta\rho = \hat{\rho}\, e^{-\chi_s\omega^2 x}\cos[\omega(x/a_s - t)]. \tag{9.2.19}$$

The damping length is

$$x_{\text{damp}} = \frac{1}{\chi_s\omega^2} \sim \frac{a_s\,\text{Re}}{\omega} \sim \frac{a_s\,\text{Pe}}{\omega}. \tag{9.2.20}$$

We see that the damping length is proportional to the wave period, $2\pi/\omega$, squared. Hence, the smaller the wave period is, the faster the wave damps.

9.3 The Structure of Shocks

In Sect. 8.6 we studied shocks. There we considered shocks as surfaces of discontinuity. However, in reality, they are narrow transitional layers with a thickness of the order of a free path of molecules or atoms constituting a gas. If we consider the shock thickness as a characteristic spatial scale, then we obtain that the Reynolds and Peclet numbers are of the order of unity. Hence, viscosity and thermal conduction are important inside a shock. The aim of this section is to describe the shock structure. We use a reference frame where the shock is at rest and the velocity is orthogonal to the shock surface at both sides. Then the shock structure is described by a stationary one-dimensional solution to the set of Eqs. (3.8.2), (7.1.4), (8.3.10), and (9.1.4) with $\boldsymbol{v} = (v, 0, 0)$. In such a solution all variables only depend on $x = x_1$. Hence, for such solutions Eqs. (3.8.2), (7.1.4), and (9.1.4) reduce to

$$\frac{d(\rho v)}{dx} = 0, \tag{9.3.1}$$

$$\rho v\frac{dv}{dx} = -\frac{dp}{dx} + \mu\frac{d^2 v}{dx^2}, \tag{9.3.2}$$

$$\rho C_v \left(v \frac{dT}{dx} + (\gamma - 1) T \frac{dv}{dx} \right) = \kappa \frac{d^2 T}{dx^2} + \mu \left(\frac{dv}{dx} \right)^2, \tag{9.3.3}$$

where $\mu = \zeta + \frac{4}{3}\eta$. We choose the direction of the x-axis opposite to the direction of shock propagation. Then the region before the shock is $x < 0$, and the region after the shock is $x > 0$, and the flow velocity in the reference frame moving together with the shock is everywhere positive. The flow parameters before the shock are v_1, ρ_1, p_1, and T_1. They can be chosen arbitrarily with the only restriction that they satisfy the Clapeyron equation and $M_1 = v_1/a_{s1} > 1$. The flow parameters after the shock, v_2, ρ_2, p_2, and T_2, are determined either by the Rankine–Hugoniot relations given by Eqs. (8.6.1)–(8.6.4), or by Eqs. (8.6.6)–(8.6.8). The solution must satisfy the boundary conditions that v, ρ, p, and T are equal to their values before and after the shock far from the shock. We can formally impose the condition that they tend to the corresponding quantities before the shock as $x \to -\infty$, and after the shock as $x \to \infty$.

Integrating Eqs. (9.3.1) and (9.3.2), and using the boundary conditions at $x \to -\infty$ we obtain

$$\rho v = \rho_1 v_1, \tag{9.3.4}$$

$$p = p_1 - \rho_1 v_1 (v - v_1) + \mu \frac{dv}{dx}, \tag{9.3.5}$$

where we used the fact that the derivatives of all quantities tend to zero as $x \to -\infty$. Using Eqs. (8.3.10) and (9.3.5) we reduce Eq. (9.3.3) to

$$C_v \rho_1 v_1 \frac{dT}{dx} + [p_1 - \rho_1 v_1 (v - v_1)] \frac{dv}{dx} = \kappa \frac{d^2 T}{dx^2}. \tag{9.3.6}$$

Integrating this equation and using Eqs. (8.3.10) and (9.2.14), and the boundary conditions at $x \to -\infty$, yields

$$\kappa \frac{dT}{dx} = C_v \rho_1 v_1 (T - \gamma T_1) + p_1 v - \frac{1}{2} \rho_1 v_1 (v - v_1)^2. \tag{9.3.7}$$

Using Eqs. (8.3.10) and (9.3.4) to eliminate p from Eq. (9.3.5) we obtain

$$\mu \frac{dv}{dx} = \rho_1 v_1 \left(C_v (\gamma - 1) \frac{T}{v} + v - v_1 \right) - p_1. \tag{9.3.8}$$

Equations (9.3.7) and (9.3.8) constitute an autonomous system of two first-order ordinary differential equations. It is straightforward to verify that $T = T_1$ and $v = v_1$, and $T = T_2$ and $v = v_2$ are critical points of this system. Hence, the shock structure is described by the separatrix connecting these two critical points.

Lemma 9.1 *The first critical point is an unstable node, and the second critical point is a saddle.*

Proof Calculating the Jacobian matrices at the two critical points we obtain

$$J_1 = \begin{pmatrix} \dfrac{\rho_1 v_1 C_v}{\kappa} & \dfrac{\rho_1 v_1^2}{\gamma\kappa M_1^2} \\ \dfrac{\rho_1 C_v(\gamma-1)}{\mu} & \dfrac{\rho_1 v_1}{\mu}\left(1 - \dfrac{1}{\gamma M_1^2}\right) \end{pmatrix}, \tag{9.3.9}$$

$$J_2 = \begin{pmatrix} \dfrac{\rho_2 v_2 C_v}{\kappa} & \dfrac{\rho_2 v_2^2}{\gamma\kappa M_2^2} \\ \dfrac{\rho_2 C_v(\gamma-1)}{\mu} & \dfrac{\rho_2 v_2}{\mu}\left(1 - \dfrac{1}{\gamma M_2^2}\right) \end{pmatrix}, \tag{9.3.10}$$

where we used Eqs. (8.6.1), (8.6.3), and (8.6.6). The characteristic equation of the matrix J_1 is

$$\lambda^2 - \rho_1 v_1 \lambda \left[\frac{C_v}{\kappa} + \frac{1}{\mu}\left(1 - \frac{1}{\gamma M_1^2}\right)\right] + \frac{\rho_1^2 v_1^2 C_v}{\kappa\mu}\left(1 - \frac{1}{M_1^2}\right) = 0. \tag{9.3.11}$$

The discriminant of this quadratic equation is

$$\rho_1^2 v_1^2 \left\{\left[\frac{C_v}{\kappa} - \frac{1}{\mu}\left(1 - \frac{1}{\gamma M_1^2}\right)\right]^2 + \frac{4C_v(\gamma-1)}{\gamma\kappa\mu M_1^2}\right\} > 0. \tag{9.3.12}$$

Since $M_1 > 1$ this result implies that Eq. (9.3.11) has two positive real roots meaning that the first critical point is an unstable node.

The characteristic equation of the matrix J_2 is

$$\lambda^2 - \rho_2 v_2 \lambda \left[\frac{C_v}{\kappa} + \frac{1}{\mu}\left(1 - \frac{1}{\gamma M_2^2}\right)\right] + \frac{\rho_2^2 v_2^2 C_v}{\kappa\mu}\left(1 - \frac{1}{M_2^2}\right) = 0. \tag{9.3.13}$$

Since $M_2 < 1$ it follows that Eq. (9.3.13) has two real roots with different signs meaning that the second critical point is a saddle. The lemma is proved. □

A typical phase portrait of the system of Eqs. (9.3.7) and (9.3.8) is shown in Fig. 9.1, and a typical dependence of v, ρ, p, and T in the shock structure is shown in Fig. 9.2. We see that, under the assumption that Pe $\sim$ Re holds for the majority of gases, a substantial variation of these quantities only occurs at a distance of a few $l_{\text{sh}} = \mu/(\rho_1 v_1)$. Hence, l_{sh} can be considered as the typical thickness of the shock structure.

We can obtain explicit expressions for v, ρ, p, and T in the case of weak shocks, that is, when $\epsilon = M_1^2 - 1 \ll 1$. In this case the deviation of the velocity and temperature from their values before the shock and after the shock are small and we

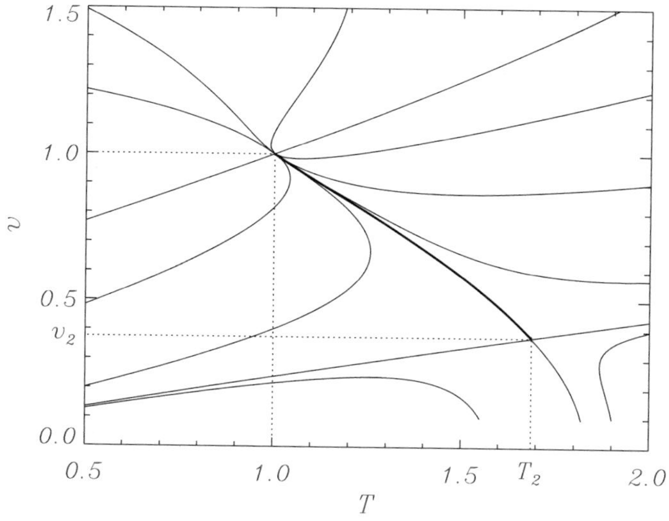

Fig. 9.1 Phase portrait of the system of Eqs. (9.3.7) and (9.3.8). The separatrix connecting the two critical points with coordinates (1, 1) and (T_2, v_2) is shown by the thick curve. T and v are given in units of T_1 and v_1, respectively

Fig. 9.2 Dependence of the velocity, density, pressure, and temperature on $X = \rho_1 v_1 x/\mu$ in the shock structure for $M_1 = 2$ and $\kappa = C_v \mu$. The solid, dashed, dotted, and dash-dotted curves correspond to v/v_1, ρ/ρ_1, p/p_1, and T/T_1, respectively

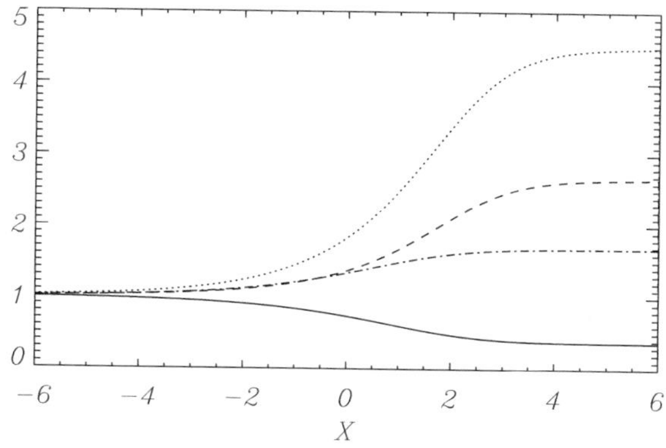

can look for an approximate solution describing the shock structure. We introduce $\delta v = v - v_2$ and $\delta T = T - T_2$. It follows from Eqs. (8.6.6) and (8.6.8) that

$$\frac{\delta v}{v_1} = \mathcal{O}(\epsilon), \quad \frac{\delta T}{T_1} = \mathcal{O}(\epsilon).$$

We obtain from Eqs. (8.6.6)–(8.6.8) and (8.6.9)

$$\frac{v_1}{v_2} = 1 + \frac{2\epsilon}{\gamma+1} - \frac{2(\gamma-1)}{(\gamma+1)^2}\epsilon^2 + \mathcal{O}(\epsilon^3),$$
$$\frac{\rho_1}{\rho_2} = 1 - \frac{2\epsilon(1-\epsilon)}{\gamma+1} + \mathcal{O}(\epsilon^3),$$
$$\frac{p_1}{p_2} = 1 - \frac{2\gamma\epsilon}{\gamma+1} + \frac{4\gamma^2\epsilon^2}{(\gamma+1)^2} + \mathcal{O}(\epsilon^3), \tag{9.3.14}$$
$$\frac{T_1}{T_2} = 1 - \frac{2(\gamma-1)\epsilon}{\gamma+1} + \frac{2(\gamma-1)(2\gamma-1)}{(\gamma+1)^2}\epsilon^2 + \mathcal{O}(\epsilon^3),$$
$$M_2^2 = 1 - \epsilon + \frac{2\gamma\epsilon^2}{\gamma+1} + \mathcal{O}(\epsilon^3).$$

Now we only keep the terms of the order of ϵ and ϵ^2 on the right-hand sides of Eqs. (9.3.7) and (9.3.8), while we drop the terms of higher orders. As a result, using Eqs. (8.6.1) and (8.3.14), we reduce Eqs. (9.3.7) and (9.3.8) to

$$\kappa\frac{d(\delta T)}{dx} = v_2 p_2\left(\frac{\delta T}{(\gamma-1)T_2} + \frac{\delta v}{v_2} - \frac{\gamma(\delta v)^2}{2v_2^2}\right), \tag{9.3.15}$$

$$\mu\frac{d(\delta v)}{dx} = p_2\left(\frac{\delta T}{T_2} + (\gamma - 1 - \gamma\epsilon)\frac{\delta v}{v_2} + \frac{(\delta v)^2}{v_2^2} - \frac{\delta v\,\delta T}{v_2 T_2}\right). \tag{9.3.16}$$

When calculating the right-hand side of Eqs. (9.3.15) and (9.3.16) we used the relation $M_2^2 \approx 1-\epsilon$. Dividing Eq. (9.3.15) by Eq. (9.3.16) yields

$$\frac{\kappa}{\mu}\frac{d(\delta T)}{d(\delta v)} = \frac{v_2\left(\dfrac{\delta T}{(\gamma-1)T_2} + \dfrac{\delta v}{v_2} - \dfrac{\gamma(\delta v)^2}{2v_2^2}\right)}{\dfrac{\delta T}{T_2} + (\gamma-1-\gamma\epsilon)\dfrac{\delta v}{v_2} + \dfrac{(\delta v)^2}{v_2^2} - \dfrac{\delta v\,\delta T}{v_2 T_2}}. \tag{9.3.17}$$

Now we look for the solution to this equation corresponding to the separatrix connecting the two critical points. The separatrix is an integral curve that starts at the first critical point, which is a node, and ends at the second critical point, which is a saddle. First we find the equation of the separatrix. The negative root to the characteristic Eq. (9.3.13) and the corresponding eigenvector of matrix J_2 are given by

$$\lambda_- = -\frac{\gamma C_v \rho_2 v_2 \epsilon}{\gamma C_v \mu + (\gamma-1)\kappa} + \mathcal{O}(\epsilon^2),$$
$$\boldsymbol{a}_- = (a_1, a_2)^T = \left(-v_2(1+\epsilon),\, \gamma C_v + \frac{\gamma^2\kappa C_v\epsilon}{\gamma C_v\mu + (\gamma-1)\kappa}\right)^T + \mathcal{O}(\epsilon^2), \tag{9.3.18}$$

where the superscript 'T' indicates transposition. Then the equation of two out of four integral curves that ends at the second critical point is $\delta T = (a_1/a_2)\delta v + \mathcal{O}(\epsilon^2)$. The equation of the first integral curve is given by this equation and the condition $\delta v > 0$, while the equation of the second integral curve is given by the same equation and the condition $\delta v < 0$. Since the velocity increases in the shock, $\delta v > 0$, we choose the first integral curve. In accordance with this we look for the solution to Eq. (9.3.17) corresponding to the separatrix in the form

$$\delta T = -\frac{v_2}{\gamma C_v}\left(1 + \frac{(\gamma C_v \mu - \kappa)\epsilon}{\gamma C_v \mu + (\gamma - 1)\kappa}\right)\delta v + A(\delta v)^2, \tag{9.3.19}$$

where the coefficient at δv is equal to $a_1/a_2 + \mathcal{O}(\epsilon^2)$, and the coefficient A is to be determined. Multiplying Eq. (9.3.17) by the denominator of its right-hand side and substituting Eq. (9.3.19) into the obtained equation yields

$$\begin{aligned}\frac{\kappa}{\gamma C_v \mu}&\left(\frac{\gamma^2 C_v \mu}{\gamma C_v \mu + (\gamma - 1)\kappa}\epsilon\,\delta v - \frac{\gamma(\delta v)^2}{v_2} - \frac{A v_2 (\delta v)^2}{T_2}\right) \\ &= \frac{\gamma\kappa\epsilon\,\delta v}{\gamma C_v \mu + (\gamma - 1)\kappa} - \frac{\gamma(\delta v)^2}{2v_2} + \frac{A v_2(\delta v)^2}{(\gamma - 1)T_2} + \mathcal{O}(\epsilon^3).\end{aligned} \tag{9.3.20}$$

The terms proportional to δv are cancelled. Collecting terms proportional to $(\delta v)^2$ we obtain

$$A = \frac{\gamma C_v \mu - 2\kappa}{2C_v[\gamma C_v \mu + (\gamma - 1)\kappa]}. \tag{9.3.21}$$

Now we substitute δT given by Eq. (9.3.19) into Eq. (9.3.16) and use Eq. (9.3.21). As a result we obtain

$$\frac{d(\delta v)}{dx} = \frac{\gamma^2 p_2 C_v}{\gamma C_v \mu + (\gamma - 1)\kappa}\left(\frac{(\gamma + 1)(\delta v)^2}{2v_2^2} - \frac{\epsilon\,\delta v}{v_2}\right). \tag{9.3.22}$$

Separating the variables yields

$$\frac{dW}{W[(\gamma + 1)W - 2\epsilon]} = \frac{dx}{\ell}, \tag{9.3.23}$$

where

$$W = \frac{\delta v}{v_2}, \quad \ell = 2\frac{\gamma C_v \mu + (\gamma - 1)\kappa}{\gamma C_v \rho_1 v_1}, \tag{9.3.24}$$

and we used the approximate relation $\gamma p_2 \approx \rho_2 v_2^2 \approx \rho_1 v_1^2$. Integrating Eq. (9.3.23) we obtain

$$\ln\frac{2\epsilon - (\gamma + 1)W}{(\gamma + 1)W} = \frac{2\epsilon x}{\ell}, \tag{9.3.25}$$

where we arbitrarily chose the integration constant. Using the approximate relation $v_1 - v_2 \approx 2\epsilon v_2/(\gamma + 1)$ we transform this equation into a very symmetric form,

$$v = \frac{v_1 + v_2}{2} - \frac{v_1 - v_2}{2} \tanh \frac{\epsilon x}{\ell}. \tag{9.3.26}$$

In this equation the first term is of the order of unity, the second of the order of ϵ, and we neglect terms of higher orders.

Using Eqs. (8.3.10), (8.3.16), (8.4.4) and Eq. (9.3.14) we obtain

$$\frac{v_2(v_1 - v_2)}{2\gamma C_v} = \frac{T_2}{2}\left(1 - \frac{T_1}{T_2}\right) + \mathcal{O}(\epsilon^2). \tag{9.3.27}$$

With the aid of these results, and Eqs. (9.3.19) and (9.3.26), we have

$$T = \frac{T_1 + T_2}{2} - \frac{T_1 - T_2}{2} \tanh \frac{\epsilon x}{\ell}, \tag{9.3.28}$$

where again we only kept terms of the order of unity and ϵ. Finally, using Eqs. (9.3.4) and (9.3.5) we obtain

$$\rho = \frac{\rho_1 + \rho_2}{2} - \frac{\rho_1 - \rho_2}{2} \tanh \frac{\epsilon x}{\ell}, \quad p = \frac{p_1 + p_2}{2} - \frac{p_1 - p_2}{2} \tanh \frac{\epsilon x}{\ell}. \tag{9.3.29}$$

The substantial variation of the hyperbolic tangent is restricted to the interval centred at zero with length of the order of unity, while it is almost constant beyond this interval. Hence, we conclude that the characteristic thickness of the shock is ℓ/ϵ. Since μ and κ/C_v are of the same order, we conclude that it is of the order of $\mu/(\rho_1 v_1 \epsilon)$. Previously we claimed that the characteristic thickness of the shock is $\mu/(\rho_1 v_1)$. Now we see that this is only true for sufficiently strong shocks with $M_1 - 1 \gtrsim 1$. For weak shocks it is multiplied by $(M_1 - 1)^{-1}$, where we took into account that $\epsilon \approx 2(M_1 - 1)$.

9.4 Burgers' Equation

In Sect. 8.8 we studied the effect of nonlinearity on acoustic waves. We showed that one-dimensional nonlinear acoustic waves are described by Eq. (8.8.1). Using this equation to study the spatial evolution of a harmonic wave driven at a particular plane perpendicular to the wave propagation direction we found that at a definite spatial position (called critical) an infinite gradient appears in the wave profile and we cannot obtain a continuous solution to Eq. (8.8.1) beyond this position. To extend the solution beyond the critical position we need to assume that there is a shock in the wave profile. Just before the critical position there are large gradients in the wave. This implies that the viscosity and thermal conduction become important even when their effect well before the critical position is negligible. In this section we aim to derive an equation that takes both nonlinearity and dissipative processes into account. This equation is a generalisation of Eq. (8.8.1).

To derive the equation generalising Eq. (8.8.1) we use a so-called reductive perturbation method. Since we are considering one-dimensional motions Eqs. (3.8.2), (7.1.4), and (8.3.5) reduce to

$$\frac{\partial \rho}{\partial t} + \frac{\partial(\rho v)}{\partial x} = 0, \tag{9.4.1}$$

$$\rho\left(\frac{\partial v}{\partial t} + v\frac{\partial v}{\partial x}\right) = -\frac{\partial p}{\partial x} + \mu\frac{\partial^2 v}{\partial x^2}, \tag{9.4.2}$$

$$\rho C_v\left(\frac{\partial T}{\partial t} + v\frac{\partial T}{\partial x} + (\gamma - 1)T\frac{\partial v}{\partial x}\right) = \kappa\frac{\partial^2 T}{\partial x^2} + \mu\left(\frac{\partial v}{\partial x}\right)^2. \tag{9.4.3}$$

We also use the Clapeyron law given by Eq. (8.3.10). Using Eqs. (8.3.12) and (8.3.14) we transform it into

$$p = (\gamma - 1)C_v\rho T. \tag{9.4.4}$$

We introduce the ratio of the characteristic velocity amplitude U to the unperturbed sound speed a_0 and assume that this ratio is small, $\epsilon = U/a_0 \ll 1$. The effects of viscosity and thermal conduction are characterised by the Reynolds and Peclet numbers, Re and Pe. We assume that these numbers are large, $\mathrm{Re}^{-1} = \mathcal{O}(\epsilon)$ and $\mathrm{Pe}^{-1} = \mathcal{O}(\epsilon)$, where we define the Reynolds and Peclet numbers as $\mathrm{Re} = \rho_0 a_0 L/\mu$ with $\mu = \zeta + \frac{4}{3}\eta$ and $\mathrm{Pe} = \rho_0 a_0 C_p L/\kappa$, L being the wavelength. Since $\mu = \rho_0 a_0 L\,\mathrm{Re}^{-1}$ and $\kappa = C_p\rho_0 a_0 L\,\mathrm{Pe}^{-1}$ it is convenient to introduce the scaled coefficients of viscosity and thermal conduction, $\bar{\mu} = \epsilon^{-1}\mu$ and $\bar{\kappa} = \epsilon^{-1}\kappa$.

As we have seen in Sect. 8.8, nonlinearity only affects the wave shape at distances of the order of the critical distance, which is, in turn, of the order of $\epsilon^{-1}L$, while on substantially smaller distances the wave propagates as predicted by the linear theory and preserves its shape. On the basis of the analysis in Sect. 9.2 we can conclude that viscosity and thermal conduction also only affect the wave at distances of the order of $\epsilon^{-1}L$. Hence, at distances substantially smaller than $\epsilon^{-1}L$ the wave propagates without changing its shape and perturbations of all variables only depend on $\tau = t - x/a_0$. On the other hand, both nonlinearity and dissipation start to affect the wave at distances of the order of $\epsilon^{-1}L$. This observation inspires us to introduce the scaled distance $\xi = \epsilon x$. Now we make the variable substitution and use τ and ξ as the independent variables instead of t and x. As a result, Eqs. (9.4.1)–(9.4.3) are transformed into

$$\frac{\partial \rho}{\partial \tau} - \frac{1}{a_0}\frac{\partial(\rho v)}{\partial \tau} + \epsilon\frac{\partial(\rho v)}{\partial \xi} = 0, \tag{9.4.5}$$

$$\rho\left(\frac{\partial v}{\partial \tau} - \frac{v}{a_0}\frac{\partial v}{\partial \tau}\right) = \frac{1}{a_0}\frac{\partial p}{\partial \tau} - \epsilon\frac{\partial p}{\partial \xi} + \frac{\epsilon\bar{\mu}}{a_0^2}\frac{\partial^2 v}{\partial \tau^2} + \mathcal{O}(\epsilon^3), \tag{9.4.6}$$

$$\frac{\partial T}{\partial \tau} - \frac{v}{a_0}\frac{\partial T}{\partial \tau} - (\gamma - 1)\frac{T}{a_0}\frac{\partial v}{\partial \tau} + \epsilon(\gamma - 1)T\frac{\partial v}{\partial \xi} = \frac{\epsilon\bar{\kappa}}{\rho_0 C_v a_0^2}\frac{\partial^2 T}{\partial \tau^2} + \mathcal{O}(\epsilon^3), \quad (9.4.7)$$

where we took into account that perturbations of all quantities are of the order of ϵ. Now we look for a solution to Eqs. (9.4.4)–(9.4.7) in the form of power series expansions with respect to ϵ,

$$\begin{aligned} v &= \epsilon v_1 + \epsilon^2 v_2 + \dots, \qquad \rho = \rho_0 + \epsilon\rho_1 + \epsilon^2\rho_2 + \dots, \\ p &= p_0 + \epsilon p_1 + \epsilon^2 p_2 + \dots, \quad T = T_0 + \epsilon T_1 + \epsilon^2 T_2 + \dots, \end{aligned} \quad (9.4.8)$$

where the variables with the subscript zero are constant. We substitute these expansions into Eqs. (9.4.4)–(9.4.7) and collect terms of the order of ϵ. This yields

$$p_1 = C_v(\gamma - 1)(\rho_0 T_1 + T_0\rho_1), \quad (9.4.9)$$

$$\frac{\partial \rho_1}{\partial \tau} - \frac{\rho_0}{a_0}\frac{\partial v_1}{\partial \tau} = 0, \quad (9.4.10)$$

$$\rho_0\frac{\partial v_1}{\partial \tau} = \frac{1}{a_0}\frac{\partial p_1}{\partial \tau}, \quad (9.4.11)$$

$$\frac{\partial T_1}{\partial \tau} - (\gamma - 1)\frac{T_0}{a_0}\frac{\partial v_1}{\partial \tau} = 0. \quad (9.4.12)$$

Assuming that all perturbations either tend to zero as $\tau \to \infty$ or that they are periodic functions of time with zero mean value over the period we can remove partial derivatives with respect to τ and obtain a system of four linear algebraic equations. It is straightforward to show with the aid of Eq. (9.4.4) that the determinant of this system is zero, which implies that it has non-trivial solutions. Using this system we can express all variables in terms of v_1,

$$\rho_1 = \frac{\rho_0 v_1}{a_0}, \quad p_1 = \rho_0 a_0 v_1, \quad T_1 = (\gamma - 1)\frac{T_0 v_1}{a_0}. \quad (9.4.13)$$

In the next step we collect terms of the order of ϵ^2 in Eqs. (9.4.4)–(9.4.7). As a result we obtain

$$\frac{\partial p_2}{\partial \tau} - C_v(\gamma - 1)\left(\rho_0\frac{\partial T_2}{\partial \tau} + T_0\frac{\partial \rho_2}{\partial \tau}\right) = C_v(\gamma - 1)\frac{\partial(\rho_1 T_1)}{\partial \tau}, \quad (9.4.14)$$

$$\frac{\partial \rho_2}{\partial \tau} - \frac{\rho_0}{a_0}\frac{\partial v_2}{\partial \tau} = \frac{1}{a_0}\frac{\partial(\rho_1 v_1)}{\partial \tau} - \rho_0\frac{\partial v_1}{\partial \xi}, \quad (9.4.15)$$

$$\rho_0\frac{\partial v_2}{\partial \tau} - \frac{1}{a_0}\frac{\partial p_2}{\partial \tau} = \frac{\rho_0 v_1}{a_0}\frac{\partial v_1}{\partial \tau} - \rho_1\frac{\partial v_1}{\partial \tau} - \frac{\partial p_1}{\partial \xi} + \frac{\bar{\mu}}{a_0^2}\frac{\partial^2 v_1}{\partial \tau^2}, \quad (9.4.16)$$

$$\frac{\partial T_2}{\partial \tau} - (\gamma - 1)\frac{T_0}{a_0}\frac{\partial v_2}{\partial \tau} = \frac{v_1}{a_0}\frac{\partial T_1}{\partial \tau} + (\gamma - 1)\left(\frac{T_1}{a_0}\frac{\partial v_1}{\partial \tau} - T_0\frac{\partial v_1}{\partial \xi}\right) + \frac{\bar{\kappa}}{\rho_0 C_v a_0^2}\frac{\partial^2 T_1}{\partial \tau^2}. \tag{9.4.17}$$

Using Eq. (9.4.13) to express the right-hand sides of Eqs. (9.4.14)–(9.4.17) in terms of v_1 we reduce these equations to

$$\frac{\partial p_2}{\partial \tau} - C_v(\gamma - 1)\left(\rho_0\frac{\partial T_2}{\partial \tau} + T_0\frac{\partial \rho_2}{\partial \tau}\right) = \frac{2(\gamma - 1)\rho_0}{\gamma}v_1\frac{\partial v_1}{\partial \tau}, \tag{9.4.18}$$

$$\frac{\partial \rho_2}{\partial \tau} - \frac{\rho_0}{a_0}\frac{\partial v_2}{\partial \tau} = \frac{2\rho_0}{a_0^2}v_1\frac{\partial v_1}{\partial \tau} - \rho_0\frac{\partial v_1}{\partial \xi}, \tag{9.4.19}$$

$$\rho_0\frac{\partial v_2}{\partial \tau} - \frac{1}{a_0}\frac{\partial p_2}{\partial \tau} = -\rho_0 a_0\frac{\partial v_1}{\partial \xi} + \frac{\bar{\mu}}{a_0^2}\frac{\partial^2 v_1}{\partial \tau^2}, \tag{9.4.20}$$

$$\frac{\partial T_2}{\partial \tau} - (\gamma - 1)\frac{T_0}{a_0}\frac{\partial v_2}{\partial \tau} = (\gamma - 1)\left(\frac{\gamma T_0 v_1}{a_0^2}\frac{\partial v_1}{\partial \tau} - T_0\frac{\partial v_1}{\partial \xi}\right) + \frac{(\gamma - 1)\bar{\kappa} T_0}{\rho_0 C_v a_0^3}\frac{\partial^2 v_1}{\partial \tau^2}. \tag{9.4.21}$$

When deriving Eq. (9.4.18) we collected terms of the order of ϵ^2 in Eq. (9.4.4) and then differentiated the result with respect to τ. The system of Eqs. (9.4.18)–(9.4.21) can be considered as a system of linear inhomogeneous algebraic equations for the derivatives of v_2, ρ_2, p_2, and T_2 with respect to τ. If we set the right-hand side of these equations equal to zero then we obtain the system of equations of the first-order approximation with Eq. (9.4.9) differentiated with respect to τ. Hence, the determinant of the system of Eqs. (9.4.18)–(9.4.21) is zero, meaning that it only has solutions if the right-hand side satisfies the compatibility condition. The simplest way to obtain this condition is to eliminate the variables of the second-order approximation from Eqs. (9.4.18)–(9.4.21). As a result, we obtain

$$\frac{\partial v_1}{\partial \xi} - \frac{\gamma + 1}{2a_0^2}v_1\frac{\partial v_1}{\partial \tau} - \frac{1}{2\rho_0 a_0^3}\left(\bar{\mu} + \frac{(\gamma - 1)\bar{\kappa}}{\gamma C_v}\right)\frac{\partial^2 v_1}{\partial \tau^2} = 0. \tag{9.4.22}$$

Returning to the original independent variables and putting $v \approx \epsilon v_1$ yields

$$\boxed{\frac{\partial v}{\partial x} + \frac{1}{a_0}\left(1 - \frac{\gamma + 1}{2a_0}v\right)\frac{\partial v}{\partial t} - \Gamma\frac{\partial^2 v}{\partial t^2} = 0,} \tag{9.4.23}$$

where

$$\Gamma = \frac{1}{2\rho_0 a_0^3}\left(\mu + \frac{(\gamma - 1)\kappa}{\gamma C_v}\right). \tag{9.4.24}$$

Equation (9.4.23) is *Burgers' equation*, named after the Dutch physicist Johannes Martinus Burgers. To clarify the relative importance of nonlinearity and dissipation in this equation we introduce the dimensionless variables

$$\tilde{x} = \frac{x}{L}, \quad \tilde{t} = \frac{a_0 t}{L}, \quad \tilde{v} = \frac{v}{U}. \tag{9.4.25}$$

Using these variables we transform Eq. (9.4.23) into

$$\frac{\partial \tilde{v}}{\partial \tilde{x}} + \left(1 - \frac{\gamma + 1}{2} M \tilde{v}\right) \frac{\partial \tilde{v}}{\partial \tilde{t}} - \frac{a_0^2 \Gamma}{L} \frac{\partial^2 \tilde{v}}{\partial \tilde{t}^2} = 0. \tag{9.4.26}$$

We recall that $M = U/a_0$. The relative importance of nonlinearity and dissipation is determined by the dimensionless parameter $ML(a_0^2\Gamma)^{-1}$, which is of the order of the minimum of the two quantities, MRe and MPe. When $ML(a_0^2\Gamma)^{-1} \gg 1$ we can neglect viscosity and thermal conduction ($\mu = 0$ and $\kappa = 0$) in Eq. (9.4.23) and reduce it to Eq. (8.8.1). On the other hand, when $ML(a_0^2\Gamma)^{-1} \ll 1$ we can drop the nonlinear term and then look for the solution to Eq. (9.4.23) proportional to $\exp[i(kx - \omega t)]$. In this case we obtain $k = \omega/a_0 + \delta k$, where δk is given by Eq. (9.2.18).

Equation (9.4.23) can be used to study the boundary value problem where v is given at $x = 0$ for $-\infty < t < \infty$ and we look for the solution in the region $x > 0$. Sometimes we need to solve the initial value problem where v is given at $t = 0$ for $-\infty < x < \infty$. In that case another form of Burgers' equation should be used. To derive it we introduce the running variable $\xi = x - a_0 t$ and the "slow" time $\tau = \epsilon t$. Then a derivation similar to that resulting in Eq. (9.4.23) gives

$$\boxed{\frac{\partial v}{\partial t} + a_0 \left(1 + \frac{\gamma + 1}{2a_0} v\right) \frac{\partial v}{\partial x} - a_0^3 \Gamma \frac{\partial^2 v}{\partial x^2} = 0.} \tag{9.4.27}$$

Let us make the variable substitution

$$\boxed{v = \frac{4a_0^2\Gamma}{\gamma + 1} \frac{\partial(\ln\phi)}{\partial t}} \tag{9.4.28}$$

in Eq. (9.4.23). This is the *Cole–Hopf substitution*, named after two American mathematicians, Julian David Cole and Eberhard Frederich Ferdinand Hopf. Substituting Eq. (9.4.28) into Eq. (9.4.23) and integrating the obtained equation with respect to t we obtain the diffusion equation

$$\boxed{\frac{\partial \phi}{\partial x} + \frac{1}{a_0} \frac{\partial \phi}{\partial t} = \Gamma \frac{\partial^2 \phi}{\partial t^2}.} \tag{9.4.29}$$

It is important that this equation is linear and, in many cases, can be solved analytically.

Example A sound wave is driven at $x = 0$ and propagates in the region $x > 0$. The boundary condition is

$$v = -U \sin(\omega t) \quad \text{at} \quad x = 0. \tag{9.4.30}$$

Find the solution to Eq. (9.4.23) describing the sound wave in the region $x > 0$.

Solution: We make the Cole–Hopf substitution defined by Eq. (9.4.28). Then it follows from Eq. (9.4.30) that

$$\phi = e^{s\cos(\omega t)} \quad \text{at} \quad x = 0, \quad s = \frac{U(\gamma + 1)}{4a_0^2 \omega \Gamma}. \tag{9.4.31}$$

The function $e^{s\cos(\omega t)}$ can be expanded in the Fourier series

$$e^{s\cos(\omega t)} = \frac{1}{2} f_0 + \sum_{n=1}^{\infty} f_n \cos(n\omega t). \tag{9.4.32}$$

The Fourier coefficients are given by

$$f_n = \frac{2}{\pi} \int_0^{\pi} e^{s\cos t} \cos(nt)\, dt = 2I_n(s), \tag{9.4.33}$$

where I_n is the modified Bessel function of the first kind (the theory of Bessel functions can be found in Riley et al., see Further Reading). The function ϕ satisfies Eq. (9.4.29). We look for the solution to this equation satisfying the boundary condition (9.4.31) in the form

$$\phi = \frac{1}{2}\phi_0(x) + \sum_{n=1}^{\infty} [\phi_n^c(x)\cos(n\omega t) + \phi_n^s(x)\sin(n\omega t)]. \tag{9.4.34}$$

Substituting this expression into equation (9.4.29) and using Eq. (9.4.31) we obtain that $\phi_0(x) = f_0 = 2I_0(s)$, and $\phi_n^c(x)$ and $\phi_n^s(x)$ are defined by the system of equations

$$\frac{d\phi_n^c}{dx} = -\Gamma n^2\omega^2\phi_n^c - \frac{n\omega}{a_0}\phi_n^s, \quad \frac{d\phi_n^s}{dx} = -\Gamma n^2\omega^2\phi_n^s + \frac{n\omega}{a_0}\phi_n^c. \tag{9.4.35}$$

They must satisfy the boundary conditions

$$\phi_n^c = 2I_n(s), \quad \phi_n^s = 0 \quad \text{at} \quad x = 0. \tag{9.4.36}$$

The solution to the system of Eq. (9.4.35) and the boundary Eq. (9.4.36) is straightforward:

$$\phi_n^c = 2I_n(s)e^{-\Gamma n^2\omega^2 x} \cos\frac{n\omega x}{a_0}, \quad \phi_n^s = 2I_n(s)e^{-\Gamma n^2\omega^2 x} \sin\frac{n\omega x}{a_0}. \tag{9.4.37}$$

Substituting these expressions into Eq. (9.4.34) yields

$$\phi = I_0(s) + 2\sum_{n=1}^{\infty} I_n(s)e^{-\Gamma n^2\omega^2 x}\cos[n\omega(t - x/a_0)]. \quad (9.4.38)$$

We now consider the case of weak nonlinearity. We have already shown that the nonlinearity is weak when $ML(a_0^2\Gamma)^{-1} \ll 1$, where $L = 2\pi a_0/\omega$ is the wavelength. This condition is equivalent to $s \ll 1$. In this case

$$I_0(s) = 1 + \mathcal{O}(s^2), \quad I_1(s) = \frac{s}{2} + \mathcal{O}(s^3), \quad I_n(s) = \mathcal{O}(s^n), \quad (9.4.39)$$

and it follows from Eq. (9.4.38) that

$$\phi = 1 + se^{-\Gamma\omega^2 x}\cos[\omega(t - x/a_0)] + \mathcal{O}(s^2). \quad (9.4.40)$$

Substituting this expression into Eq. (9.4.28) we obtain

$$v = -Ue^{-\Gamma\omega^2 x}\sin[\omega(t - x/a_0)] + \mathcal{O}(s^2). \quad (9.4.41)$$

We note that the first term in this expression is of the order of s. It is straightforward to verify that $\Gamma = \chi_s$. Then if we use Eqs. (9.2.10) and (9.2.19) to calculate the dependence of v on x in the sound wave driven at $x = 0$ then we obtain an expression similar to that given by Eq. (9.4.41). The only difference is that its phase is shifted by $\pi/2$. Hence, using the expansion with respect to the small parameter s we reproduce the result obtained in the linear approximation.

Problems

9.1: Suppose there is a cavity between two rigid walls placed at $x = 0$ and $x = L$. This cavity is filled with a thermal conducting but inviscid gas. At the initial time the gas in the cavity is at rest and its density, pressure, and temperature are equal to ρ_0, p_0, and T_0, respectively. The cavity is thermally insulated, implying that the heat flux at the cavity walls is zero. At $t = 0$ the left boundary of the cavity starts to oscillate harmonically and its position is given by $x = -l\cos(\omega t)$. Describe the gas motion in the cavity for $t > 0$. You can assume that $l \ll L$ and use the linear approximation. You also can assume that the dissipation is weak, $\text{Pe} \gg 1$.

9.2: Consider a thermal conducting gas without viscosity. Find the solution describing the structure of shock in this gas and determine the conditions when this solution exists. You can take $\gamma = 1.4$.

9.3: Use Burgers' Eq. (9.4.27) to obtain the solution describing the structure of a weak shock given by Eq. (9.3.26).
[Hint: Use the substitution $x' = -x$ and $v' = -v - C$ in Eq. (9.4.27), where C is the propagation speed of the shock-like structure, and then drop the prime. Then look for the solution to the obtained equation that only depends on $y = x + Ct$.]

9.4: At the initial time a perturbation is launched at $x = 0$. The perturbation velocity is given by

$$v = \begin{cases} 0, & t < 0, \\ -v_0, & t > 0, \end{cases}$$

where $v_0 > 0$ is a constant and $v_0 \ll a_0$, where a_0 is the sound speed in the unperturbed gas. Use Burgers' equation and the Cole–Hopf substitution to calculate the spatial dependence of the perturbation for $x > 0$. Find the limiting form of the solution for large x.

9.5: Calculate the next term giving the first nonlinear correction in the expansion of v with respect to $s \ll 1$ in the last example.

Appendix
Solutions to Problems

A.2 Solutions for Chap. 2

2.1: We have

$$\hat{\mathbf{A}}\hat{\mathbf{A}}^T = \begin{pmatrix} \frac{1}{2} & -\frac{\sqrt{3}}{2} & 0 \\ \frac{\sqrt{3}}{2} & \frac{1}{2} & 0 \\ 0 & 0 & 1 \end{pmatrix} \begin{pmatrix} \frac{1}{2} & \frac{\sqrt{3}}{2} & 0 \\ -\frac{\sqrt{3}}{2} & \frac{1}{2} & 0 \\ 0 & 0 & 1 \end{pmatrix} = \begin{pmatrix} 1 & 0 & 0 \\ 0 & 1 & 0 \\ 0 & 0 & 1 \end{pmatrix}.$$

Hence, $\hat{\mathbf{A}}$ is orthogonal. Next,

$$\det(\hat{\mathbf{A}}) = \begin{vmatrix} \frac{1}{2} & -\frac{\sqrt{3}}{2} & 0 \\ \frac{\sqrt{3}}{2} & \frac{1}{2} & 0 \\ 0 & 0 & 1 \end{vmatrix} = 1,$$

which implies that $\hat{\mathbf{A}}$ is proper orthogonal.

2.2: The new coordinates are given by

$$\begin{pmatrix} \frac{1}{2} & -\frac{\sqrt{3}}{2} & 0 \\ \frac{\sqrt{3}}{2} & \frac{1}{2} & 0 \\ 0 & 0 & 1 \end{pmatrix} \begin{pmatrix} 1 \\ -1 \\ 2 \end{pmatrix} = \begin{pmatrix} \frac{\sqrt{3}+1}{2} \\ \frac{\sqrt{3}-1}{2} \\ 2 \end{pmatrix}.$$

2.3:
(a)

$$x_i y_i = x_1 y_1 + x_2 y_2 + x_3 y_3.$$

© Springer Nature Switzerland AG 2019
M. S. Ruderman, *Fluid Dynamics and Linear Elasticity*, Springer Undergraduate Mathematics Series, https://doi.org/10.1007/978-3-030-19297-6

(b)
$$\begin{cases} t_1 = n_j\sigma_{j1} = n_1\sigma_{11} + n_2\sigma_{21} + n_3\sigma_{31} \\ t_2 = n_j\sigma_{j2} = n_1\sigma_{12} + n_2\sigma_{22} + n_3\sigma_{32} \\ t_3 = n_j\sigma_{j3} = n_1\sigma_{13} + n_2\sigma_{23} + n_3\sigma_{33}. \end{cases}$$

(c)
$$T_{ii} = T_{11} + T_{22} + T_{33}$$

(d)
$$B_{ii}C_{jj} = (B_{11} + B_{22} + B_{33})(C_{11} + C_{22} + C_{33}).$$

(e)
$$\delta_{ij}B_iC_j = B_iC_i = B_1C_1 + B_2C_2 + B_3C_3.$$

(f)
$$\delta_{ij}T_{ij} = T_{ii} = T_{11} + T_{22} + T_{33}.$$

2.4: (i)

$$\begin{aligned} \mathrm{tr}\left(\hat{\mathbf{T}}^2\right) &= T_{ij}T_{ji} \\ &= T_{11}^2 + T_{22}^2 + T_{33}^2 + T_{12}T_{21} + T_{13}T_{31} + T_{21}T_{12} + T_{23}T_{32} + T_{31}T_{13} + T_{32}T_{23} \\ &= T_{11}^2 + T_{22}^2 + T_{33}^2 + 2T_{12}T_{21} + 2T_{13}T_{31} + 2T_{23}T_{32}. \end{aligned}$$

(ii)

$$\begin{aligned} I_2 &= \frac{1}{2}(I_1^2 - T_{ij}T_{ji}) \\ &= \frac{1}{2}(T_{11} + T_{22} + T_{33})^2 - \frac{1}{2}(T_{11}^2 + T_{22}^2 + T_{33}^2 + 2T_{12}T_{21} + 2T_{13}T_{31} + 2T_{23}T_{32}) \\ &= T_{11}T_{22} + T_{11}T_{33} + T_{22}T_{33} - T_{12}T_{21} - T_{13}T_{31} - T_{23}T_{32}. \end{aligned}$$

(iii)

$$\begin{aligned} \det(\hat{\mathbf{T}} - \lambda\hat{\mathbf{I}}) &= \begin{vmatrix} T_{11}-\lambda & T_{12} & T_{13} \\ T_{21} & T_{22}-\lambda & T_{23} \\ T_{31} & T_{32} & T_{33}-\lambda \end{vmatrix} \\ &= (T_{11}-\lambda)\begin{vmatrix} T_{22}-\lambda & T_{23} \\ T_{32} & T_{33}-\lambda \end{vmatrix} - T_{12}\begin{vmatrix} T_{21} & T_{23} \\ T_{31} & T_{33}-\lambda \end{vmatrix} + T_{13}\begin{vmatrix} T_{21} & T_{22}-\lambda \\ T_{31} & T_{32} \end{vmatrix} \\ &= (T_{11}-\lambda)[\lambda^2 + T_{22}T_{33} - \lambda(T_{22}+T_{33}) - T_{23}T_{32}] \\ &\quad - T_{12}(T_{21}T_{33} - \lambda T_{21} - T_{31}T_{23}) + T_{13}(T_{21}T_{32} + \lambda T_{31} - T_{31}T_{22}) \\ &= -\lambda^3 + \lambda^2(T_{11}+T_{22}+T_{33}) - \lambda(T_{11}T_{22} + T_{11}T_{33} + T_{22}T_{33} - T_{23}T_{32}) \\ &\quad + T_{11}\begin{vmatrix} T_{22} & T_{23} \\ T_{32} & T_{33} \end{vmatrix} + \lambda T_{12}T_{21} - T_{12}\begin{vmatrix} T_{21} & T_{23} \\ T_{31} & T_{33} \end{vmatrix} + \lambda T_{13}T_{31} + T_{13}\begin{vmatrix} T_{21} & T_{22} \\ T_{31} & T_{32} \end{vmatrix} \\ &= -\lambda^3 + I_1\lambda^2 - \lambda(T_{11}T_{22} + T_{11}T_{33} + T_{22}T_{33} - T_{12}T_{21} - T_{23}T_{32} - T_{13}T_{31}) \\ &\quad + \begin{vmatrix} T_{11} & T_{12} & T_{13} \\ T_{21} & T_{22} & T_{23} \\ T_{31} & T_{32} & T_{33} \end{vmatrix} \\ &= -\lambda^3 + I_1\lambda^2 - I_2\lambda + I_3. \end{aligned}$$

Hence, the equation $\det(\hat{\mathbf{T}} - \lambda\hat{\mathbf{I}}) = 0$ can be written as

$$\lambda^3 - I_1\lambda^2 + I_2\lambda - I_3 = 0.$$

2.5: $x_i' = a_{ij}x_j \Longrightarrow x_j' = a_{jk}x_k \Longrightarrow a_{ji}x_j' = a_{ji}a_{jk}x_k \Longrightarrow a_{ji}x_j' = \delta_{ik}x_k$ (because the matrix $\hat{\mathbf{A}}$ is orthogonal) $\Longrightarrow x_i = a_{ji}x_j'$.

2.6:

$$\det(\hat{\mathbf{T}}) = \begin{vmatrix} T_{11} & T_{12} & T_{13} \\ T_{21} & T_{22} & T_{23} \\ T_{31} & T_{32} & T_{33} \end{vmatrix} = T_{11}\begin{vmatrix} T_{22} & T_{23} \\ T_{32} & T_{33} \end{vmatrix} - T_{12}\begin{vmatrix} T_{21} & T_{23} \\ T_{31} & T_{33} \end{vmatrix} + T_{13}\begin{vmatrix} T_{21} & T_{22} \\ T_{31} & T_{32} \end{vmatrix}$$

$$= T_{11}(T_{22}T_{33} - T_{23}T_{32}) + T_{12}(T_{23}T_{31} - T_{21}T_{33}) + T_{13}(T_{21}T_{32} - T_{31}T_{22})$$

$$\varepsilon_{ijk}T_{1i}T_{2j}T_{3k} = \varepsilon_{1jk}T_{11}T_{2j}T_{3k} + \varepsilon_{2jk}T_{12}T_{2j}T_{3k} + \varepsilon_{3jk}T_{13}T_{2j}T_{3k}.$$

But $\varepsilon_{1jk} = \begin{cases} 1 \text{ if } (jk) = (2,3) \\ -1 \text{ if } (jk) = (3,2) \\ 0 \text{ all others } (jk) \end{cases}$ and similar results for ε_{2jk} and ε_{3jk}

$$\Longrightarrow \varepsilon_{ijk}T_{1i}T_{2j}T_{3k} = T_{11}(T_{22}T_{33} - T_{23}T_{32}) + T_{12}(T_{23}T_{31} - T_{21}T_{33})$$

$$+T_{13}(T_{21}T_{32} - T_{31}T_{22}) = \det(\hat{\mathbf{T}}).$$

2.7:

$$\varepsilon_{ijk}\frac{\partial^2 v_k}{\partial x_i \partial x_j} = \varepsilon_{ijk}\frac{\partial^2 v_k}{\partial x_j \partial x_i} = \varepsilon_{jik}\frac{\partial^2 v_k}{\partial x_i \partial x_j} = -\varepsilon_{ijk}\frac{\partial^2 v_k}{\partial x_i \partial x_j} \Longrightarrow \varepsilon_{ijk}\frac{\partial^2 v_k}{\partial x_i \partial x_j} = 0.$$

2.8: (i) Using Fig. A.1 we easily find that the components of the unit vectors of the new coordinate system are given in the old coordinate system by

$$\boldsymbol{e}_1' = (\cos\theta, \sin\theta, 0), \quad \boldsymbol{e}_2' = (-\sin\theta, \cos\theta, 0), \quad \boldsymbol{e}_3' = (0, 0, 1)$$

This implies that the transformation matrix is

$$\hat{\mathbf{A}} = \begin{pmatrix} \cos\theta & \sin\theta & 0 \\ -\sin\theta & \cos\theta & 0 \\ 0 & 0 & 1 \end{pmatrix}.$$

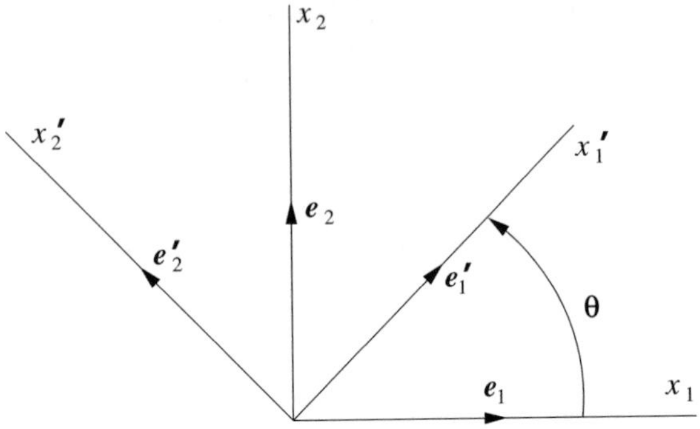

Fig. A.1 New Cartesian coordinates are obtained by rotating the old ones by angle θ about the x_3-axis

(ii) Let T_{ij} be the components of the tensor $\mathbf{T}$ in the old coordinates, and T'_{ij} be its components in the new coordinates obtained from the old ones by rotating the coordinate axes about the x_3-axis by the angle θ. Then the matrices of the old and new coordinates are related by $\hat{\mathbf{T}}' = \hat{\mathbf{A}}\hat{\mathbf{T}}\hat{\mathbf{A}}^T$. Multiplying this identity by $\hat{\mathbf{A}}$ from the right and taking into account that $\hat{\mathbf{A}}^T = \hat{\mathbf{A}}^{-1}$ and $\hat{\mathbf{T}}' = \hat{\mathbf{T}}$ we obtain

$$\begin{pmatrix} T_{11} & T_{12} & T_{13} \\ T_{21} & T_{22} & T_{23} \\ T_{31} & T_{32} & T_{33} \end{pmatrix} \begin{pmatrix} \cos\theta & \sin\theta & 0 \\ -\sin\theta & \cos\theta & 0 \\ 0 & 0 & 1 \end{pmatrix} = \begin{pmatrix} \cos\theta & \sin\theta & 0 \\ -\sin\theta & \cos\theta & 0 \\ 0 & 0 & 1 \end{pmatrix} \begin{pmatrix} T_{11} & T_{12} & T_{13} \\ T_{21} & T_{22} & T_{23} \\ T_{31} & T_{32} & T_{33} \end{pmatrix}.$$

Let us take $\theta = \frac{\pi}{2}$. Then we have

$$\begin{pmatrix} T_{11} & T_{12} & T_{13} \\ T_{21} & T_{22} & T_{23} \\ T_{31} & T_{32} & T_{33} \end{pmatrix} \begin{pmatrix} 0 & 1 & 0 \\ -1 & 0 & 0 \\ 0 & 0 & 1 \end{pmatrix} = \begin{pmatrix} 0 & 1 & 0 \\ -1 & 0 & 0 \\ 0 & 0 & 1 \end{pmatrix} \begin{pmatrix} T_{11} & T_{12} & T_{13} \\ T_{21} & T_{22} & T_{23} \\ T_{31} & T_{32} & T_{33} \end{pmatrix}.$$

This equation can be transformed to

$$\begin{pmatrix} -T_{12} & T_{11} & T_{13} \\ -T_{22} & T_{21} & T_{23} \\ -T_{32} & T_{31} & T_{33} \end{pmatrix} = \begin{pmatrix} T_{21} & T_{22} & T_{23} \\ -T_{11} & -T_{12} & -T_{13} \\ T_{31} & T_{32} & T_{33} \end{pmatrix}.$$

It follows from this equation that

$$T_{11} = T_{22}, \quad T_{12} = -T_{21}, \quad T_{13} = T_{23}, \quad T_{13} = -T_{23}, \quad T_{31} = -T_{32}, \quad T_{31} = T_{32}$$

It follows from the last four equations that

$$T_{13} = T_{23} = T_{31} = T_{32} = 0.$$

Then, introducing the notation $T_{11} = T_{22} = T_1$, $T_{33} = T_3$ and $T_{12} = T_2$ we obtain

$$\hat{\mathbf{T}} = \begin{pmatrix} T_1 & T_2 & 0 \\ -T_2 & T_1 & 0 \\ 0 & 0 & T_3 \end{pmatrix}.$$

Now it is straightforward to verify by direct calculation that $\hat{\mathbf{T}} = \hat{\mathbf{A}}\hat{\mathbf{T}}\hat{\mathbf{A}}^T$.

(iii) $\mathrm{tr}(\hat{\mathbf{T}}) = 2T_1 + T_3$,

$$\hat{\mathbf{T}}^2 = \begin{pmatrix} T_1 & T_2 & 0 \\ -T_2 & T_1 & 0 \\ 0 & 0 & T_3 \end{pmatrix}\begin{pmatrix} T_1 & T_2 & 0 \\ -T_2 & T_1 & 0 \\ 0 & 0 & T_3 \end{pmatrix} = \begin{pmatrix} T_1^2 - T_2^2 & 2T_1T_2 & 0 \\ -2T_1T_2 & T_1^2 - T_2^2 & 0 \\ 0 & 0 & T_3^2 \end{pmatrix},$$

$\mathrm{tr}(\hat{\mathbf{T}}^2) = 2T_1^2 - 2T_2^2 + T_3^2$.

(iv) Using the previous results we obtain

$$I_2 = \tfrac{1}{2}[(2T_1 + T_3)^2 - 2T_1^2 + 2T_2^2 - T_3^2] = T_1^2 + T_2^2 + 2T_1T_3, \quad I_3 = T_3(T_1^2 + T_2^2).$$

Then we have the system of equations

$$\begin{cases} 2T_1 + T_3 = 0, \\ T_1^2 + T_2^2 + 2T_1T_3 = 0, \\ T_3(T_1^2 + T_2^2) = 8. \end{cases}$$

It follows from the second equation that $T_1^2 + T_2^2 = -2T_1T_3$. Substituting this result into the third equation we obtain $T_1T_3^2 = -4$. Using the first equation to eliminate T_1 from this equation we obtain $T_3^3 = 8$, so that $T_3 = 2$. Then $T_1 = -1$. Substituting these results into the second equation we obtain $T_2^2 = 3$, so that $T_2 = \sqrt{3}$ (recall that $T_2 > 0$).

2.9: Since $\hat{\mathbf{A}}$ is orthogonal, we have $\hat{\mathbf{A}}\hat{\mathbf{A}}^T = \hat{\mathbf{I}}$, where $\hat{\mathbf{I}}$ is the unit matrix. Then

$$\begin{pmatrix} \frac{1}{2} & \frac{\sqrt{3}}{2} & 0 \\ -\frac{\sqrt{3}}{2} & \mu & 0 \\ 0 & 0 & \lambda \end{pmatrix}\begin{pmatrix} \frac{1}{2} & -\frac{\sqrt{3}}{2} & 0 \\ \frac{\sqrt{3}}{2} & \mu & 0 \\ 0 & 0 & \lambda \end{pmatrix} = \begin{pmatrix} 1 & \frac{(2\mu-1)\sqrt{3}}{4} & 0 \\ \frac{(2\mu-1)\sqrt{3}}{4} & \mu^2 + \frac{3}{4} & 0 \\ 0 & 0 & \lambda^2 \end{pmatrix} = \begin{pmatrix} 1 & 0 & 0 \\ 0 & 1 & 0 \\ 0 & 0 & 1 \end{pmatrix}.$$

This implies that

$$\mu = \frac{1}{2}, \quad \lambda^2 = 1.$$

Since $\hat{\mathbf{A}}$ is proper orthogonal, $\det(\hat{\mathbf{A}}) = 1$. Substituting the expression for μ we obtain

$$\det(\hat{\mathbf{A}}) = \begin{vmatrix} \frac{1}{2} & \frac{\sqrt{3}}{2} & 0 \\ -\frac{\sqrt{3}}{2} & \frac{1}{2} & 0 \\ 0 & 0 & \lambda \end{vmatrix} = \lambda,$$

which implies that $\lambda = 1$.

2.10:

$$\hat{\mathbf{A}}\hat{\mathbf{A}}^T = \begin{pmatrix} \frac{1}{\sqrt{2}} & \frac{-1}{\sqrt{2}} & 0 \\ \frac{1}{\sqrt{2}} & \frac{1}{\sqrt{2}} & 0 \\ 0 & 0 & 1 \end{pmatrix} \begin{pmatrix} \frac{1}{\sqrt{2}} & \frac{1}{\sqrt{2}} & 0 \\ \frac{-1}{\sqrt{2}} & \frac{1}{\sqrt{2}} & 0 \\ 0 & 0 & 1 \end{pmatrix} = \begin{pmatrix} 1 & 0 & 0 \\ 0 & 1 & 0 \\ 0 & 0 & 1 \end{pmatrix} = \hat{\mathbf{I}}$$

$\Longrightarrow \hat{\mathbf{A}}$ is orthogonal.

$$\det(\hat{\mathbf{A}}) = \begin{vmatrix} \frac{1}{\sqrt{2}} & \frac{-1}{\sqrt{2}} \\ \frac{1}{\sqrt{2}} & \frac{1}{\sqrt{2}} \end{vmatrix} = 1$$

$\Longrightarrow \hat{\mathbf{A}}$ is proper orthogonal.

$$\hat{\mathbf{T}}' = \hat{\mathbf{A}}\hat{\mathbf{T}}\hat{\mathbf{A}}^T = \begin{pmatrix} \frac{1}{\sqrt{2}} & \frac{-1}{\sqrt{2}} & 0 \\ \frac{1}{\sqrt{2}} & \frac{1}{\sqrt{2}} & 0 \\ 0 & 0 & 1 \end{pmatrix} \begin{pmatrix} 0 & 1 & 0 \\ 1 & 0 & 0 \\ 0 & 0 & 0 \end{pmatrix} \begin{pmatrix} \frac{1}{\sqrt{2}} & \frac{1}{\sqrt{2}} & 0 \\ \frac{-1}{\sqrt{2}} & \frac{1}{\sqrt{2}} & 0 \\ 0 & 0 & 1 \end{pmatrix} = \begin{pmatrix} -1 & 0 & 0 \\ 0 & 1 & 0 \\ 0 & 0 & 0 \end{pmatrix}.$$

2.11: (i) $\|\boldsymbol{a}\| = \sqrt{4+1+1} = \sqrt{6}$, $\|\boldsymbol{b}\| = \sqrt{1+1+1} = \sqrt{3} \Longrightarrow$

$$\boldsymbol{e}_1' = \frac{\boldsymbol{a}}{\sqrt{6}} = \left(\frac{2}{\sqrt{6}}, \frac{1}{\sqrt{6}}, \frac{-1}{\sqrt{6}}\right), \quad \boldsymbol{e}_2' = \frac{\boldsymbol{b}}{\sqrt{3}} = \left(\frac{1}{\sqrt{3}}, \frac{-1}{\sqrt{3}}, \frac{1}{\sqrt{3}}\right).$$

(ii)

$$\boldsymbol{e}_3' = \boldsymbol{e}_1' \times \boldsymbol{e}_2' = \frac{\boldsymbol{a} \times \boldsymbol{b}}{3\sqrt{2}} = \frac{1}{3\sqrt{2}} \begin{vmatrix} \boldsymbol{e}_1 & \boldsymbol{e}_2 & \boldsymbol{e}_3 \\ 2 & 1 & -1 \\ 1 & -1 & 1 \end{vmatrix} = \frac{(0, -3, -3)}{3\sqrt{2}} = \left(0, \frac{-1}{\sqrt{2}}, \frac{-1}{\sqrt{2}}\right).$$

(iii)

$$\hat{\mathbf{A}} = \begin{pmatrix} \frac{2}{\sqrt{6}} & \frac{1}{\sqrt{6}} & \frac{-1}{\sqrt{6}} \\ \frac{1}{\sqrt{3}} & \frac{-1}{\sqrt{3}} & \frac{1}{\sqrt{3}} \\ 0 & \frac{-1}{\sqrt{2}} & \frac{-1}{\sqrt{2}} \end{pmatrix}.$$

(iv)

$$\det(\hat{\mathbf{A}}) = \begin{vmatrix} \frac{2}{\sqrt{6}} & \frac{1}{\sqrt{6}} & \frac{-1}{\sqrt{6}} \\ \frac{1}{\sqrt{3}} & \frac{-1}{\sqrt{3}} & \frac{1}{\sqrt{3}} \\ 0 & \frac{-1}{\sqrt{2}} & \frac{-1}{\sqrt{2}} \end{vmatrix} = \frac{2}{\sqrt{6}}\left(\frac{1}{\sqrt{6}} + \frac{1}{\sqrt{6}}\right) - \frac{1}{\sqrt{6}}\frac{1}{\sqrt{3}}\frac{-1}{\sqrt{2}} + \frac{-1}{\sqrt{6}}\frac{1}{\sqrt{3}}\frac{-1}{\sqrt{2}} = 1.$$

2.12:

(i) $$\hat{\mathbf{T}}' = \hat{\mathbf{A}}\hat{\mathbf{T}}\hat{\mathbf{A}}^T = \begin{pmatrix} \cos\theta & \sin\theta & 0 \\ -\sin\theta & \cos\theta & 0 \\ 0 & 0 & 1 \end{pmatrix} \begin{pmatrix} a & b & 0 \\ b & c & 0 \\ 0 & 0 & 0 \end{pmatrix} \begin{pmatrix} \cos\theta & -\sin\theta & 0 \\ \sin\theta & \cos\theta & 0 \\ 0 & 0 & 1 \end{pmatrix}$$

$$= \begin{pmatrix} \cos\theta & \sin\theta & 0 \\ -\sin\theta & \cos\theta & 0 \\ 0 & 0 & 1 \end{pmatrix} \begin{pmatrix} a\cos\theta + b\sin\theta & -a\sin\theta + b\cos\theta & 0 \\ b\cos\theta + c\sin\theta & -b\sin\theta + c\cos\theta & 0 \\ 0 & 0 & 0 \end{pmatrix}$$

$$= \begin{pmatrix} a\cos^2\theta + 2b\cos\theta\sin\theta + c\sin^2\theta & h & 0 \\ h & a\sin^2\theta - 2b\cos\theta\sin\theta + c\cos^2\theta & 0 \\ 0 & 0 & 0 \end{pmatrix},$$

where

$$h = -a\cos\theta\sin\theta + b(\cos^2\theta - \sin^2\theta) + c\cos\theta\sin\theta = b\cos 2\theta + \frac{1}{2}(c-a)\sin 2\theta.$$

Eventually,

$$\hat{\mathbf{T}}' = \begin{pmatrix} a\cos^2\theta + b\sin 2\theta + c\sin^2\theta & b\cos 2\theta + \frac{1}{2}(c-a)\sin 2\theta & 0 \\ b\cos 2\theta + \frac{1}{2}(c-a)\sin 2\theta & a\sin^2\theta - b\sin 2\theta + c\cos^2\theta & 0 \\ 0 & 0 & 0 \end{pmatrix}.$$

(ii) If $\tan 2\theta = 2b/(a-c)$ then

(a)

$$T'_{12} = T'_{21} = b\cos 2\theta + \frac{1}{2}(c-a)\sin 2\theta = \cos 2\theta\left[b + \frac{1}{2}(c-a)\frac{2b}{(a-c)}\right] = 0.$$

(b)

$$T'_{11} = a\cos^2\theta + b\sin 2\theta + c\sin^2\theta = \frac{a}{2}(1+\cos 2\theta) + b\sin 2\theta + \frac{c}{2}(1-\cos 2\theta)$$

$$= \frac{1}{2}(a+c) + \frac{1}{2}\cos 2\theta\left[a - c + 2b\frac{2b}{(a-c)}\right] = \frac{1}{2}(a+c) + \frac{1}{2}\frac{\cos 2\theta}{a-c}[(a-c)^2 + 4b^2].$$

Since $0 < \theta < \pi/4$ it follows that $a > c$. Then

$$\frac{1}{\cos^2 2\theta} = 1 + \tan^2 2\theta \quad \Longrightarrow \cos 2\theta = \frac{1}{(1+\tan^2 2\theta)^{1/2}} = \frac{a-c}{[(a-c)^2+4b^2]^{1/2}}.$$

Then

$$T'_{11} = \frac{a+c}{2} + \frac{a-c}{2[(a-c)^2+4b^2]^{1/2}} \frac{[(a-c)^2+4b^2]}{a-c} = \frac{1}{2}\left\{a+c+[(a-c)^2+4b^2]^{1/2}\right\}.$$

2.13: (i)

$$\hat{\mathbf{A}}\hat{\mathbf{A}}^T = \begin{pmatrix} \frac{1}{\sqrt{3}} & \frac{-1}{\sqrt{3}} & \frac{1}{\sqrt{3}} \\ \frac{1}{\sqrt{2}} & \frac{1}{\sqrt{2}} & 0 \\ \frac{1}{\sqrt{6}} & \frac{-1}{\sqrt{6}} & \frac{-2}{\sqrt{6}} \end{pmatrix} \begin{pmatrix} \frac{1}{\sqrt{3}} & \frac{1}{\sqrt{2}} & \frac{1}{\sqrt{6}} \\ \frac{-1}{\sqrt{3}} & \frac{1}{\sqrt{2}} & \frac{-1}{\sqrt{6}} \\ \frac{1}{\sqrt{3}} & 0 & \frac{-2}{\sqrt{6}} \end{pmatrix} = \begin{pmatrix} 1 & 0 & 0 \\ 0 & 1 & 0 \\ 0 & 0 & 1 \end{pmatrix} = \hat{\mathbf{I}}.$$

(ii)

$$\hat{\mathbf{T}}' = \hat{\mathbf{A}}\hat{\mathbf{T}}\hat{\mathbf{A}}^T = \begin{pmatrix} \frac{1}{\sqrt{3}} & \frac{-1}{\sqrt{3}} & \frac{1}{\sqrt{3}} \\ \frac{1}{\sqrt{2}} & \frac{1}{\sqrt{2}} & 0 \\ \frac{1}{\sqrt{6}} & \frac{-1}{\sqrt{6}} & \frac{-2}{\sqrt{6}} \end{pmatrix} \begin{pmatrix} 2 & 0 & c \\ 0 & a & 0 \\ 0 & 0 & b \end{pmatrix} \begin{pmatrix} \frac{1}{\sqrt{3}} & \frac{1}{\sqrt{2}} & \frac{1}{\sqrt{6}} \\ \frac{-1}{\sqrt{3}} & \frac{1}{\sqrt{2}} & \frac{-1}{\sqrt{6}} \\ \frac{1}{\sqrt{3}} & 0 & \frac{-2}{\sqrt{6}} \end{pmatrix}$$

$$= \begin{pmatrix} \frac{2}{\sqrt{3}} & \frac{-a}{\sqrt{3}} & \frac{b+c}{\sqrt{3}} \\ \frac{2}{\sqrt{2}} & \frac{a}{\sqrt{2}} & \frac{c}{\sqrt{2}} \\ \frac{2}{\sqrt{6}} & \frac{-a}{\sqrt{6}} & \frac{c-2b}{\sqrt{6}} \end{pmatrix} \begin{pmatrix} \frac{1}{\sqrt{3}} & \frac{1}{\sqrt{2}} & \frac{1}{\sqrt{6}} \\ \frac{-1}{\sqrt{3}} & \frac{1}{\sqrt{2}} & \frac{-1}{\sqrt{6}} \\ \frac{1}{\sqrt{3}} & 0 & \frac{-2}{\sqrt{6}} \end{pmatrix} = \begin{pmatrix} \frac{2+a+b+c}{3} & \frac{2-a}{\sqrt{6}} & \frac{2+a-2b-2c}{3\sqrt{2}} \\ \frac{2-a+c}{\sqrt{6}} & \frac{2+a}{2} & \frac{2-a-2c}{2\sqrt{3}} \\ \frac{2+a-2b+c}{3\sqrt{2}} & \frac{2-a}{2\sqrt{3}} & \frac{2+a+4b-2c}{6} \end{pmatrix}.$$

(iii) Since $\hat{\mathbf{T}}' = \hat{\mathbf{T}}$, we, in particular, have

$$\frac{2+a+b+c}{3} = 2, \quad \frac{2-a}{\sqrt{6}} = 0, \quad \frac{2-a+c}{\sqrt{6}} = 0,$$

and we obtain $a = 2$, $b = 2$, $c = 0$, so that $\hat{\mathbf{T}} = 2\hat{\mathbf{I}}$. It is not necessary to verify that all other components of $\hat{\mathbf{T}}'$ and $\hat{\mathbf{T}}$ are the same, because the unit tensor $\mathbf{I}$ has the same components δ_{ij} in any coordinate system.

2.14: We obtain for $\hat{\mathbf{U}}$:

$$\hat{\mathbf{U}} = \hat{\mathbf{B}}^{-1}\hat{\mathbf{T}} = \hat{\mathbf{B}}^T\hat{\mathbf{T}}$$

$$= \frac{1}{\sqrt{30}} \begin{pmatrix} 2\sqrt{5} & 2 & \sqrt{6} \\ -\sqrt{5} & 5 & 0 \\ \sqrt{5} & 1 & -2\sqrt{6} \end{pmatrix} \begin{pmatrix} 6\sqrt{5} & -\sqrt{5} & 17\sqrt{5} \\ 12 & 11 & 11 \\ -4\sqrt{6} & 3\sqrt{6} & -17\sqrt{6} \end{pmatrix} = \sqrt{30} \begin{pmatrix} 2 & 1 & 3 \\ 1 & 2 & -1 \\ 3 & -1 & 10 \end{pmatrix}.$$

Then

$$\boldsymbol{u}\cdot\mathbf{U}\cdot\boldsymbol{u} = \sqrt{30}(u_1,\ u_2,\ u_3)\begin{pmatrix}2 & 1 & 3\\ 1 & 2 & -1\\ 3 & -1 & 10\end{pmatrix}\begin{pmatrix}u_1\\ u_2\\ u_3\end{pmatrix}$$

$$= \sqrt{30}(2u_1^2 + 2u_2^2 + 10u_3^2 + 2u_1u_2 + 6u_1u_3 - 2u_2u_3)$$

$$= \sqrt{30}\big[(u_1 + u_2)^2 + (u_1 + 3u_3)^2 + (u_2 - u_3)^2\big].$$

We see that the condition $\boldsymbol{u}\cdot\mathbf{U}\cdot\boldsymbol{u} = 0$ is equivalent to

$$u_1 + u_2 = 0, \quad u_1 + 3u_3 = 0, \quad u_2 - u_3 = 0,$$

which gives $u_1 = u_2 = u_3 = 0$. This implies that $\mathbf{U}$ is positive-definite.

2.15: The condition $\nabla\cdot\mathbf{U} = 0$ gives

$$\frac{\partial f}{\partial x_1} + 2x_1 = 0, \quad \frac{\partial g}{\partial x_2} + 2x_2 + \cos x_3 = 0, \quad \frac{\partial h}{\partial x_3} - \sin x_1 = 0.$$

Taking into account the conditions $f = 0$ at $x_1 = 0$, $g = 0$ at $x_2 = 0$, and $h = 0$ at $x_3 = 0$, we obtain from these equations

$$f = -x_1^2, \quad g = -x_2^2 - x_2\cos x_3, \quad h = x_3\sin x_1.$$

2.16:
$$[\boldsymbol{v}\times(\nabla\times\boldsymbol{v})]_i = \varepsilon_{ijk}v_j(\nabla\times\boldsymbol{v})_k = \varepsilon_{ijk}v_j\varepsilon_{klm}\frac{\partial v_m}{\partial x_l} = \varepsilon_{ijk}\varepsilon_{lmk}v_j\frac{\partial v_m}{\partial x_l}$$

$$= (\delta_{il}\delta_{jm} - \delta_{im}\delta_{jl})v_j\frac{\partial v_m}{\partial x_l} = v_j\frac{\partial v_j}{\partial x_i} - v_j\frac{\partial v_i}{\partial x_j}$$

$$= \frac{1}{2}\frac{\partial(v_jv_j)}{\partial x_i} - v_j\frac{\partial v_i}{\partial x_j} = \frac{1}{2}\frac{\partial(\|\boldsymbol{v}\|^2)}{\partial x_i} - (\boldsymbol{v}\cdot\nabla)v_i.$$

2.17: Using the expression for the operator ∇ in spherical coordinates given by Eq. (2.13.12) we obtain

$$\nabla\boldsymbol{e}_r = \frac{1}{r}\,\boldsymbol{e}_\theta(\cos\theta\cos\phi, \cos\theta\sin\phi, -\sin\theta)$$
$$+\frac{\boldsymbol{e}_\phi}{r\sin\theta}(-\sin\theta\sin\phi, \sin\theta\cos\phi, 0) = \frac{1}{r}\,\boldsymbol{e}_\theta\boldsymbol{e}_\theta + \frac{1}{r}\,\boldsymbol{e}_\phi\boldsymbol{e}_\phi,$$

$$\nabla \boldsymbol{e}_\theta = -\frac{1}{r}\,\boldsymbol{e}_\theta(\sin\theta\cos\phi, \sin\theta\sin\phi, \cos\theta)$$
$$+\frac{\boldsymbol{e}_\phi}{r\sin\theta}(-\cos\theta\sin\phi, \cos\theta\cos\phi, 0) = -\frac{1}{r}\,\boldsymbol{e}_\theta\boldsymbol{e}_r + \frac{\cot\theta}{r}\,\boldsymbol{e}_\phi\boldsymbol{e}_\phi,$$

$$\nabla \boldsymbol{e}_\phi = -\frac{\boldsymbol{e}_\phi}{r\sin\theta}(\cos\phi, \sin\phi, 0)$$
$$= -\frac{\boldsymbol{e}_\phi}{r\sin\theta}(\boldsymbol{e}_r\sin\theta + \boldsymbol{e}_\theta\cos\theta) = -\frac{1}{r}\,\boldsymbol{e}_\phi\boldsymbol{e}_r - \frac{\cot\theta}{r}\,\boldsymbol{e}_\phi\boldsymbol{e}_\theta.$$

2.18: (i) Using the expression for $\nabla\cdot\boldsymbol{v}$ in spherical coordinates we obtain

$$\nabla\cdot\boldsymbol{v} = \frac{1}{r^2}\frac{\partial(r^2 v_r)}{\partial r} + \frac{1}{r\sin\theta}\frac{\partial(v_\theta\sin\theta)}{\partial\theta} = \frac{1}{r^2}\frac{\partial}{\partial r}\left(\frac{1}{\sin\theta}\frac{\partial\psi}{\partial\theta}\right)$$
$$-\frac{1}{r\sin\theta}\frac{\partial}{\partial\theta}\left(\frac{1}{r}\frac{\partial\psi}{\partial r}\right) = \frac{1}{r^2\sin\theta}\left(\frac{\partial^2\psi}{\partial r\partial\theta} - \frac{\partial^2\psi}{\partial\theta\partial r}\right) = 0.$$

(ii) We start by calculating $\frac{1}{2}\nabla(\|\boldsymbol{v}\|^2)$:

$$\frac{1}{2}\nabla(\|\boldsymbol{v}\|^2) = \left(v_r\frac{\partial v_r}{\partial r} + v_\theta\frac{\partial v_\theta}{\partial r}\right)\boldsymbol{e}_r + \frac{1}{r}\left(v_r\frac{\partial v_r}{\partial\theta} + v_\theta\frac{\partial v_\theta}{\partial\theta}\right)\boldsymbol{e}_\theta$$
$$= \frac{1}{r\sin^2\theta}\left[\frac{\partial\psi}{\partial r}\frac{\partial}{\partial r}\left(\frac{1}{r}\frac{\partial\psi}{\partial r}\right) + \frac{1}{r}\frac{\partial\psi}{\partial\theta}\frac{\partial}{\partial r}\left(\frac{1}{r^2}\frac{\partial\psi}{\partial\theta}\right)\right]\boldsymbol{e}_r$$
$$+\frac{1}{r^3\sin\theta}\left[\frac{\partial\psi}{\partial r}\frac{\partial}{\partial\theta}\left(\frac{1}{\sin\theta}\frac{\partial\psi}{\partial r}\right) + \frac{1}{r^2}\frac{\partial\psi}{\partial\theta}\frac{\partial}{\partial\theta}\left(\frac{1}{\sin\theta}\frac{\partial\psi}{\partial\theta}\right)\right]\boldsymbol{e}_\theta.$$

Next we calculate $\nabla\times\boldsymbol{v}$:

$$\nabla\times\boldsymbol{v} = \frac{1}{r}\left(\frac{\partial(rv_\theta)}{\partial r} - \frac{\partial v_r}{\partial\theta}\right)\boldsymbol{e}_\phi = -\frac{1}{r}\left[\frac{1}{\sin\theta}\frac{\partial^2\psi}{\partial r^2} + \frac{1}{r^2}\frac{\partial}{\partial\theta}\left(\frac{1}{\sin\theta}\frac{\partial\psi}{\partial\theta}\right)\right]\boldsymbol{e}_\phi.$$

Now we calculate $\boldsymbol{v}\times(\nabla\times\boldsymbol{v})$:

$$\boldsymbol{v}\times(\nabla\times\boldsymbol{v}) = (\nabla\times\boldsymbol{v})_\phi(v_\theta\boldsymbol{e}_r - v_r\boldsymbol{e}_\theta)$$
$$= \frac{1}{r^2\sin\theta}\left[\frac{1}{\sin\theta}\frac{\partial^2\psi}{\partial r^2} + \frac{1}{r^2}\frac{\partial}{\partial\theta}\left(\frac{1}{\sin\theta}\frac{\partial\psi}{\partial\theta}\right)\right]\left(\frac{\partial\psi}{\partial r}\boldsymbol{e}_r + \frac{1}{r}\frac{\partial\psi}{\partial\theta}\boldsymbol{e}_\theta\right).$$

Using these results we finally obtain

$$
\begin{aligned}
(\boldsymbol{v}\cdot\nabla)\boldsymbol{v} &= \frac{1}{2}\nabla(\|\boldsymbol{v}\|^2) - \boldsymbol{v}\times(\nabla\times\boldsymbol{v}) \\
&= \frac{1}{r\sin^2\theta}\left[\frac{\partial\psi}{\partial r}\frac{\partial}{\partial r}\left(\frac{1}{r}\frac{\partial\psi}{\partial r}\right) + \frac{1}{r}\frac{\partial\psi}{\partial\theta}\frac{\partial}{\partial r}\left(\frac{1}{r^2}\frac{\partial\psi}{\partial\theta}\right)\right]\boldsymbol{e}_r \\
&\quad + \frac{1}{r^3\sin\theta}\left[\frac{\partial\psi}{\partial r}\frac{\partial}{\partial\theta}\left(\frac{1}{\sin\theta}\frac{\partial\psi}{\partial r}\right) + \frac{1}{r^2}\frac{\partial\psi}{\partial\theta}\frac{\partial}{\partial\theta}\left(\frac{1}{\sin\theta}\frac{\partial\psi}{\partial\theta}\right)\right]\boldsymbol{e}_\theta \\
&\quad - \frac{1}{r^2\sin\theta}\left[\frac{1}{\sin\theta}\frac{\partial^2\psi}{\partial r^2} + \frac{1}{r^2}\frac{\partial}{\partial\theta}\left(\frac{1}{\sin\theta}\frac{\partial\psi}{\partial\theta}\right)\right]\left(\frac{\partial\psi}{\partial r}\boldsymbol{e}_r + \frac{1}{r}\frac{\partial\psi}{\partial\theta}\boldsymbol{e}_\theta\right) \\
&= \frac{1}{r^2\sin^2\theta}\left[\frac{\partial\psi}{\partial\theta}\frac{\partial}{\partial r}\left(\frac{1}{r^2}\frac{\partial\psi}{\partial\theta}\right) - \frac{\sin\theta}{r^2}\frac{\partial\psi}{\partial r}\frac{\partial}{\partial\theta}\left(\frac{1}{\sin\theta}\frac{\partial\psi}{\partial\theta}\right) - \frac{1}{r}\left(\frac{\partial\psi}{\partial r}\right)^2\right]\boldsymbol{e}_r \\
&\quad + \frac{1}{r^3\sin\theta}\left[\frac{\partial\psi}{\partial r}\frac{\partial}{\partial\theta}\left(\frac{1}{\sin\theta}\frac{\partial\psi}{\partial r}\right) - \frac{1}{\sin\theta}\frac{\partial\psi}{\partial\theta}\frac{\partial^2\psi}{\partial r^2}\right]\boldsymbol{e}_\theta.
\end{aligned}
$$

2.19: (i) Let $C_{ijk}\boldsymbol{e}_i\boldsymbol{e}_j\boldsymbol{e}_k = 0$. We calculate the value of the tensor on the left-hand side of this equality for three vector arguments, $\boldsymbol{e}_l$, $\boldsymbol{e}_m$ and $\boldsymbol{e}_n$. Then

$$0 = C_{ijk}\boldsymbol{e}_i\boldsymbol{e}_j\boldsymbol{e}_k(\boldsymbol{e}_l, \boldsymbol{e}_m, \boldsymbol{e}_n) = C_{ijk}(\boldsymbol{e}_i\cdot\boldsymbol{e}_l)(\boldsymbol{e}_j\cdot\boldsymbol{e}_m)(\boldsymbol{e}_k\cdot\boldsymbol{e}_n) = C_{ijk}\delta_{il}\delta_{jm}\delta_{kn} = C_{lmn}.$$

(ii) Let us take three arbitrary vectors $\boldsymbol{x}$, $\boldsymbol{y}$ and $\boldsymbol{z}$. They can be written as $\boldsymbol{x} = x_i\boldsymbol{e}_i$, $\boldsymbol{y} = y_j\boldsymbol{e}_j$ and $\boldsymbol{z} = z_k\boldsymbol{e}_k$. Then

$$\mathbf{T}(\boldsymbol{x}, \boldsymbol{y}, \boldsymbol{z}) = \mathbf{T}(x_i\boldsymbol{e}_i, y_j\boldsymbol{e}_j, z_k\boldsymbol{e}_k) = x_i y_j z_k\mathbf{T}(\boldsymbol{e}_i, \boldsymbol{e}_j, \boldsymbol{e}_k) = x_i y_j z_k T_{ijk},$$

where $T_{ijk} = \mathbf{T}(\boldsymbol{e}_i, \boldsymbol{e}_j, \boldsymbol{e}_k)$. Now we notice that

$$x_i y_j z_k = (\boldsymbol{e}_i\cdot\boldsymbol{x})(\boldsymbol{e}_j\cdot\boldsymbol{y})(\boldsymbol{e}_k\cdot\boldsymbol{z}) = \boldsymbol{e}_i\boldsymbol{e}_j\boldsymbol{e}_k(\boldsymbol{x}, \boldsymbol{y}, \boldsymbol{z})$$

and consequently

$$\mathbf{T}(\boldsymbol{x}, \boldsymbol{y}, \boldsymbol{z}) = T_{ijk}\boldsymbol{e}_i\boldsymbol{e}_j\boldsymbol{e}_k(\boldsymbol{x}, \boldsymbol{y}, \boldsymbol{z}).$$

Since this relation is valid for any $\boldsymbol{x}$, $\boldsymbol{y}$ and $\boldsymbol{z}$, we conclude that

$$\mathbf{T} = T_{ijk}\boldsymbol{e}_i\boldsymbol{e}_j\boldsymbol{e}_k,$$

which means that the tensors $\boldsymbol{e}_i\boldsymbol{e}_j\boldsymbol{e}_k$, $i = 1, 2, 3$, $j = 1, 2, 3$, $k = 1, 2, 3$, constitute a basis in the linear space of the third-order tensors. Since the basis consists of 27 tensors, the dimension of the linear space of the third-order tensors is 27.

A.3 Solutions for Chap. 3

3.1: The motion is a *stretch* in the x_1-direction (see Fig. A.2), at a rate determined by α.

At $t = 0$, the length of the material line AB is $(b - a)$, and at time t it is $(b - a)(1 + \alpha t) > (b - a)$.

3.2: Let us calculate the Jacobian:

$$J = \frac{D(x_1, x_2, x_3)}{D(\xi_1, \xi_2, \xi_3)} = \det\left(\frac{\partial x_i}{\partial \xi_j}\right) = \begin{vmatrix} 1 & \omega t & 0 \\ -\omega t & 1 & 0 \\ 0 & 0 & 1 \end{vmatrix} = 1 + (\omega t)^2.$$

The velocity and acceleration in the Lagrangian description are given by

$$\boldsymbol{v}(\boldsymbol{\xi}, t) = \frac{D\boldsymbol{x}}{Dt} = \omega(\xi_2, -\xi_1, 0), \quad \boldsymbol{a}(\boldsymbol{\xi}, t) = \frac{D\boldsymbol{v}}{Dt} = 0.$$

To obtain the expression for the velocity in the Eulerian description we first have to express $\boldsymbol{\xi}$ in terms of $\boldsymbol{x}$:

$$\begin{cases} \xi_1 + \xi_2 \omega t = x_1, \\ -\xi_1 \omega t + \xi_2 = x_2. \end{cases}$$

Solving this system of equations we obtain

$$\xi_1 = \frac{1}{J}(x_1 - x_2 \omega t), \quad \xi_2 = \frac{1}{J}(x_1 \omega t + x_2).$$

Then

$$\boldsymbol{v}(\boldsymbol{x}, t) = \boldsymbol{v}(\boldsymbol{\xi}(\boldsymbol{x}, t), t) = \frac{\omega}{J}(x_1 \omega t + x_2, \ -x_1 + x_2 \omega t, \ 0).$$

Using this result we obtain

$$\begin{aligned} \frac{\partial \boldsymbol{v}}{\partial t} &= -\frac{\omega}{J^2}\frac{dJ}{dt}(x_1 \omega t + x_2, \ -x_1 + x_2 \omega t, \ 0) + \frac{\omega^2}{J}(x_1, \ x_2, \ 0) \\ &= \frac{\omega^2}{J^2}\left((1 - \omega^2 t^2)x_1 - 2x_2 \omega t, \ 2x_1 \omega t + (1 - \omega^2 t^2)x_2, \ 0\right), \end{aligned}$$

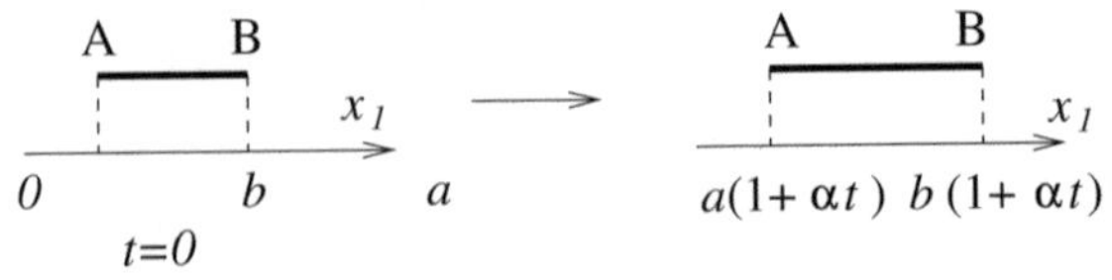

Fig. A.2 Schematic picture of deformation

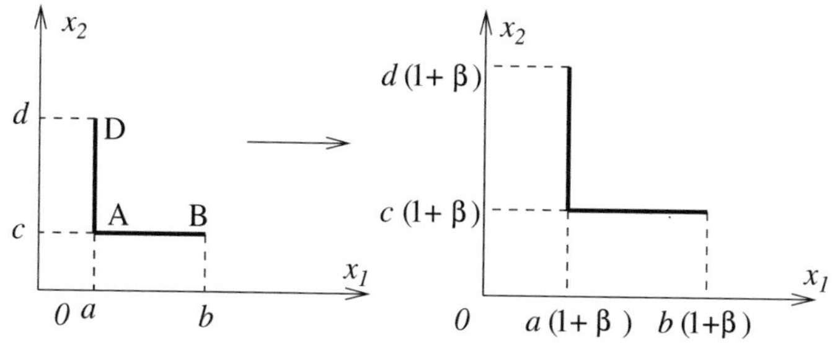

Fig. A.3 Schematic picture of stretching in the x_1- and x_2-direction

$$(\boldsymbol{v}\cdot\nabla)\boldsymbol{v} = \left(v_1\frac{\partial}{\partial x_1} + v_2\frac{\partial}{\partial x_2}\right)\boldsymbol{v} = \frac{\omega}{J}(v_1\omega t + v_2,\ -v_1 + v_2\omega t,\ 0)$$

$$= \frac{\omega^2}{J^2}\left((\omega^2t^2 - 1)x_1 + 2x_2\omega t,\ -2x_1\omega t + (\omega^2t^2 - 1)x_2,\ 0\right).$$

Then we obtain

$$\frac{\partial \boldsymbol{v}}{\partial t} + (\boldsymbol{v}\cdot\nabla)\boldsymbol{v} = 0 = \frac{D\boldsymbol{v}}{Dt}.$$

3.3: (i) When $\kappa = 0$ we have $x_1 = (1+\beta)\xi_1,\ x_2 = (1+\beta)\xi_2,\ x_3 = \xi_3$. This is a stretch in the x_1- and x_2-direction (see Fig. A.3).

At $t = 0$ the length of the material element AB is $(b-a)$, and AD is $(d-c)$. At time t their lengths are $(1+\beta)(b-a) > (b-a)$ and $(1+\beta)(d-c) > (d-c)$ respectively.

(ii)

$$\mathcal{J} = \frac{D(x_1, x_2, x_3)}{D(\xi_1, \xi_2, \xi_3)} = \det\left(\frac{\partial x_i}{\partial \xi_j}\right) = \begin{vmatrix} 1+\beta & \kappa & 0 \\ 0 & 1+\beta & 0 \\ 0 & 0 & 1 \end{vmatrix} = (1+\beta)^2 > 1$$

(recall that $\beta > 0$). $\mathcal{J} > 1$ means that there is an increase in volume.

(iii) Using the relation between the differentials of the initial and current coordinates we obtain

$$dx_1 = \frac{\partial x_1}{\partial \xi_j}\,d\xi_j = (1+\beta)\,d\xi_1 + \kappa\,d\xi_2 = [(1+\beta)\cos\theta + \kappa\sin\theta]\,dl_0$$

$$dx_2 = \frac{\partial x_2}{\partial \xi_j}\,d\xi_j = (1+\beta)\,d\xi_2 = (1+\beta)\,dl_0\sin\theta, \quad dx_3 = 0.$$

Then

$$dl = \sqrt{\boldsymbol{x} \cdot \boldsymbol{x}} = dl_0 \sqrt{[(1+\beta)\cos\theta + \kappa \sin\theta]^2 + (1+\beta)^2 \sin^2\theta}$$

$$= dl_0 \sqrt{(1+\beta)^2 + \kappa(1+\beta)\sin 2\theta + \kappa^2 \sin^2\theta}.$$

For θ acute $\sin\theta > 0$ and $\sin 2\theta > 0$, so that $dl > dl_0$, i.e. the element is stretched.

3.4: (i) Let us calculate the Jacobian:

$$\mathcal{J} = \frac{D(x_1, x_2, x_3)}{D(\xi_1, \xi_2, \xi_3)} = \det\left(\frac{\partial x_i}{\partial \xi_j}\right) = \begin{vmatrix} f' \cos(k\xi_2) & -kf \sin(k\xi_2) & 0 \\ f' \sin(k\xi_2) & kf \cos(k\xi_2) & 0 \\ 0 & 0 & 1 \end{vmatrix}$$

$$= kff'[\cos^2(k\xi_2) + \sin^2(k\xi_2)] = kff',$$

where the prime indicates the derivative. Then, using $\rho\mathcal{J} = \rho_0$, we obtain

$$\rho = \frac{\rho_0}{kf(\xi_1)f'(\xi_1)}.$$

(ii) On the planes initially normal to the ξ_1-axis $\xi_1 = \text{const}$, so that $f(\xi_1)$ is also constant. Hence,

$$x_1^2 + x_2^2 = [f(\xi_1)]^2[\cos^2(k\xi_2) + \sin^2(k\xi_2)] = [f(\xi_1)]^2 = \text{const}.$$

Thus the images of the planes $\xi_1 = \text{const}$ under the mapping $\boldsymbol{\xi} \longrightarrow \boldsymbol{x}$ are portions of surfaces $x_1^2 + x_2^2 = \text{const}$, which are cylinders.

(iii) On a plane initially normal to the ξ_2-axis $\xi_2 = a = \text{const}$. If the point (x_1, x_2, x_3) belongs to such a plane, then

$$\sin(ka)x_1 - \cos(ka)x_2 = 0. \qquad (*)$$

This is the equation of a plane containing the coordinate origin and having the unit normal vector $\boldsymbol{n} = (\sin(ka), -\cos(ka), 0)$. To show that the whole plane Π_t determined by $(*)$ is the image of the plane Π_0 determined by $\xi_2 = a$, we have to show that $\forall\ (x_1, x_2, x_3) \in \Pi_t\ \exists\ (\xi_1, \xi_2, \xi_3) \in \Pi_0$ such that $(\xi_1, \xi_2, \xi_3) \longrightarrow (x_1, x_2, x_3)$. It is straightforward to verify that, for $\cos(ka) \neq 0$, if we take

$$\xi_1 = f^{-1}(x_1/\cos(ka)), \quad \xi_2 = a, \quad \xi_3 = x_3,$$

where f^{-1} is the function inverse to f, then $(\xi_1, \xi_2, \xi_3) \longrightarrow (x_1, x_2, x_3)$. The inverse function f^{-1} exists because f is monotonic.

If $\cos(ka) = 0$, then we take

$$\xi_1 = f^{-1}(x_2/\sin(ka)), \quad \xi_2 = a, \quad \xi_3 = x_3.$$

Note that the mapping $(\xi_1, \xi_2, \xi_3) \longrightarrow (x_1, x_2, x_3)$ is one-to-one, i.e. the images of different points do not coincide. Indeed, assume that $(\xi_1, a, \xi_3) \longrightarrow (x_1, x_2, x_3)$ and $(\xi_1', a, \xi_3') \longrightarrow (x_1', x_2', x_3')$. If $\xi_3' \neq \xi_3$, then $x_3' \neq x_3$ and $(x_1', x_2', x_3') \neq (x_1, x_2, x_3)$. If $\xi_3' = \xi_3$, but $\xi_1' \neq \xi_1$, then $x_1' \neq x_1$ because $f(\xi_1)$ is monotonic, and again $(x_1', x_2', x_3') \neq (x_1, x_2, x_3)$.

(iv) The points that are in the plane $x_3 = 0$ in the current configuration were in the plane $\xi_3 = 0$ in the initial configuration, where they occupied the rectangle $a \leq \xi_1 \leq b, \ -c \leq \xi_2 \leq c$.

The image of the line $\xi_1 = h = \text{const}$ is the circle $x_1^2 + x_2^2 = [f(h)]^2$. Hence, the images of the boundaries $\xi_1 = a, \ -c \leq \xi_2 \leq c$, and $\xi_1 = b, \ -c \leq \xi_2 \leq c$ lie on the circles $x_1^2 + x_2^2 = [f(a)]^2$ and $x_1^2 + x_2^2 = [f(b)]^2$, respectively (note that $f(a) < f(b)$ because $f(\xi_1)$ is monotonically increasing).

The image of the line $\xi_2 = q = \text{const}$ is the line $x_2 = x_1 \tan(kq)$. Hence, the images of the boundaries $a \leq \xi_1 \leq b, \ \xi_2 = -c$ and $a \leq \xi_1 \leq b, \ \xi_2 = c$ lie on the lines $x_2 = -x_1 \tan(kc)$ and $x_2 = x_1 \tan(kc)$, respectively. As a result, we obtain that the image of the rectangle $a \leq \xi_1 \leq b, \ -c \leq \xi_2 \leq c$ has the shape shown in Fig. A.4.

(v) If there is no change in volume, then $\mathcal{J} = 1 \implies kf(\xi_1)f'(\xi_1) = 1$

$$\implies \frac{d}{d\xi_1}\left\{\frac{1}{2}[f(\xi_1)]^2\right\} = \frac{1}{k} \implies [f(\xi_1)]^2 = \frac{2\xi_1}{k} + A \implies f(\xi_1) = \sqrt{\frac{2\xi_1}{k} + A},$$

where A is an arbitrary constant (but such that $2\xi_1/k + A > 0$ in the interval of variation of ξ_1). We have chosen the positive square root because $f(\xi_1) > 0$.

3.5: We take an arbitrary point A in the plane Π_0. Its coordinates are (a, ξ_2, ξ_3), where ξ_2 and ξ_3 are arbitrary. Then $A \longrightarrow \text{B} = (x_1, x_2, x_3)$, where $x_1 = a + \omega^2 t^2 \xi_2$, $x_2 = (1 + \omega^2 t^2)\xi_2$, $x_3 = \xi_3$. Since

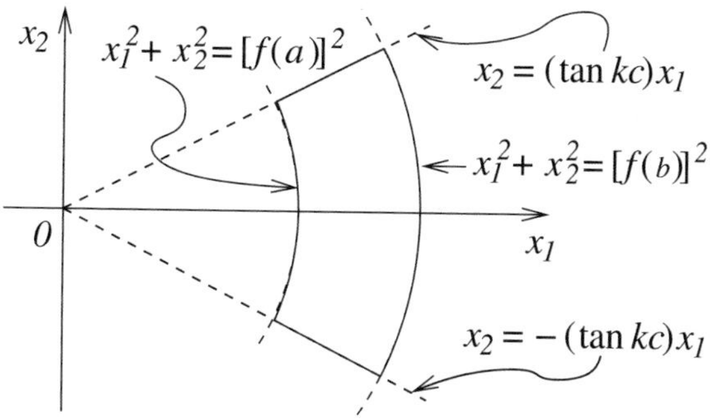

Fig. A.4 Image of the rectangle

$$x_1 = a + \frac{\omega^2 t^2 x_2}{(1+\omega^2 t^2)}, \qquad (*)$$

we conclude that B $\in \Pi_t$. Hence, Π_0 is mapped into Π_t.

Let us now take an arbitrary B $\in \Pi_t$, B $= (x_1, x_2, x_3)$. Since B $\in \Pi_t$, x_1 and x_2 are related by $(*)$. Then it is straightforward to verify that the point A $= (a, \xi_2, \xi_3)$ with $\xi_2 = x_2(1+\omega^2 t^2)^{-1}$, $\xi_3 = x_3$ is mapped to B. Hence, each point in Π_t is the image of a point in Π_0, which means that Π_0 is mapped onto Π_t. It is easy to show that this mapping is one-to-one.

The unit normal vector to the plane Π_0 is $\boldsymbol{e}_1$, and the normal (not unit) vector to Π_t is $\boldsymbol{n} = (1,\ -\omega^2 t^2(1+\omega^2 t^2)^{-1},\ 0)$. Let θ be the acute angle between Π_0 and Π_t. Then

$$\cos\theta = \frac{|\boldsymbol{n}\cdot\boldsymbol{e}_1|}{\|\boldsymbol{n}\|} = \frac{1}{\sqrt{1+\omega^4 t^4(1+\omega^2 t^2)^{-2}}} = \frac{1+\omega^2 t^2}{\sqrt{\omega^4 t^4 + (1+\omega^2 t^2)^2}}.$$

Then $\lim_{t\to\infty}\cos\theta = 1/\sqrt{2}$ and $\lim_{t\to\infty}\theta = \pi/4$.

3.6: (i) The equations of streamlines are:

$$\frac{dx_1}{d\lambda} = \frac{-x_2(x_1^2+x_2^2)}{T(a^2+x_1^2+x_2^2)} + \frac{t x_1(a^2+x_1^2+x_2^2)}{a^2(t^2+T^2)},$$
$$\frac{dx_2}{d\lambda} = \frac{x_1(x_1^2+x_2^2)}{T(a^2+x_1^2+x_2^2)} + \frac{t x_2(a^2+x_1^2+x_2^2)}{a^2(t^2+T^2)}.$$

Using the substitution $x_1 = r\cos\phi,\ x_2 = r\sin\phi$ we reduce this system of equations to

$$\frac{dr}{d\lambda} = \frac{tr(a^2+r^2)}{a^2(t^2+T^2)}, \quad \frac{d\phi}{d\lambda} = \frac{r^2}{T(a^2+r^2)}. \qquad (*)$$

(ii) Separating variables in the first equation $(*)$ we obtain

$$\frac{a^2\,dr}{r(a^2+r^2)} = \frac{t\,d\lambda}{t^2+T^2}.$$

Now

$$\int\frac{a^2\,dr}{r(a^2+r^2)} = \frac{a^2}{2}\int\frac{dr^2}{r^2(a^2+r^2)} = \frac{1}{2}\int\left(\frac{1}{r^2} - \frac{1}{a^2+r^2}\right)dr^2 = \frac{1}{2}\ln\frac{r^2}{a^2+r^2} + C_1,$$

where C_1 is an arbitrary constant. Hence, we obtain

$$\frac{1}{2}\ln\frac{r^2}{a^2+r^2} + C_1 = \frac{t\lambda}{t^2+T^2} \quad\Longrightarrow\quad \frac{r^2}{a^2+r^2} = \widetilde{C}_1\exp\left(\frac{2t\lambda}{t^2+T^2}\right) \quad\Longrightarrow$$

$$\frac{a^2+r^2}{r^2} = \frac{1}{\widetilde{C}_1}\exp\left(-\frac{2t\lambda}{t^2+T^2}\right) \implies r^2 = \frac{a^2}{C_2\exp\left(-\frac{2t\lambda}{t^2+T^2}\right)-1},$$

where $\widetilde{C}_1$ is an arbitrary positive constant and $C_2 = 1/\widetilde{C}_1$. Substituting this result into the second equation (∗) we obtain

$$\frac{d\phi}{d\lambda} = \frac{1}{TC_2}\exp\left(\frac{2t\lambda}{t^2+T^2}\right),$$

so that

$$\phi = \frac{t^2+T^2}{2tTC_2}\exp\left(\frac{2t\lambda}{t^2+T^2}\right) + C_3,$$

where C_3 is an arbitrary constant. Using this equation to express λ in terms of ϕ and substituting the result into the expression for r^2 we obtain

$$r^2 = \frac{a^2}{\dfrac{t^2+T^2}{2tT(\phi - C_3)} - 1}. \qquad (\dagger)$$

Using the condition that $r = r_0$ at $\phi = \phi_0$ we obtain

$$r_0^2 = \frac{a^2}{\dfrac{t^2+T^2}{2tT(\phi_0 - C_3)} - 1} \implies C_3 = \phi_0 - \frac{r_0^2(t^2+T^2)}{2tT(a^2+r_0^2)}.$$

Substituting C_3 into (†) we eventually arrive at

$$r = a\sqrt{\frac{2tT(a^2+r_0^2)(\phi-\phi_0) + r_0^2(t^2+T^2)}{a^2(t^2+T^2) - 2tT(a^2+r_0^2)(\phi-\phi_0)}}.$$

(iii) The equations of the trajectories are

$$\frac{dx_1}{dt} = \frac{-x_2(x_1^2+x_2^2)}{T(a^2+x_1^2+x_2^2)} + \frac{tx_1(a^2+x_1^2+x_2^2)}{a^2(t^2+T^2)},$$

$$\frac{dx_2}{dt} = \frac{x_1(x_1^2+x_2^2)}{T(a^2+x_1^2+x_2^2)} + \frac{tx_2(a^2+x_1^2+x_2^2)}{a^2(t^2+T^2)}.$$

Using the substitution $x_1 = r\cos\phi,\ x_2 = r\sin\phi$ we reduce this system of equations to

$$\frac{dr}{dt} = \frac{tr(a^2+r^2)}{a^2(t^2+T^2)}, \quad \frac{d\phi}{dt} = \frac{r^2}{T(a^2+r^2)}. \qquad (\ddagger)$$

(iv) Separating variables in the first equation (‡) we obtain

$$\frac{a^2\,dr}{r(a^2+r^2)} = \frac{t\,dt}{t^2+T^2}.$$

Integration of this equation yields

$$\frac{1}{2}\ln\frac{r^2}{a^2+r^2} = \frac{1}{2}\ln(t^2+T^2) + C_1, \qquad (*\dagger)$$

where C_1 is an arbitrary constant. Hence, we obtain

$$r^2 = \frac{C_2 a^2(t^2+T^2)}{1-C_2(t^2+T^2)}, \qquad (*\ddagger)$$

where C_2 is an arbitrary positive constant. Substituting these results into the second equation (‡) we obtain

$$\frac{d\phi}{dt} = \frac{C_2(t^2+T^2)}{T} \quad\Longrightarrow\quad \phi = \frac{C_2 t}{3T}(t^2+3T^2) + C_3, \qquad (\dagger\dagger)$$

where C_3 is an arbitrary constant. Using the condition that $r = r_0$ and $\phi = \phi_0$ at $t = t_0$ we obtain

$$r_0^2 = \frac{C_2 a^2(t_0^2+T^2)}{1-C_2(t_0^2+T^2)}, \quad \phi_0 = \frac{C_2 t_0}{3T}(t_0^2+3T^2) + C_3 \quad\Longrightarrow$$

$$C_2 = \frac{r_0^2}{(a^2+r_0^2)(t_0^2+T^2)}, \quad C_3 = \phi_0 - \frac{r_0^2 t_0(t_0^2+3T^2)}{3T(a^2+r_0^2)(t_0^2+T^2)}.$$

Substituting the expressions for C_2 and C_3 into $(*\dagger)$ and $(\dagger\dagger)$ we obtain the equations of the trajectory in parametric form with t as a parameter:

$$r = ar_0\sqrt{\frac{t^2+T^2}{a^2(t_0^2+T^2)+r_0^2(t_0^2-t^2)}}, \quad \phi = \phi_0 + \frac{r_0^2[t^3-t_0^3+3T^2(t-t_0)]}{3T(a^2+r_0^2)(t_0^2+T^2)}.$$

We note that $r \to \infty$ as $t \to \sqrt{t_0^2 + (a/r_0)^2(t_0^2+T^2)}$.

(v) Our starting point is equation $(*\dagger)$. Substituting $r = a$ and $t = 0$ into this equation, we obtain

$$-\ln 2 = 2\ln T + 2C_1,$$

so that $2C_1 = -\ln(2T^2)$. Equation ($*\dagger$) can be rewritten as

$$\frac{r^2}{a^2 + r^2} = \frac{t^2 + T^2}{2T^2}.$$

Now the moment of time when $r = 2a$ is determined by equation

$$\frac{t^2 + T^2}{2T^2} = \frac{4}{5} \quad \Longrightarrow \quad 5t^2 = 3T^2 \quad \Longrightarrow \quad t = T\sqrt{\frac{3}{5}}.$$

3.7: The differential equations determining the streamlines are

$$\frac{dx}{d\lambda} = v_0 \sin(\omega t)\cos(kx)\cosh[k(y+h)], \quad \frac{dy}{d\lambda} = v_0 \sin(\omega t)\sin(kx)\sinh[k(y+h)].$$

Recall that t here is considered as a parameter. Dividing the second equacion by the first one we obtain

$$\frac{dy}{dx} = \tan(kx)\tanh[k(y+h)].$$

Separating the variables we obtain

$$\frac{\cosh[k(y+h)]\,dy}{\sinh[k(y+h)]} = \frac{\sin(kx)\,dx}{\cos(kx)}.$$

Integrating this equation we obtain

$$\ln\sinh[k(y+h)] = -\ln|\cos(kx)| + \text{const}.$$

This equation can be rewritten as

$$\cos(kx)\sinh[k(y+h)] = C,$$

where C is a constant.

3.8: (i) If $X(t)$ is the radius of the sphere at time t then

$$\frac{dX}{dt} = \frac{a^2 V}{a^2 + t^2}.$$

Integrating this equation and using the initial condition, $X(0) = R$, we obtain

$$X = aV\arctan(t/a) + R.$$

(ii) Since v is proportional to r, it follows that

$$v = \frac{a^2 V r}{(a^2 + t^2) X(t)}. \qquad (*)$$

(iii) We use the equation of mass conservation in spherical coordinates when the motion is spherically symmetric:

$$\frac{\partial \rho}{\partial t} + \frac{1}{r^2} \frac{\partial (r^2 \rho v)}{\partial r} = 0.$$

Substituting $\rho = r^2 f(t)$ and $(*)$ into this equation we obtain

$$\frac{df}{dt} = -\frac{5a^2 V f}{(a^2 + t^2)[aV \arctan(t/a) + R]}.$$

Integrating this equation and using the condition that $f = \rho_0/R^2$ at $t = 0$ yields

$$\ln f = -5 \ln[\arctan(t/a) + R/aV] + \ln(\rho_0/R^2) + 5 \ln(R/aV).$$

It follows from this equation that

$$f = \frac{\rho_0 R^3}{[aV \arctan(t/a) + R]^5}.$$

Then we obtain

$$\rho \to \frac{\rho_0 r^2 R^3}{(\pi a V/2 + R)^5} \quad \text{as} \quad t \to \infty.$$

A.4 Solutions for Chap. 4

4.1: Consider an infinitesimal cube with the sides parallel to coordinate axes and the side length equal to l (see Fig. A.5).

Using the relations

$$\boldsymbol{t}(x_1, x_2, x_3, -\boldsymbol{e}_j) = -\boldsymbol{t}(x_1, x_2, x_3, \boldsymbol{e}_j), \quad j = 1, 2, 3,$$

we obtain

$$\begin{aligned} t_i(x_1, x_2, x_3, -\boldsymbol{e}_1) &= -T_{i1}(x_1, x_2, x_3), \\ t_i(x_1 + l, x_2, x_3, \boldsymbol{e}_1) &= T_{i1}(x_1 + l, x_2, x_3), \\ t_i(x_1, x_2, x_3, -\boldsymbol{e}_2) &= -T_{i2}(x_1, x_2, x_3), \end{aligned}$$

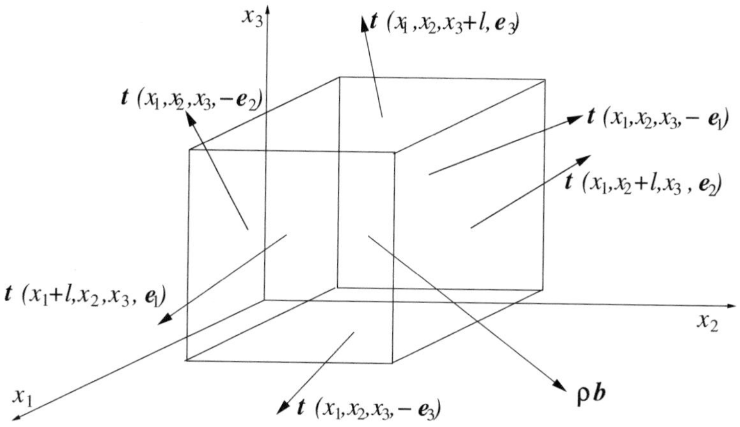

Fig. A.5 Illustration of the derivation of the equilibrium equations

$$
\begin{aligned}
t_i(x_1, x_2 + l, x_3, \boldsymbol{e}_2) &= T_{i2}(x_1, x_2 + l, x_3), \\
t_i(x_1, x_2, x_3, -\boldsymbol{e}_3) &= -T_{i3}(x_1, x_2, x_3), \\
t_i(x_1, x_2, x_3 + l, \boldsymbol{e}_3) &= T_{i3}(x_1, x_2, x_3 + l).
\end{aligned}
$$

Taking the projection of forces on the coordinate axes we obtain

$$
\begin{aligned}
[T_{i1}(x_1 + l, x_2, x_3) - T_{i1}(x_1, x_2, x_3) + T_{i2}(x_1, x_2 + l, x_3) - T_{i2}(x_1, x_2, x_3) \\
+ T_{i3}(x_1, x_2, x_3 + l) - T_{i3}(x_1, x_2, x_3)]l^2 + \rho b_i l^3 = 0.
\end{aligned}
$$

Using the relations

$$
\begin{aligned}
\lim_{l \to 0} \frac{T_{i1}(x_1 + l, x_2, x_3) - T_{i1}(x_1, x_2, x_3)}{l} &= \frac{\partial T_{i1}}{\partial x_1}, \\
\lim_{l \to 0} \frac{T_{i2}(x_1, x + l_2, x_3) - T_{i2}(x_1, x_2, x_3)}{l} &= \frac{\partial T_{i2}}{\partial x_2}, \\
\lim_{l \to 0} \frac{T_{i3}(x_1, x_2, x_3 + l) - T_{i3}(x_1, x_2, x_3)}{l} &= \frac{\partial T_{i3}}{\partial x_3},
\end{aligned}
$$

we reduce this equation to

$$
\frac{\partial T_{i1}}{\partial x_1} + \frac{\partial T_{i2}}{\partial x_2} + \frac{\partial T_{i3}}{\partial x_3} + \rho b_i = 0,
$$

which can be rewritten as

$$
\frac{\partial T_{ij}}{\partial x_j} + \rho b_i = 0.
$$

4.2: (i) The normal unit vector to the plane normal to the x_1-axis is $\boldsymbol{e}_1$, so that the stress vector (surface traction) on this plane is

$$\boldsymbol{t} = \mathbf{T} \cdot \boldsymbol{e}_1 = \begin{pmatrix} 3 & 1 & 1 \\ 1 & 0 & 2 \\ 1 & 2 & 0 \end{pmatrix} \begin{pmatrix} 1 \\ 0 \\ 0 \end{pmatrix} = \begin{pmatrix} 3 \\ 1 \\ 1 \end{pmatrix},$$

which can be rewritten as $\boldsymbol{t} = 3\boldsymbol{e}_1 + \boldsymbol{e}_2 + \boldsymbol{e}_3$.

(ii) The normal vector to the plane with normal in the direction of the vector $\boldsymbol{a} = (0, 1, 1)$ is

$$\boldsymbol{n} = \frac{\boldsymbol{a}}{\|\boldsymbol{a}\|} = \left(0, \frac{1}{\sqrt{2}}, \frac{1}{\sqrt{2}}\right).$$

Then the stress on this plane is

$$\boldsymbol{t} = \mathbf{T} \cdot \boldsymbol{n} = \begin{pmatrix} 3 & 1 & 1 \\ 1 & 0 & 2 \\ 1 & 2 & 0 \end{pmatrix} \begin{pmatrix} 0 \\ \frac{1}{\sqrt{2}} \\ \frac{1}{\sqrt{2}} \end{pmatrix} = \begin{pmatrix} \sqrt{2} \\ \sqrt{2} \\ \sqrt{2} \end{pmatrix}.$$

For the normal stress we obtain $t_n = \boldsymbol{t} \cdot \boldsymbol{n} = 2$, and the shear component is given by $\boldsymbol{t}_\tau = \boldsymbol{t} - t_n \boldsymbol{n} = (\sqrt{2}, 0, 0)$.

(iii) A unit vector $\boldsymbol{n}$ is in the principal direction of the stress tensor $\mathbf{T}$ if $\boldsymbol{t}(\boldsymbol{n}) \| \boldsymbol{n}$. Since $\boldsymbol{t} = \mathbf{T} \cdot \boldsymbol{n}$, the equation defining the principal directions and stresses is

$$\mathbf{T} \cdot \boldsymbol{n} = T\boldsymbol{n}. \qquad (*)$$

Equation $(*)$ can be written in the matrix form as

$$(\mathbf{T} - T\mathbf{I}) \cdot \boldsymbol{n} = \begin{pmatrix} 3 - T & 1 & 1 \\ 1 & -T & 2 \\ 1 & 2 & -T \end{pmatrix} \begin{pmatrix} n_1 \\ n_2 \\ n_3 \end{pmatrix} = 0.$$

The components of the unit vector $\boldsymbol{n}$ which is in the principal direction are defined by the system of equations

$$\begin{cases} (3 - T)n_1 + n_2 + n_3 = 0, \\ n_1 - Tn_2 + 2n_3 = 0, \\ n_1 + 2n_2 - Tn_3 = 0, \end{cases} \qquad (\dagger)$$

and the condition $n_1^2 + n_2^2 + n_3^2 = 1$. The system (†) has a non-trivial solution only when its determinant is zero. Calculating this determinant we obtain

$$\begin{vmatrix} 3-T & 1 & 1 \\ 1 & -T & 2 \\ 1 & 2 & -T \end{vmatrix} = -(T^3 - 3T^2 - 6T + 8),$$

so that the principal stresses are the roots of the equation

$$T^3 - 3T^2 - 6T + 8 = 0.$$

Firstly we note that $T = 1$ is the root of this equation. Keeping this in mind we rewrite it as

$$T^3 - T^2 - 2T^2 + 2T - 8T + 8 = (T-1)(T^2 - 2T - 8) = 0.$$

Then the two other roots are the roots of the quadratic $T^2 - 2T - 8 = 0$, and they are equal to -2 and 4. Hence, the principal stresses are $-2,\ 1,\ 4$.

For $T = -2$ equation (†) takes the form

$$\begin{cases} 5n_1 + n_2 + n_3 = 0, \\ n_1 + 2n_2 + 2n_3 = 0, \\ n_1 + 2n_2 + 2n_3 = 0, \end{cases}$$

so that $n_1 = 0$ and $n_3 = -n_2$. Then $\left(1+1\right)n_2^2 = 1$, $n_2 = 1/\sqrt{2}$ and

$$\boldsymbol{n}_1 = \frac{(0,\, 1,\, -1)}{\sqrt{2}}.$$

For $T = 1$ equation (†) takes the form

$$\begin{cases} 2n_1 + n_2 + n_3 = 0, \\ n_1 - n_2 + 2n_3 = 0, \\ n_1 + 2n_2 - n_3 = 0. \end{cases}$$

It follows from the first equation that $n_3 = -2n_1 - n_2$. Substituting this into the second and third equation we obtain

$$\begin{cases} 3n_1 + 3n_2 = 0, \\ 3n_1 + 3n_2 = 0, \end{cases}$$

so that $n_2 = -n_1$ and $n_3 = -n_1 \Longrightarrow$

$$\boldsymbol{n}_2 = \frac{(1, -1, -1)}{\sqrt{3}}.$$

For $T = 4$ equation (†) takes the form

$$\begin{cases} -n_1 + n_2 + n_3 = 0, \\ n_1 - 4n_2 + 2n_3 = 0, \\ n_1 + 2n_2 - 4n_3 = 0. \end{cases}$$

It follows from the first equation that $n_3 = n_1 - n_2$. Substituting this into the second and third equation we obtain

$$\begin{cases} 3n_1 - 6n_2 = 0, \\ 3n_1 - 6n_2 = 0, \end{cases}$$

so that $n_1 = 2n_2$ and $n_3 = n_2 \Longrightarrow$

$$\boldsymbol{n}_3 = \frac{(2, 1, 1)}{\sqrt{6}}.$$

(iv) We obtain for the surface stress

$$\boldsymbol{t} = \mathbf{T} \cdot \boldsymbol{n} = \begin{pmatrix} 3 & 1 & 1 \\ 1 & 0 & 2 \\ 1 & 2 & 0 \end{pmatrix} \begin{pmatrix} 0 \\ n_2 \\ n_3 \end{pmatrix} = \begin{pmatrix} n_2 + n_3 \\ 2n_3 \\ 2n_2 \end{pmatrix}.$$

The condition $\boldsymbol{t} \perp \boldsymbol{n}$ means that

$$0 = \boldsymbol{t} \cdot \boldsymbol{n} = 2n_2n_3 + 2n_2n_3 = 4n_2n_3,$$

so that either $n_2 = 0$ or $n_3 = 0$. Hence, we have four possible vectors: $\boldsymbol{n}_{1,2} = (0, \pm 1, 0)$ and $\boldsymbol{n}_{3,4} = (0, 0, \pm 1)$.

4.3: (i) We need to show that $\partial T_{ij}/\partial x_j = 0$ for $i = 1, 2, 3$.

$$\begin{aligned} \frac{\partial T_{1j}}{\partial x_j} &= \frac{\partial T_{11}}{\partial x_1} + \frac{\partial T_{12}}{\partial x_2} + \frac{\partial T_{13}}{\partial x_3} \\ &= -\frac{\pi m^2 W}{2L} \sin\frac{\pi x_1}{2L} \sinh(mx_2) + \frac{\pi m^2 W}{2L} \sin\frac{\pi x_1}{2L} \sinh(mx_2) = 0, \end{aligned}$$

$$\begin{aligned}\frac{\partial T_{2j}}{\partial x_j} &= \frac{\partial T_{21}}{\partial x_1} + \frac{\partial T_{22}}{\partial x_2} + \frac{\partial T_{23}}{\partial x_3} \\ &= \frac{\pi^2 m W}{4L^2} \cos \frac{\pi x_1}{2L} \cosh(m x_2) - \frac{\pi^2 m W}{4L^2} \cos \frac{\pi x_1}{2L} \cosh(m x_2) = 0, \\ \frac{\partial T_{3j}}{\partial x_j} &= \frac{\partial T_{31}}{\partial x_1} + \frac{\partial T_{32}}{\partial x_2} + \frac{\partial T_{33}}{\partial x_3} = 0,\end{aligned}$$

$\Longrightarrow$ the plate is in equilibrium with no body force.

(ii) The outward unit normal on the edge $x_2 = h$ is $\boldsymbol{e}_2$, and the stress vector on $x_2 = h$ is

$$\begin{aligned}\boldsymbol{t}(\boldsymbol{e}_2) =& \mathbf{T} \cdot \boldsymbol{e}_2 = T_{ij} \boldsymbol{e}_i (\boldsymbol{e}_j \cdot \boldsymbol{e}_2) = T_{ij} \boldsymbol{e}_i \delta_{j2} = T_{12} \boldsymbol{e}_1 + T_{22} \boldsymbol{e}_2 \\ =& \frac{\pi m W}{2L} \sin\left(\frac{\pi x_1}{2L}\right) \cosh(mh) \boldsymbol{e}_1 - \frac{\pi^2 W}{4L^2} \cos\left(\frac{\pi x_1}{2L}\right) \sinh(mh) \boldsymbol{e}_2,\end{aligned}$$

where T_{12} and T_{22} have been evaluated at $x_2 = h$.

The outward unit normal on the edge $x_1 = -L$ is $-\boldsymbol{e}_1$, and the stress vector on $x_1 = -L$ is

$$\begin{aligned}\boldsymbol{t}(-\boldsymbol{e}_1) =& - \boldsymbol{t}(\boldsymbol{e}_1) = -\mathbf{T} \cdot \boldsymbol{e}_1 = -T_{ij} \boldsymbol{e}_i (\boldsymbol{e}_j \cdot \boldsymbol{e}_1) = -T_{ij} \boldsymbol{e}_i \delta_{j1} \\ =& - T_{11} \boldsymbol{e}_1 - T_{21} \boldsymbol{e}_2 = \frac{\pi m W}{2L} \cosh(m x_2) \boldsymbol{e}_2,\end{aligned}$$

where T_{12} and T_{21} have been evaluated at $x_1 = -L$.

4.4: (i) Using Eq. (4.4.5) we obtain

$$\frac{2 f a x_1}{l^4} + \rho_0 b_1 = 0, \quad \frac{4 f x_2}{l^4} + \rho_0 b_2 = 0, \quad \frac{4 f x_3}{l^4} + \rho_0 b_3 = 0.$$

Then it follows that

$$\boldsymbol{b} = \frac{2f}{\rho_0 l^4} \exp\left[\left(x_1^2 + x_2^2 + x_3^2\right) / l^2\right] (a x_1, 2 x_2, 2 x_3).$$

(ii) A simple calculation gives

$$\nabla \times \boldsymbol{b} = \frac{4f}{\rho_0 l^6} \exp\left[\left(x_1^2 + x_2^2 + x_3^2\right) / l^2\right] (0, 2 x_1 x_3 - a x_1 x_3, a x_1 x_2 - 2 x_1 x_2) = 0.$$

It follows from this equation that $a = 2$.

4.5: Since $\boldsymbol{v} = 0$ the momentum equation takes the form

$$\nabla \cdot \mathbf{T} + \rho \boldsymbol{b} = 0.$$

Using $\nabla \cdot \mathbf{T} = -\nabla p$ we rewrite this equation as

$$\nabla p = \rho \boldsymbol{b}.$$

Using the expression for $\boldsymbol{b}$ and the relation between p and ρ we obtain three scalar equations for p from this vector equation,

$$\frac{\partial p}{\partial r} = -\frac{g p R^2}{\alpha r^2}, \quad \frac{\partial p}{\partial \theta} = 0, \quad \frac{\partial p}{\partial \varphi} = 0.$$

It follows from the second and third equations that p only depends on r. Then, using the variable separation to integrate the first equation, and taking the condition $p = p_0$ at $r = R$ into account, we finally obtain

$$\ln p = \frac{g R^2}{\alpha}\left(\frac{1}{r} - \frac{1}{R}\right) + \ln p_0.$$

This equation can be transformed to

$$p = p_0 \exp\left(\frac{g R^2}{\alpha}\left(\frac{1}{r} - \frac{1}{R}\right)\right).$$

4.6: We have for the components of the body force

$$\rho b_1 = 0, \quad \rho b_2 = \rho_e E, \quad \rho b_3 = -\rho g.$$

Then the equilibrium equations are written as

$$\frac{\partial T_1}{\partial x_1} = 0, \qquad (*)$$

$$\frac{\partial T_1}{\partial x_2} + \frac{\partial T_3}{\partial x_3} + \rho_e E = 0, \qquad (\dagger)$$

$$\frac{\partial T_3}{\partial x_2} + \frac{\partial T_2}{\partial x_3} - \rho g = 0. \qquad (\ddagger)$$

Equation $(*)$ is satisfied automatically. It follows from $(\dagger)$ that

$$\frac{\partial T_3}{\partial x_3} = -2q x_2 x_3 - \rho_e E.$$

Integrating this equation and taking into account that $T_3 = 0$ at $x_3 = 0$ we obtain

$$T_3 = -q x_2 x_3^2 - \rho_e E x_3.$$

Finally, it follows from (‡) that

$$\frac{\partial T_2}{\partial x_3} = q x_3^2 + \rho g.$$

Integrating this equation and taking into account that $T_2 = 0$ at $x_3 = 0$ we obtain

$$T_2 = \frac{q}{3}\, x_3^3 + \rho g x_3.$$

Summarising,

$$T_2 = \frac{q}{3}\, x_3^3 + \rho g x_3, \quad T_3 = -q x_2 x_3^2 - \rho_e E x_3.$$

A.5 Solutions for Chap. 5

5.1: Let the contour C enclose the cylinder and the contour C_1 be the cylinder boundary. We consider the contour $\widetilde{C}$ that consists of the contour C traversed in the positive (that is counterclockwise) direction, the contour C_1 traversed in the negative direction, and two coinciding straight lines connecting points on each of the two contours (see Fig. A.6). Since $\boldsymbol{v}$ is potential, it follows that $\oint_{\widetilde{C}} \boldsymbol{v} \cdot d\boldsymbol{l} = 0$. The integrals along the straight lines cancel each other. The integral over the contour C_1 traversed in the negative direction is equal to minus the integral over the contour C_1 traversed in the positive direction. As a result we have $\oint_C \boldsymbol{v} \cdot d\boldsymbol{l} = \oint_{C_1} \boldsymbol{v} \cdot d\boldsymbol{l}$.

5.2: The y-component of the total force acting on the cylinder is

$$\begin{aligned} f_y &= -a \int_0^{2\pi} p \sin\phi \, d\phi = -a \int_0^{2\pi} \left[\left(p_0 - \frac{\rho}{2}\, V^2\right) \sin\phi + \rho V^2 \sin\phi \cos 2\phi\right] d\phi \\ &= -a \int_0^{2\pi} \left[\left(p_0 - \rho V^2\right) \sin\phi + \frac{\rho}{2}\, V^2 \sin 3\phi\right] d\phi \\ &= a \left(p_0 - \rho V^2\right) \cos\phi \Big|_0^{2\pi} + \frac{\rho}{6}\, V^2 \cos 3\phi \Big|_0^{2\pi} = 0. \end{aligned}$$

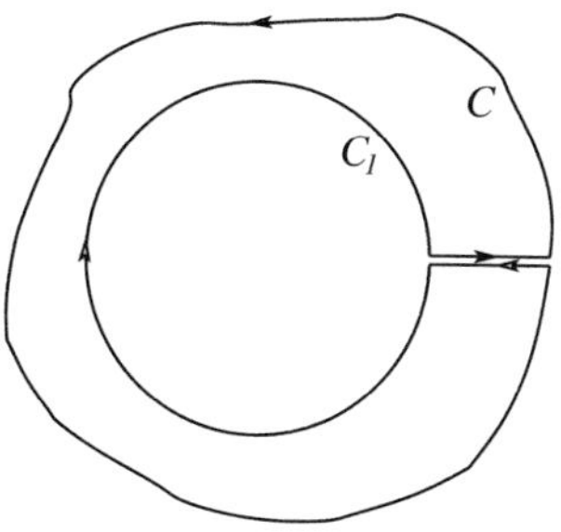

Fig. A.6 The contour $\widetilde{C}$. The arrows show the direction of traversing the contour

5.3: Let us substitute $\boldsymbol{v} = \nabla\Phi$ into Euler's equation written in the Gromeka–Lamb form. Taking into account that $\nabla \times \nabla\Phi = 0$, we obtain

$$\nabla\left(\rho\frac{\partial\Phi}{\partial t} + p + \frac{\rho}{2}\|\mathbf{v}\|^2 + \rho\varphi\right) = 0.$$

It follows from this equation that the expression in brackets is independent of the spatial variables. However, it still can depend on time. As a result, we obtain

$$\frac{\partial\Phi}{\partial t} + \frac{p}{\rho} + \frac{1}{2}\|\nabla\Phi\|^2 + \varphi = C(t).$$

5.4: Since the bathysphere's diameter is much smaller than the depth at which it is immersed we can neglect the pressure variation on the sphere's surface and take it to be the same at any point of the bathysphere's surface. The surface of the bathysphere is $S = 4\pi r^2 \approx 12.57\,\mathrm{m}^2$. Using Eq. (5.2.4) we obtain that the water pressure at the bathysphere is $p \approx 201p_a \approx 2.01 \times 10^7\,\mathrm{Nm}^{-2}$. The pressure force imposed on the bathysphere is approximately equal to $201Sp_a \approx 2.53 \times 10^8\,\mathrm{N}$.

5.5: (ia) In Fig. A.7 a vertical cross-section of the tank by a plane parallel to one of its sides is shown. The curved solid lines are the streamlines, and the curved dashed lines show the water jet leaking from the tank.

Let us introduce the z-axis in the vertical direction with the origin at the tank bottom. The potential of the gravity field is $\varphi = gz$ (up to an arbitrary additive constant). The pressure at the beginning of a streamline, which is at the water surface, is equal to the atmospheric pressure p_a, and it is the same at the end of this streamline, which is at the hole. The water speed at the beginning of the streamline is zero, and it is v at the end of the streamline. Then Bernoulli's integral gives

$$\frac{1}{2}v^2 + \frac{p_a}{\rho} = gz + \frac{p_a}{\rho},$$

where z is the water depth in the tank. This equation reduces to

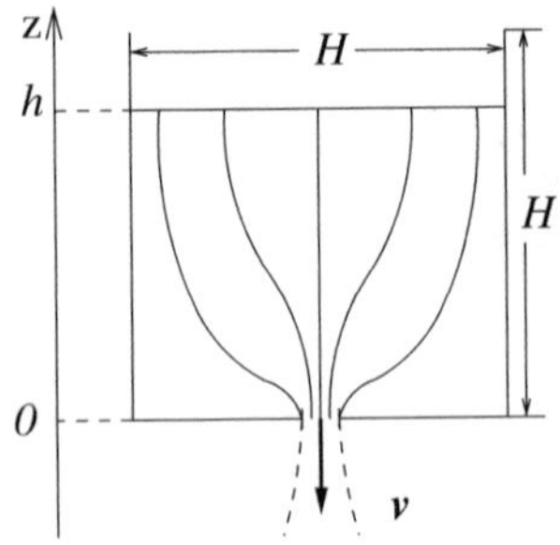

Fig. A.7 Sketch of the tank and streamlines

$$v^2 = 2gh,$$

where $h = z$ is the water depth in the tank.

(ib) The pressure at the water surface is equal to the atmospheric pressure p_a, and it is the same at the hole. The water speed at the surface is zero, and it is v at the hole. Then the Lagrange–Cauchy integral gives

$$\frac{1}{2}v^2 + \frac{p_a}{\rho} = gz + \frac{p_a}{\rho},$$

which leads to

$$v^2 = 2gh.$$

(ii) The amount of water that leaks from the tank in one second is $\pi r^2 v$, and the amount of water in the tank is hH^2. Hence,

$$\frac{d(hH^2)}{dt} = -\pi r^2 v = -\pi r^2 \sqrt{2gh}.$$

Separating the variables we obtain

$$\frac{dh}{\sqrt{h}} = -\frac{\pi r^2 \sqrt{2g}}{H^2}\, dt.$$

Integration yields

$$2\sqrt{h} = C - \frac{\pi r^2 t \sqrt{2g}}{H^2}.$$

Using the initial condition $h = H$ at $t = 0$ we obtain $C = 2\sqrt{H}$, so that

$$h = \left(\sqrt{H} - \frac{\pi r^2 t \sqrt{2g}}{2H^2}\right)^2.$$

The time when $h = H/2$ is given by the equation

$$\left(\sqrt{H} - \frac{\pi r^2 t \sqrt{2g}}{2H^2}\right)^2 = \frac{H}{2},$$

which results in

$$t = \frac{H^2\sqrt{H}(\sqrt{2}-1)}{\pi r^2 \sqrt{g}} \approx 420\ \text{s} = 7\ \text{min}.$$

5.6: Consider an arbitrary body immersed in water see (Fig. A.8). The pressure force acting on the body is given by

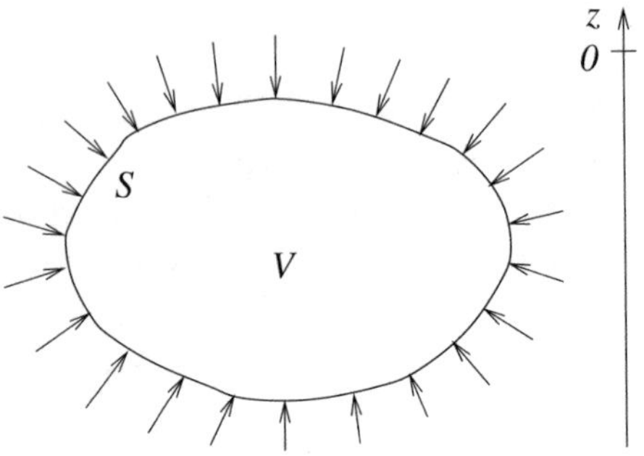

Fig. A.8 Body immersed in water. The arrows show the pressure force

$$F = -\int_S p\boldsymbol{n}\,dS = -\boldsymbol{e}_i \int_S p n_i\,dS.$$

Note that $pn_i = (p\boldsymbol{e}_i)\cdot\boldsymbol{n}$, so that

$$\int_S pn_i\,dS = \int_S (p\boldsymbol{e}_i)\cdot\boldsymbol{n}\,dS = \int_S (p\boldsymbol{e}_i)\cdot d\boldsymbol{S},$$

where $d\boldsymbol{S} = \boldsymbol{n}\,dS$. Using Gauss' divergence theorem we obtain

$$\int_S (p\boldsymbol{e}_i)\cdot d\boldsymbol{S} = \int_V \nabla\cdot(p\boldsymbol{e}_i)\,dV. = \int_V \frac{\partial p}{\partial x_i}\,dV$$

Hence,

$$\boldsymbol{F} = -\boldsymbol{e}_i \int_V \frac{\partial p}{\partial x_i}\,dV = -\int_V \nabla p\,dV.$$

It follows from the equilibrium equation that

$$\nabla p = \rho\boldsymbol{g} = -\rho g\boldsymbol{e}_z,$$

where $\boldsymbol{e}_z$ is the unit vector of the z-axis directed upward. Then

$$\boldsymbol{F} = \rho g\boldsymbol{e}_z \int_V dV = \rho V g\boldsymbol{e}_z = gM\boldsymbol{e}_z,$$

where $M = \rho V$ is the mass of water displaced by the body, and gM is its weight.

5.7: A bathysphere is in equilibrium under action of three forces: Archimedes' force, gravity, and the rope tension. The force of gravity is equal to 6×10^4 N. Archimedes' force is equal to $\frac{4}{3}\pi r^2\rho g$, where $\rho \approx 10^3$ kg/m^3 is the water density, and $g \approx 9.8$ m/s^2 is the acceleration due to gravity. Then the equilibrium equation is

$$6 \times 10^4 = \frac{4}{3}\pi r^2 \rho g + T,$$

where T is the rope tension. It follows from this equation that

$$T = 6 \times 10^4 - \frac{4}{3}\pi r^2 \rho g \approx 6 \times 10^4 - \frac{4}{3}\pi \times 10^3 \times 9.8 \approx 1.81 \times 10^4 \text{ N}.$$

5.8: (i) We introduce the function $\tilde{\Psi}$ as

$$\tilde{\Psi} = \int^y v_x \, dy'.$$

Then

$$v_x = \frac{\partial \tilde{\Psi}}{\partial y}.$$

It follows from this equation and the equation $\nabla \cdot \boldsymbol{v} = 0$ that

$$\frac{\partial v_y}{\partial y} = -\frac{\partial v_x}{\partial x} = -\frac{\partial^2 \tilde{\Psi}}{\partial x \partial y}.$$

Integrating this equation we obtain

$$v_y = -\frac{\partial \tilde{\Psi}}{\partial x} + f(x),$$

where $f(x)$ is an arbitrary function. Introducing $\Psi = \tilde{\Psi} - \int f(x)\, dx$ we eventually obtain

$$v_x = \frac{\partial \Psi}{\partial y}, \quad v_y = -\frac{\partial \Psi}{\partial x}. \qquad (*)$$

(ii) For a planar motion the equation of a streamline is

$$\frac{dx}{v_x} = \frac{dy}{v_y}.$$

Using equation $(*)$ we rewrite this equation as

$$\frac{\partial \Psi}{\partial x} dx + \frac{\partial \Psi}{\partial y} dy = 0,$$

which implies that

$$\Psi(x, y) = \text{const.}$$

(iii) Since the motion is potential we have

$$v_x = \frac{\partial \Phi}{\partial x}, \quad v_y = \frac{\partial \Phi}{\partial y}.$$

Comparing this equation with equation (∗) we obtain

$$\frac{\partial \Phi}{\partial x} = \frac{\partial \Psi}{\partial y}, \quad \frac{\partial \Phi}{\partial y} = -\frac{\partial \Psi}{\partial x}.$$

Differentiating these relations yields

$$\frac{\partial^2 \Phi}{\partial x^2} = \frac{\partial \Psi}{\partial x \partial y}, \quad \frac{\partial^2 \Phi}{\partial y^2} = -\frac{\partial \Psi}{\partial x \partial y}.$$

Adding these two equations we obtain

$$\frac{\partial^2 \Phi}{\partial x^2} + \frac{\partial^2 \Phi}{\partial y^2} \equiv \nabla^2 \Phi = 0.$$

5.9: (i) The boundary condition at the surface of the sphere is $v_n = 0$, where v_n is the velocity component normal to the surface. Since $v_n = \dfrac{\partial \Phi}{\partial r}$, we obtain

$$\frac{\partial \Phi}{\partial r} = 0 \quad \text{at} \quad r = a.$$

(ii) Substituting $\boldsymbol{v} = \nabla \Phi$ into equation $\nabla \cdot \boldsymbol{v} = 0$ we obtain $\nabla^2 \Phi = 0$. Using Eqs. (2.13.11) and (2.13.17) and taking into account that Φ is independent of ϕ we write this equation as

$$\frac{\partial}{\partial r} r^2 \frac{\partial \Phi}{\partial r} + \frac{1}{\sin\theta} \frac{\partial}{\partial \theta} \sin\theta \frac{\partial \Phi}{\partial \theta} = 0.$$

Substituting $\Phi = R(r)\cos\theta$ into this equation yields

$$\frac{d}{dr} r^2 \frac{dR}{\partial r} - 2R = 0. \tag{$*$}$$

We look for the solution in the form $R = r^\alpha$. Substituting this expression into equation (∗) we obtain $\alpha^2 + \alpha - 2 = 0$, so $\alpha_1 = 1$ and $\alpha_2 = -2$, and the general solution to equation (∗) is

$$R(r) = Ar + Br^{-2}, \tag{$\dagger$}$$

where A and B are arbitrary constants. Since $\boldsymbol{v} \approx -V\boldsymbol{e}_z$ when $r \gg a$, where $\boldsymbol{e}_z$ is the unit vector of the z-axis, we have $\Phi \approx -Vz = -Vr\cos\theta$ when $r \gg a$. This implies that $R \approx -Vr$ when $r \gg a$. Comparing this expression with equation (†) we obtain $A = -V$, so $R(r) = -Vr + Br^{-2}$. Now the boundary condition at the sphere gives

$$\left.\frac{\partial\Phi}{\partial r}\right|_{r=a} = -\left(V + \frac{2B}{a^3}\right)\cos\theta = 0.$$

It follows from this equation that $B = -a^3V/2$, so

$$\Phi = -V\left(r + \frac{a^3}{2r^2}\right)\cos\theta.$$

(iii) The velocity is given by

$$\boldsymbol{v} = \nabla\Phi = -V\left(1 - \frac{a^3}{r^3}\right)\cos\theta\,\boldsymbol{e}_r + V\left(1 + \frac{a^3}{2r^3}\right)\sin\theta\,\boldsymbol{e}_\theta.$$

Now, using the Bernoulli integral, we obtain

$$p\big|_{r=a} = p_0 + \frac{\rho V^2}{2} - \frac{9\rho V^2}{8}\sin^2\theta.$$

The force acting on the sphere is

$$\boldsymbol{F} = -\int_S p\boldsymbol{n}\,dS,$$

where S is the surface of the sphere and $\boldsymbol{n}$ is the normal vector to this surface. Using the expression for the components of $\boldsymbol{n}$ in Cartesian coordinates,

$$\boldsymbol{n} = (\sin\theta\cos\phi,\ \sin\theta\sin\phi,\ \cos\theta),$$

we obtain

$$\begin{aligned}
\boldsymbol{F} &= -a^2\int_0^\pi\left(p_0 + \frac{\rho V^2}{2} - \frac{9\rho V^2}{8}\sin^2\theta\right)\sin\theta\,d\theta\int_0^{2\pi}(\sin\theta\cos\phi,\ \sin\theta\sin\phi,\ \cos\theta)\,d\phi \\
&= -a^2\int_0^\pi\left(p_0 + \frac{\rho V^2}{2} - \frac{9\rho V^2}{8}\sin^2\theta\right)\left(\sin\theta\sin\phi\Big|_0^{2\pi},\ -\sin\theta\cos\phi\Big|_0^{2\pi},\ \cos\theta\,\phi\Big|_0^{2\pi}\right)\sin\theta\,d\theta \\
&= -a^2\int_0^\pi\left(p_0 + \frac{\rho V^2}{2} - \frac{9\rho V^2}{8}\sin^2\theta\right)(0,\ 0,\ 2\pi\cos\theta)\sin\theta\,d\theta \\
&= -2\pi a^2\left(0,\ 0,\ \frac{2p_0 + \rho V^2}{4}\sin^2\theta\Big|_0^\pi - \frac{9\rho V^2}{32}\sin^4\theta\Big|_0^\pi\right) = (0,\ 0,\ 0).
\end{aligned}$$

5.10: (i) We use the spherical coordinate system with the origin at the sphere's centre. Since the fluid is incompressible, $\nabla \cdot \boldsymbol{v} = 0$. Since the motion is spherically symmetric, $\boldsymbol{v} = v\boldsymbol{e}_r$ and we have

$$\frac{1}{r^2}\frac{\partial(r^2 v)}{\partial r} = 0.$$

It follows from this equation that

$$r^2 v = C, \tag{$*$}$$

where C is a constant. At the surface of the sphere the fluid velocity coincides with the velocity of the surface,

$$v = \frac{dR}{dt} \quad \text{at} \quad r = R(t).$$

It follows from this equation and equation $(*)$ that

$$C = R^2 \frac{dR}{dt}.$$

Hence,

$$v = \frac{R^2}{r^2}\frac{dR}{dt}.$$

Since the motion is potential we obtain

$$v = \frac{\partial \Phi}{\partial r}.$$

Then it follows that

$$\frac{\partial \Phi}{\partial r} = \frac{R^2}{r^2}\frac{dR}{dt}.$$

Integrating this equation we obtain

$$\Phi = -\frac{R^2}{r}\frac{dR}{dt} + g(t),$$

where $g(t)$ is an arbitrary function.

(ii) Substituting the expressions for Φ and v into the Lagrange–Cauchy integral yields

$$-\frac{\rho}{r}\frac{d}{dt}R^2\frac{dR}{dt} + p + \frac{\rho R^4}{2r^4}\left(\frac{dR}{dt}\right)^2 = f(t) - \rho g'(t). \tag{$\dagger$}$$

Taking $r \to \infty$ and using the condition that $p \to p_0$ as $r \to \infty$ we obtain

$$f(t) - \rho g'(t) = p_0.$$

Using this result and substituting $r = R(t)$ into equation (†) yields

$$p = p_0 + \frac{\rho}{R}\frac{d}{dt}R^2\frac{dR}{dt} - \frac{\rho}{2}\left(\frac{dR}{dt}\right)^2 = p_0 + \rho R\frac{d^2R}{dt^2} + \frac{3\rho}{2}\left(\frac{dR}{dt}\right)^2.$$

Then the force acting on the sphere is

$$F = 4\pi R^2 p = 4\pi\rho R^2\left[\frac{p_0}{\rho} + R\frac{d^2R}{dt^2} + \frac{3}{2}\left(\frac{dR}{dt}\right)^2\right].$$

5.11: (i) Let the current radius of the shell be r, and the shell be at the depth h. Since the volume of the shell is proportional to r^3, the air pressure in the shell is $p = p_0(r_0/r)^3$. The water pressure outside the shell is $p_a + g\rho h$, so that the difference between the pressure inside and outside the shell is $p_0(r_0/r)^3 - (p_a + g\rho h)$. At the initial time the difference between the pressure inside and outside the shell is $p_0 - (p_a + g\rho h_0)$. Since the volume of the shell is proportional to the pressure difference, we have

$$\frac{p_0(r_0/r)^3 - (p_a + g\rho h)}{p_0 - (p_a + g\rho h_0)} = \frac{r^3}{r_0^3}.$$

After simple algebra we reduce this equation to

$$(p_0 - p_a - g\rho h_0)\left(\frac{r}{r_0}\right)^6 + (p_a + g\rho h)\left(\frac{r}{r_0}\right)^3 - p_0 = 0.$$

Substituting the numerical values we arrive at

$$\left(\frac{r}{r_0}\right)^6 + (1 + 0.1h)\left(\frac{r}{r_0}\right)^3 - 6 = 0.$$

The positive solution to this equation is

$$r = r_0\left(\frac{1}{2}\sqrt{(1 + 0.1h)^2 + 24} - \frac{1 + 0.1h}{2}\right)^{1/3}. \qquad (*)$$

(ii) The shell is moving under the action of two forces: the force of gravity, equal to mg, which is directed downward, and Archimedes' force directed upward. The volume of the shell at the depth h is $\frac{4}{3}\pi r^3$, so that Archimedes' force is equal to $\frac{4}{3}\pi r^3 g\rho$. Then Newton's second law for the shell is

$$m\frac{d^2h}{dt^2} = gm - \frac{4}{3}\pi r^3 g\rho.$$

Substituting the numerical values we reduce this equation to the approximate form

$$\frac{d^2h}{dt^2} = 10 - 11.2\left(\frac{r}{r_0}\right)^3. \qquad (\dagger)$$

Substituting $h = 0$ into equation $(*)$ we obtain $(r/r_0)^3 = 2$. Substituting this result into equation $(\dagger)$ we obtain $\dfrac{d^2h}{dt^2} \approx -12.4$ m/s^2. The minus sign indicates that the acceleration is directed upward, i.e. the depth is decreasing.

5.12: We use the reference frame moving together with the sphere and put the origin of the Cartesian coordinates at the centre of the sphere. This reference frame is non-inertial because it is moving with acceleration $\boldsymbol{a}$. This implies that there is a fictitious force equal to $-\boldsymbol{a}$ per unit mass. This force can be written as $-\nabla(az)$. Since we assume that the flow is potential, it follows that its velocity is given by $\boldsymbol{v} = \nabla\Phi$. Since the liquid is incompressible we obtain $\nabla^2\Phi = 0$. The boundary condition at the sphere's boundary is $\partial\Phi/\partial r = 0$. The flow velocity at infinity is $-V\boldsymbol{e}_z$. Hence, the boundary problem for Φ is the same as in problem **5.9** and we can use the expression for Φ obtained when solving that problem. Assuming that the sphere was at rest at the initial time we obtain $V = at$. Then in spherical coordinates with the origin at the centre of the sphere and the polar axis coinciding with the z-axis of Cartesian coordinates we have

$$\Phi = -at\left(r + \frac{R^3}{2r^2}\right)\cos\theta.$$

Now we use the Lagrange–Cauchy integral (see problem **5.3**) and take into account that the body force potential is $\varphi = az = ar\cos\theta$ to obtain

$$p = \rho\left(C(t) - \frac{\partial\Phi}{\partial t} - \frac{1}{2}\|\nabla\Phi\|^2 - ar\cos\theta\right).$$

Substituting the expression for Φ we obtain that the pressure at the sphere boundary is given by

$$p = \rho\left(C(t) + \frac{1}{2}aR\cos\theta - \frac{9}{8}a^2t^2\sin^2\theta\right).$$

The force acting on an elementary surface dS with the unit normal $\boldsymbol{n}$ is $-p\boldsymbol{n}\,dS$. In Cartesian coordinates the components of $\boldsymbol{n}$ are given by

$$n_x = \sin\theta\cos\phi, \quad n_y = \sin\theta\sin\phi, \quad n_z = \cos\theta.$$

Now we can calculate the Cartesian components of the force $\boldsymbol{f}$ acting on the sphere:

$$\begin{aligned}
f_x &= -\int_S p n_x \, dS = -R^2 \int_0^\pi p(\theta) \sin^2\theta \, d\theta \int_0^{2\pi} \cos\phi \, d\phi = 0, \\
f_y &= -\int_S p n_y \, dS = -R^2 \int_0^\pi p(\theta) \sin^2\theta \, d\theta \int_0^{2\pi} \sin\phi \, d\phi = 0, \\
f_z &= -\int_S p n_z \, dS = -R^2 \int_0^\pi p(\theta) \sin\theta \cos\theta \, d\theta \int_0^{2\pi} d\phi \\
&= -2\pi R^2 \rho \int_0^\pi \left(C(t) + \frac{1}{2} a R \cos\theta - \frac{9}{8} a^2 t^2 \sin^2\theta \right) \sin\theta \cos\theta \, d\theta \\
&= -\pi R^2 \rho \left(C(t) \sin^2\theta - \frac{1}{3} a R \cos^3\theta - \frac{9}{16} a^2 t^2 \sin^4\theta \right) \Big|_0^\pi = -\frac{2\pi}{3} a \rho R^3.
\end{aligned}$$

There is also the fictitious gravity force $-aM\boldsymbol{e}_z$ acting on the sphere. Since the sphere is immovable in the non-inertial reference frame we conclude that $F = aM + \frac{2}{3}\pi a \rho R^3$. Then the added mass is

$$M_{\text{add}} = \frac{F}{a} - M = \frac{2\pi}{3} \rho R^3.$$

Note that M_{add} is equal to half of the volume of the sphere times the fluid density.

A.6 Solutions for Chap. 6

6.1: (i) Using Eq. (2.13.9) we obtain $\nabla \times \boldsymbol{u} = 0$. Then it follows from Eq. (6.3.12) that $\nabla^2 \boldsymbol{u} = \nabla\nabla \cdot \boldsymbol{u}$. Using this result, the expression for $\nabla \cdot \boldsymbol{u}$ given by Eq. (2.13.8), and taking $\boldsymbol{b} = 0$, we obtain from Eq. (6.3.8)

$$(\lambda + \mu) \frac{d}{dr} \left(\frac{1}{r} \frac{d(ru)}{dr} \right) + \mu \frac{d}{dr} \left(\frac{1}{r} \frac{d(ru)}{dr} \right) = 0 \quad \Longrightarrow \quad \frac{d}{dr} \left(\frac{1}{r} \frac{d(ru)}{dr} \right) = 0.$$

It follows from this equation that

$$\frac{1}{r} \frac{d(ru)}{dr} = 2A,$$

where A is a constant. Multiplying this equation by r and integrating we obtain

$$u = Ar + \frac{B}{r}, \qquad (*)$$

where B is a constant.

(ii) It follows from Eq. (2.13.7) that

$$\nabla \boldsymbol{u} = \frac{du_r}{dr}\boldsymbol{e}_r\boldsymbol{e}_r + \frac{u_r}{r}\boldsymbol{e}_\phi\boldsymbol{e}_\phi.$$

In particular, it follows from this expression that $(\nabla \boldsymbol{u})^T = \nabla \boldsymbol{u}$. Then, using this result and the identity $\mathbf{I} = \boldsymbol{e}_r\boldsymbol{e}_r + \boldsymbol{e}_\phi\boldsymbol{e}_\phi + \boldsymbol{e}_z\boldsymbol{e}_z$ we obtain from Eq. (6.3.16)

$$\mathbf{T} = \left[(\lambda + 2\mu)\frac{du}{dr} + \lambda\frac{u}{r}\right]\boldsymbol{e}_r\boldsymbol{e}_r + \left[\lambda\frac{du}{dr} + (\lambda + 2\mu)\frac{u}{r}\right]\boldsymbol{e}_\phi\boldsymbol{e}_\phi + \frac{\lambda}{r}\frac{d(ru)}{dr}\boldsymbol{e}_z\boldsymbol{e}_z.$$

Now we obtain with the aid of (∗)

$$\boldsymbol{t}(r_0) = \mathbf{T}\cdot\boldsymbol{n} = 2\boldsymbol{n}\left[\lambda A + \mu\left(A - \frac{B}{r_0^2}\right)\right],$$

where $\boldsymbol{n} = \pm\boldsymbol{e}_r$. Then it follows that

$$\boldsymbol{t}(a) = -2\boldsymbol{e}_r\left[\lambda A + \mu\left(A - \frac{B}{a^2}\right)\right], \quad \boldsymbol{t}(b) = 2\boldsymbol{e}_r\left[\lambda A + \mu\left(A - \frac{B}{b^2}\right)\right].$$

Substituting these expressions into the boundary conditions yields

$$2(\lambda + \mu)A - \frac{2\mu B}{a^2} = -p_0, \quad 2(\lambda + \mu)A - \frac{2\mu B}{b^2} = -p_a.$$

Solving this system of equations we obtain

$$A = \frac{a^2 p_0 - b^2 p_a}{2(\lambda + \mu)(b^2 - a^2)}, \quad B = \frac{a^2 b^2 (p_0 - p_a)}{2\mu(b^2 - a^2)}.$$

The increase in the internal radius is equal to the displacement at the internal boundary, $\delta = u(a)$. Substituting the expressions for A and B, and $r = a$, into (∗) we obtain

$$\delta = \frac{a\{[\mu a^2 + (\lambda + \mu)b^2]p_0 - b^2(\lambda + 2\mu)p_a\}}{2\mu(\lambda + \mu)(b^2 - a^2)}.$$

Substituting the numerical values into this expression we arrive at

$$\delta \approx 0.056a \approx 6.15 \text{ mm}.$$

6.2: (i) The volume of the water at 0° C is $\frac{4}{3}\pi a^3$, and its mass is $\frac{4}{3}\pi\rho_1 a^3$. Let the internal radius of the shall after the water was heated to 80° C be R. Then the volume of water is $\frac{4}{3}\pi R^3$, and its mass is $\frac{4}{3}\pi\rho_2 R^3$. Since the masses of water at 0° and at 80° are equal, we have

$$\frac{4}{3}\pi\rho_2 R^3 = \frac{4}{3}\pi\rho_1 a^3.$$

From this equation we find that

$$R = a(\rho_1/\rho_2)^{1/3} \approx 1.005a.$$

(ii) The water expansion causes the displacement of all points in the shell in the radial direction in spherical coordinates r, θ, ϕ with the origin at the shell centre. Hence, $\boldsymbol{u} = u\boldsymbol{e}_r$. Since the problem is spherically symmetric, $u = u(r)$. We neglect the body force, which is the force of gravity, because it is small in comparison with the forces due to the water expansion. Using Eq. (2.13.18) we obtain $\nabla \times \boldsymbol{u} = 0$. Then it follows from Eq. (6.3.12) that $\nabla^2\boldsymbol{u} = \nabla(\nabla \cdot \boldsymbol{u})$. Now, with the aid of Eq. (2.13.17) we reduce Eq. (6.3.8) to

$$\frac{d}{dr}\left(\frac{1}{r^2}\frac{d(r^2u)}{dr}\right) = 0.$$

It follows from this equation that

$$\frac{d(r^2u)}{dr} = 3Ar^2,$$

where A is a constant. Integrating this equation, we eventually obtain

$$u = Ar + \frac{B}{r^2}, \qquad (*)$$

where B is a constant. The displacement of the internal boundary is equal to $R - a = 0.005a$, so that it is small in comparison with a. The same is true for the displacement of the external boundary. This implies that we can neglect these displacements and impose the boundary conditions not at the perturbed boundaries, but at the unperturbed boundaries. The kinematic boundary condition at the internal boundary is

$$u = R - a \quad \text{at} \quad r = a. \qquad (\dagger)$$

At the positive side of the external boundary the surface traction is equal to $-p_a\boldsymbol{e}_r$, where $p_a \approx 10^5$ N m^{-2}. It follows from Eq. (2.13.15) that

$$\nabla\boldsymbol{u} = \frac{du}{dr}\boldsymbol{e}_r\boldsymbol{e}_r + \frac{u}{r}(\boldsymbol{e}_\theta\boldsymbol{e}_\theta + \boldsymbol{e}_\phi\boldsymbol{e}_\phi).$$

We see that $\nabla\boldsymbol{u}$ is symmetric, $\nabla\boldsymbol{u} = (\nabla\boldsymbol{u})^T$. Using this result and the identity $\mathbf{I} = \boldsymbol{e}_r\boldsymbol{e}_r + \boldsymbol{e}_\phi\boldsymbol{e}_\phi + \boldsymbol{e}_z\boldsymbol{e}_z$ we obtain from Eq. (6.3.16) that the stress tensor is given by

$$\mathbf{T} = \left(\frac{\lambda}{r^2}\frac{d(r^2u)}{dr} + 2\mu\frac{du}{dr}\right)\boldsymbol{e}_r\boldsymbol{e}_r + \left(\frac{\lambda}{r^2}\frac{d(r^2u)}{dr} + 2\mu\frac{u}{r}\right)(\boldsymbol{e}_\theta\boldsymbol{e}_\theta + \boldsymbol{e}_\phi\boldsymbol{e}_\phi).$$

The surface traction at the inner side of the external boundary is

$$\mathsf{T}\cdot\boldsymbol{e}_r\Big|_{r=b} = \left(\frac{\lambda}{r^2}\frac{d(r^2u)}{dr}\Big|_{r=b} + 2\mu\frac{du}{dr}\Big|_{r=b}\right)\boldsymbol{e}_r.$$

Using (∗) we eventually obtain

$$\mathsf{T}\cdot\boldsymbol{e}_r\Big|_{r=b} = \left(3\lambda A + 2\mu A - 4\mu B b^{-3}\right)\boldsymbol{e}_r.$$

Since the surface traction has to be continuous at any boundary, we arrive at the dynamic boundary condition at the external boundary:

$$(3\lambda + 2\mu)A - 4\mu B b^{-3} = -p_a.$$

Using this equation and the kinematic boundary condition (†) at the internal boundary we find

$$A = \frac{4\mu a^2(R-a) - p_a b^3}{3\lambda b^3 + 2\mu(2a^3 + b^3)}, \quad B = \frac{a^2 b^3 (R-a)(3\lambda + 2\mu) + p_a a^3 b^3}{3\lambda b^3 + 2\mu(2a^3 + b^3)}.$$

Substituting these expressions into (∗) we obtain

$$u = \frac{a^2(R-a)[4\mu r + (3\lambda + 2\mu)b^3 r^{-2}] - p_a b^3(r - a^3 r^{-2})}{3\lambda b^3 + 2\mu(2a^3 + b^3)}.$$

(iii) Substituting the expression for u into the expression for T we find

$$\begin{aligned}\mathsf{T} &= \left[(3\lambda + 2\mu)A - 4\mu B r^{-3}\right]\boldsymbol{e}_r\boldsymbol{e}_r + \left[(3\lambda + 2\mu)A + 2\mu B r^{-3}\right](\boldsymbol{e}_\theta\boldsymbol{e}_\theta + \boldsymbol{e}_\phi\boldsymbol{e}_\phi)\\ &= \left[3\lambda b^3 + 2\mu(2a^3 + b^3)\right]^{-1}\big\{\left[4\mu a^2(R-a)(3\lambda + 2\mu)(1 - b^3 r^{-3})\right.\\ &\quad \left. - p_a b^3(3\lambda + 2\mu + 4\mu a^3 r^{-3})\right]\boldsymbol{e}_r\boldsymbol{e}_r + \left[2\mu a^2(R-a)(3\lambda + 2\mu)(2 + b^3 r^{-3})\right.\\ &\quad \left. - p_a b^3(3\lambda + 2\mu - 2\mu a^3 r^{-3})\right](\boldsymbol{e}_\theta\boldsymbol{e}_\theta + \boldsymbol{e}_\phi\boldsymbol{e}_\phi)\big\}.\end{aligned}$$

6.3: The stress vector at the surface with the unit normal vector $\boldsymbol{n}$ is given by $\boldsymbol{t} = \mathsf{T}\cdot\boldsymbol{n}$. Substituting $\lambda = \mu$ into the expression for T, we obtain

$$\begin{aligned}\boldsymbol{t} &= \left(4a^3 + 5b^3\right)^{-1}\big\{\left[20\mu a^2(R-a)(1 - b^3 r^{-3}) - p_a b^3(5 + 4a^3 r^{-3})\right]n_r\boldsymbol{e}_r\\ &\quad + \left[10\mu a^2(R-a)(2 + b^3 r^{-3}) - p_a b^3(5 - 2a^3 r^{-3})\right](n_\theta\boldsymbol{e}_\theta + n_\phi\boldsymbol{e}_\phi)\big\}.\end{aligned}$$

Then, using the identity $n_r^2 + n_\theta^2 + n_\phi^2 = 1$ we obtain

$$\begin{aligned}\|\boldsymbol{t}\|^2 &= \left(4a^3+5b^3\right)^{-2}\left[10\mu a^2(R-a)(2+b^3r^{-3}) - p_a b^3(5-2a^3r^{-3})\right]^2 \\ &\quad + \left(4a^3+5b^3\right)^{-2}\left\{\left[20\mu a^2(R-a)(1-b^3r^{-3}) - p_a b^3(5+4a^3r^{-3})\right]^2 \right.\\ &\quad \left. - \left[10\mu a^2(R-a)(2+b^3r^{-3}) - p_a b^3(5-2a^3r^{-3})\right]^2\right\}n_r^2 \\ &= \left(4a^3+5b^3\right)^{-2}\left\{\left[10\mu a^2(R-a)(2+b^3r^{-3}) - p_a b^3(5-2a^3r^{-3})\right]^2\right. \\ &\quad \left. -12a^2b^3r^{-3}[5\mu(R-a)+p_a a]\left[5\mu a^2(R-a)(4-b^3r^{-3}) - p_a b^3(5+a^3r^{-3})\right]n_r^2\right\}.\end{aligned}$$

Let us look for μ satisfying

$$5\mu a^2(R-a)(4-b^3r^{-3}) > p_a b^3(5+a^3r^{-3}). \qquad (\ddagger)$$

Then the second term in the expression for $\|\boldsymbol{t}\|^2$ is negative unless $n_r = 0$. This implies that, at fixed r, $\|\boldsymbol{t}\|$ takes its maximum value $\tilde{t}(r)$ when $n_r = 0$, and this value is given by

$$\tilde{t}(r) = \left(4a^3+5b^3\right)^{-1}\left|10\mu a^2(R-a)(2+b^3r^{-3}) - p_a b^3(5-2a^3r^{-3})\right|.$$

It follows from ($\ddagger$) that the expression between the signs of the absolute value is positive, so that we can rewrite the expression for $\tilde{t}(r)$ as

$$\tilde{t}(r) = \left(4a^3+5b^3\right)^{-1}\left\{5\left[4\mu a^2(R-a) - p_a b^3\right] + 2a^2b^3r^{-3}\left[5\mu(R-a)+p_a a\right]\right\}.$$

Obviously, $\tilde{t}(r)$ is a monotonically decreasing function of r. Then it takes its maximum value, $t_{\max}$, at $r = a$, i.e.

$$t_{\max} = \tilde{t}(a) = \left(4a^3+5b^3\right)^{-1}\left[10\mu(R/a-1)(2a^3+b^3) - 3p_a b^3\right].$$

Substituting the numerical values we obtain

$$t_{\max} \approx 0.0144\mu - 4.25\times10^4.$$

Then $t_{\max} < t_{cr}$ when $0.0144\mu - 4.25\times10^4 < 10^8$, i.e. $\mu \le 6.93\times10^9$, so that $\mu_{\max} \approx 6.93\times10^9$. It is straightforward to verify that $\mu_{\max}$ satisfies ($\ddagger$). Since $\mu_{\max} < 4\times10^{10}$, the shell will not endure the pressure of heated water if it is made of cast iron, and it will crack.

6.4: (i) The equilibrium of an elastic isotropic medium is described by Eq. (6.3.8). Using Eq. (2.13.8) we obtain that $\nabla\cdot\boldsymbol{u} = 0$ for $\boldsymbol{u} = (0, u(r,z), 0)$. Then Eq. (6.3.8) reduces to

$$\nabla^2\boldsymbol{u} = 0, \qquad (*)$$

while Eq. (6.3.12) reduces to

$$\nabla^2 \boldsymbol{u} = -\nabla \times \nabla \times \boldsymbol{u}.$$

Using Eq. (2.13.9) we obtain

$$\nabla \times \boldsymbol{u} = -\frac{\partial u}{\partial z}\boldsymbol{e}_r + \frac{1}{r}\frac{\partial (ru)}{\partial r}\boldsymbol{e}_z.$$

Again using Eq. (2.13.9) yields

$$\nabla \times \nabla \times \boldsymbol{u} = -\left(\frac{\partial}{\partial r}\frac{1}{r}\frac{\partial (ru)}{\partial r} + \frac{\partial^2 u}{\partial z^2}\right)\boldsymbol{e}_\phi.$$

Then equation (∗) reduces to

$$\frac{\partial}{\partial r}\frac{1}{r}\frac{\partial (ru)}{\partial r} + \frac{\partial^2 u}{\partial z^2} = 0.$$

Substituting $u = rf(z)$ into this equation we obtain

$$\frac{d^2 f}{dz^2} = 0 \quad \Longrightarrow \quad f(z) = Az + B,$$

where A and B are constants. Since $u = 0$ at $z = 0$ and $u = \alpha r$ at $z = L$ we have

$$B = 0, \quad A = \frac{\alpha}{L} \quad \Longrightarrow \quad f(z) = \frac{\alpha z}{L}.$$

(ii) The expression for the stress tensor in terms of the displacement is given by Eq. (6.2.12). In our case $I_1(\mathbf{E}) = \nabla \cdot \boldsymbol{u} = 0$, so $\mathbf{T} = 2\mu\mathbf{E}$. The expression for $\mathbf{E}$ is given by Eqs. (3.4.2) and (3.4.5). It follows from these equations that

$$2\mathbf{E} = \nabla \boldsymbol{u} + (\nabla \boldsymbol{u})^T.$$

The expression for $\nabla \boldsymbol{u}$ is given by Eq. (2.13.7). It follows from this equation that

$$\nabla \boldsymbol{u} = \frac{\partial u}{\partial r}\boldsymbol{e}_r\boldsymbol{e}_\phi - \frac{u}{r}\boldsymbol{e}_\phi\boldsymbol{e}_r + \frac{\partial u}{\partial z}\boldsymbol{e}_z\boldsymbol{e}_\phi.$$

To obtain the expression for $(\nabla \boldsymbol{u})^T$ we have to swap the unit vectors in the dyadic products in the expression for $\nabla \boldsymbol{u}$. As a result we obtain

$$(\nabla \boldsymbol{u})^T = \frac{\partial u}{\partial r}\boldsymbol{e}_\phi\boldsymbol{e}_r - \frac{u}{r}\boldsymbol{e}_r\boldsymbol{e}_\phi + \frac{\partial u}{\partial z}\boldsymbol{e}_\phi\boldsymbol{e}_z.$$

Then

$$\mathbf{T} = 2\mu\mathbf{E} = \mu\big[\nabla\boldsymbol{u} + (\nabla\boldsymbol{u})^T\big] = \mu\left(\frac{\partial u}{\partial r} - \frac{u}{r}\right)(\boldsymbol{e}_r\boldsymbol{e}_\phi + \boldsymbol{e}_\phi\boldsymbol{e}_r) + \mu\frac{\partial u}{\partial z}(\boldsymbol{e}_\phi\boldsymbol{e}_z + \boldsymbol{e}_z\boldsymbol{e}_\phi).$$

Let us now calculate the surface traction at the upper edge of the rod. Since the unit normal vector at this surface is $\boldsymbol{e}_z$, the surface traction is

$$\boldsymbol{t} = \mathbf{T}\cdot\boldsymbol{e}_z = \mu\frac{\partial u}{\partial z}\boldsymbol{e}_\phi = \frac{\mu\alpha r}{L}\boldsymbol{e}_\phi,$$

where we have taken into account that $u = \alpha r z/L$. Then the momentum of the force needed to rotate the upper surface by the angle α is

$$M = \int_0^{2\pi} d\phi \int_0^R (r\|\boldsymbol{t}\|)r\,dr = \frac{2\pi\mu\alpha}{L}\int_0^R r^3 dr = \frac{\pi\mu\alpha R^4}{2L}.$$

It follows from this equation that

$$\alpha = \frac{2LM}{\pi\mu R^4}.$$

6.5: (i) We use Eq. (2.13.9) to obtain $\nabla\times\boldsymbol{u} = 0$. Then, using the identity $\nabla^2\boldsymbol{u} = \nabla(\nabla\cdot\boldsymbol{u}) - \nabla\times\nabla\times\boldsymbol{u}$ yields $\nabla^2\boldsymbol{u} = \nabla(\nabla\cdot\boldsymbol{u})$. Hence, Eq. (6.3.8) reduces to

$$\nabla(\nabla\cdot\boldsymbol{u}) = 0. \qquad (*)$$

Using Eq. (2.13.8) and recalling that $\boldsymbol{u} = u_r(r)\boldsymbol{e}_r + u_z(z)\boldsymbol{e}_z$ we obtain

$$\nabla\cdot\boldsymbol{u} = \frac{1}{r}\frac{d(ru_r)}{dr} + \frac{du_z}{dz}. \qquad (\dagger)$$

Then the vector equation $(*)$ reduces to the system of two scalar equations:

$$\frac{d}{dr}\left(\frac{1}{r}\frac{d(ru_r)}{dr}\right) = 0, \quad \frac{d^2u_z}{dz^2} = 0. \qquad (\ddagger)$$

Integrating the first equation in $(\ddagger)$ we obtain

$$\frac{\partial(ru_r)}{\partial r} = 2Ar,$$

where A is a constant. Integrating this equation yields

$$u_r = Ar + \frac{C}{r},$$

where C is a constant. Since u_r must be regular at $r = 0$, we conclude that $C = 0$ and $u_r = Ar$.

Since we can neglect the deformation of the horizontal surface, it follows that u_z is zero at the base of the cylinder. Then, integrating the second equation in (‡) gives

$$u_z = Bz.$$

(ii) Using Eq. (2.13.7) we obtain

$$(\nabla \boldsymbol{u})_{rr} = \frac{du_r}{dr} = A, \quad (\nabla \boldsymbol{u})_{\phi\phi} = \frac{u_r}{r} = A, \quad (\nabla \boldsymbol{u})_{zz} = \frac{du_z}{dz} = B,$$

while all other components of the tensor $\nabla \boldsymbol{u}$ are equal to zero. It follows from these results that

$$\nabla \boldsymbol{u} = A\boldsymbol{e}_r\boldsymbol{e}_r + A\boldsymbol{e}_\phi\boldsymbol{e}_\phi + B\boldsymbol{e}_z\boldsymbol{e}_z.$$

Using equation (†) yields $\nabla \cdot \boldsymbol{u} = 2A + B$. Substituting this expression and the expression for $\nabla \boldsymbol{u}$ into Eq. (6.3.16), and noting that $(\nabla \boldsymbol{u})^T = \nabla \boldsymbol{u}$ we obtain

$$\begin{aligned}\mathsf{T} &= \lambda(2A + B)(\boldsymbol{e}_r\boldsymbol{e}_r + \boldsymbol{e}_\phi\boldsymbol{e}_\phi + \boldsymbol{e}_z\boldsymbol{e}_z) + 2\mu(A\boldsymbol{e}_r\boldsymbol{e}_r + A\boldsymbol{e}_\phi\boldsymbol{e}_\phi + B\boldsymbol{e}_z\boldsymbol{e}_z) \\ &= [2A(\lambda + \mu) + \lambda B](\boldsymbol{e}_r\boldsymbol{e}_r + \boldsymbol{e}_\phi\boldsymbol{e}_\phi) + [2A\lambda + B(\lambda + 2\mu)]\boldsymbol{e}_z\boldsymbol{e}_z.\end{aligned}$$

(iii) The normal vector at the flat top of the cylinder is $\boldsymbol{e}_z$. Then the surface traction at the flat top is

$$\boldsymbol{t}_{\text{upper}} = \mathsf{T} \cdot \boldsymbol{e}_z = [2A\lambda + B(\lambda + 2\mu)]\boldsymbol{e}_z.$$

The normal vector at the side surface of the cylinder is $\boldsymbol{e}_r$. Then the surface traction at the side surface is

$$\boldsymbol{t}_{\text{side}} = \mathsf{T} \cdot \boldsymbol{e}_r = [2A(\lambda + \mu) + \lambda B]\boldsymbol{e}_r.$$

There is a pressure equal to $F/\pi R^2$ at the flat top, which corresponds to the surface traction equal to $-(F/\pi R^2)\boldsymbol{e}_z$. Then it follows from the condition of continuity of the surface traction that

$$2A\lambda + B(\lambda + 2\mu) = -\frac{F}{\pi R^2}.$$

Since we can neglect the air pressure, the surface traction from outside of the side surface is zero. Then it follows from the condition of continuity of the surface traction that

$$2A(\lambda + \mu) + \lambda B = 0.$$

As a result, we obtain a system of two linear algebraic equations for A and B. The solution to this system is straightforward:

$$A = \frac{\lambda F}{2\pi R^2 \mu(3\lambda + 2\mu)}, \quad B = -\frac{(\lambda + \mu)F}{\pi R^2 \mu(3\lambda + 2\mu)}.$$

The decrease in the cylinder height is equal to $|u_z|$ at $z = H$. We obtain

$$|u_z| = |B|H = \frac{(\lambda + \mu)FH}{\pi R^2 \mu(3\lambda + 2\mu)}.$$

The increase in the cylinder radius is equal to u_r at $r = R$. Hence, it is given by

$$u_r = AR = \frac{\lambda F}{2\pi R \mu(3\lambda + 2\mu)}.$$

The ratio of the relative increase in the cylinder radius and the relative decrease in the cylinder height is

$$\frac{u_r/R}{|u_z|/H} = \frac{\lambda}{2(\lambda + \mu)} = \nu.$$

The stress at the cylinder top is $F/\pi R^2$, and the relative change in the cylinder length is $|u_z|/H$. Then we obtain

$$\frac{F/\pi R^2}{|u_z|/H} = \mu \frac{3\lambda + 2\mu}{\lambda + \mu} = E.$$

(iv) Substituting the numerical values into the expressions for $|u_z|$ at u_r we obtain

$$|u_z| \approx 1.27 \times 10^{-3} \text{ m} = 1.27 \text{ mm}, \quad u_r \approx 0.318 \times 10^{-3} \text{ m} = 0.318 \text{ mm}.$$

6.6: (i) Repeating the solution to the previous problem and using the condition that the displacement is zero at the coordinate origin we obtain that

$$u_r = Ar, \quad u_z = Bz, \tag{$*$}$$

where A and B are constants.

(ii) In the same way as in the solution to the previous problem we obtain

$$\mathbf{T} = [2A(\lambda + \mu) + \lambda B](\boldsymbol{e}_r \boldsymbol{e}_r + \boldsymbol{e}_\phi \boldsymbol{e}_\phi) + [2A\lambda + B(\lambda + 2\mu)]\boldsymbol{e}_z \boldsymbol{e}_z.$$

The normal vectors at the flat surfaces of the rod are $\pm \boldsymbol{e}_z$. Then the surface tractions at these flat surfaces are

$$\boldsymbol{t}_{\text{flat}} = \pm \mathbf{T} \cdot \boldsymbol{e}_z = \pm[2A\lambda + B(\lambda + 2\mu)]\boldsymbol{e}_z.$$

The normal vector at the side surface of the rod is $\boldsymbol{e}_r$. Then the surface traction at the side surface is

$$\boldsymbol{t}_{\text{side}} = \mathbf{T} \cdot \boldsymbol{e}_r = [2A(\lambda + \mu) + \lambda B]\boldsymbol{e}_r.$$

The surface tractions from outside of the flat surfaces are equal to $\pm F/\pi R^2$. Then it follows from the condition of continuity of the surface traction that

$$2A\lambda + B(\lambda + 2\mu) = \frac{F}{\pi R^2}.$$

Since we can neglect the air pressure, the surface traction from outside of the side surface is zero. Then it follows from the condition of continuity of the surface traction that

$$2A(\lambda + \mu) + \lambda B = 0.$$

As a result, we obtain a system of two linear algebraic equations for A and B. The solution to this system is straightforward:

$$A = \frac{-\lambda F}{2\pi R^2 \mu(3\lambda + 2\mu)}, \qquad B = \frac{(\lambda + \mu)F}{\pi R^2 \mu(3\lambda + 2\mu)}.$$

The rod extension is equal to $2u_z$ at $z = L/2$. We obtain

$$l = 2u_z = BL = \frac{(\lambda + \mu)FL}{\pi R^2 \mu(3\lambda + 2\mu)},$$

which implies that

$$F = \frac{\pi l R^2 \mu(3\lambda + 2\mu)}{(\lambda + \mu)L}.$$

(iii) The decrease in the cylinder radius is equal to $|u_r|$ calculated at $r = R$. We obtain

$$|u_r| = |A|R = \frac{\lambda F}{2\pi R \mu(3\lambda + 2\mu)}.$$

Then the ratio of the relative decrease in the cylinder radius to its relative axial extension is

$$\frac{|u_r|/R}{l/L} = \frac{\lambda}{2(\lambda + \mu)} = \nu.$$

The ratio of stress at the flat surface of the rod to its relative extension is

$$\frac{F/\pi R^2}{l/L} = \mu\frac{3\lambda + 2\mu}{\lambda + \mu} = E.$$

6.7: (i) It follows from Eq. (2.13.9) that $\nabla \times \boldsymbol{u} = 0$. Then, using Eq. (6.3.12) we obtain $\nabla^2(u(t, r)\boldsymbol{e}_r) = \nabla[\nabla \cdot (u(t, r)\boldsymbol{e}_r)]$ and reduce Eq. (6.3.7) to

$$\frac{\partial^2 \boldsymbol{u}}{\partial t^2} = c_p^2 \nabla(\nabla \cdot \boldsymbol{u}).$$

Using the relation $\nabla \cdot (u(t,r)\boldsymbol{e}_r) = \dfrac{1}{r}\dfrac{d(ru)}{dr}$ we transform this equation to

$$\frac{\partial^2 u}{\partial t^2} = c_p^2 \left(\frac{\partial^2 u}{\partial r^2} + \frac{1}{r}\frac{\partial u}{\partial r} - \frac{u}{r^2} \right). \tag{$*$}$$

(ii) Substituting $u = \sin(\omega t)U(r)$ into equation ($*$) yields

$$\frac{d^2 U}{dr^2} + \frac{1}{r}\frac{dU}{dr} + \left(\frac{\omega^2}{c_p^2} - \frac{1}{r^2} \right) U = 0. \tag{$\dagger$}$$

The variable substitution $x = \omega r/c_p$ reduces this equation to the Bessel equation

$$\frac{d^2 U}{dx^2} + \frac{1}{x}\frac{dU}{dx} + \left(1 - \frac{1}{x^2} \right) U = 0.$$

Then the general solution to equation ($\dagger$) is

$$U(r) = AJ_1(kr) + BY_1(kr), \tag{$\ddagger$}$$

where A and B are arbitrary constants, $k = \omega/c_p$, and $J_1(x)$ and $Y_1(x)$ are the Bessel functions of the first and second kind and the first order. It is shown in the solution to problem **6.1** that for $\boldsymbol{u} = u(r)\boldsymbol{e}_r$ the stress tensor is given by

$$\mathbf{T} = \left[(\lambda + 2\mu)\frac{du}{dr} + \lambda\frac{u}{r} \right] \boldsymbol{e}_r\boldsymbol{e}_r + \left[\lambda\frac{du}{dr} + (\lambda + 2\mu)\frac{u}{r} \right] \boldsymbol{e}_\phi\boldsymbol{e}_\phi + \frac{\lambda}{r}\frac{d(ru)}{dr}\boldsymbol{e}_z\boldsymbol{e}_z.$$

Using this expression and neglecting the air pressure we obtain from the condition of stress continuity at the tube boundaries that

$$(\lambda + 2\mu)\frac{du}{dr} + \lambda\frac{u}{r} = 0 \quad \text{at } r = R,\ R + l,$$

where the prime indicates the derivative. Substituting $u = \sin(\omega t)U(r)$ into this equation and using equation ($\ddagger$) we obtain

$$A\left[(\lambda + 2\mu)J_1'(kr) + \frac{\lambda}{kr}J(kr) \right] + B\left[(\lambda + 2\mu)Y_1'(kr) + \frac{\lambda}{kr}Y(kr) \right] = 0$$
$$\text{at } r = R,\ R + l.$$

As a result we have a system of two linear homogeneous algebraic equations for two variables, A and B. It only has nontrivial solutions when its determinant is zero,

$$\left[(\lambda+2\mu)J_1'(kR)+\frac{\lambda}{kR}J(kR)\right]\left[(\lambda+2\mu)Y_1'(k(R+l))+\frac{\lambda Y(k(R+l))}{k(R+l)}\right]$$
$$-\left[(\lambda+2\mu)J_1'(k(R+l))+\frac{\lambda J(k(R+l))}{k(R+l)}\right]\left[(\lambda+2\mu)Y_1'(kR)+\frac{\lambda}{kR}Y(kR)\right]=0. \tag{$\natural$}$$

This equation determines k and, consequently, $\omega = kc_p$.

(iii) We now consider the case where $\epsilon = l/R \ll 1$ and look for solutions satisfying the condition $kl = \mathcal{O}(1)$. This condition implies that $kR = \mathcal{O}(\epsilon^{-1})$. Then, using the asymptotic relations

$$J_1(x)=\sqrt{\frac{2}{\pi x}}\cos\left(x-\frac{3\pi}{4}\right)+\mathcal{O}(x^{-1}),\quad Y_1(x)=\sqrt{\frac{2}{\pi x}}\sin\left(x-\frac{3\pi}{4}\right)+\mathcal{O}(x^{-1}),$$

valid for $x \gg 1$, we reduce equation ($\natural$) in the leading order approximation with respect to ϵ to

$$\sin(kl)=0.$$

It follows from this equation that

$$\omega=\frac{\pi c_p n}{l},\quad n=1,2,\ldots$$

We see that $kl = \omega l/c_p = \pi n = \mathcal{O}(1)$.

6.8: (i) The components of $\boldsymbol{e}_1$ are equal to δ_{1j}. Also $u_i = u\delta_{1i}$ and $\partial/\partial x_j = \delta_{1j}\partial/\partial x$, so that $\nabla\cdot\boldsymbol{u} = \partial u/\partial x$. Then, using the expression

$$T_{ij}=\lambda\delta_{ij}\nabla\cdot\boldsymbol{u}+\mu\left(\frac{\partial u_i}{\partial x_j}+\frac{\partial u_j}{\partial x_i}\right),$$

we obtain

$$t_i=T_{ij}\delta_{1j}=T_{i1}=\lambda\delta_{i1}\frac{\partial u}{\partial x}+\mu\frac{\partial u}{\partial x}(\delta_{i1}+\delta_{1i})=\delta_{1i}(\lambda+2\mu)\frac{\partial u}{\partial x}.$$

Hence,

$$t_1=(\lambda+2\mu)\frac{\partial u}{\partial x},\quad t_2=t_3=0.$$

Since $\boldsymbol{t}$ has to be continuous at $x = 0$, we obtain

$$(\lambda_- + 2\mu_-)\left(\frac{\partial u}{\partial x}\right)_- = (\lambda_+ + 2\mu_+)\left(\frac{\partial u}{\partial x}\right)_+, \tag{$*$}$$

where, for any function f,

$$f_\pm = \lim_{\epsilon \to +0} f(\pm\epsilon).$$

Since $\boldsymbol{u}$ is continuous at $x = 0$, we obtain

$$u_- = u_+. \tag{$\dagger$}$$

(ii) The incoming wave drives the transmitted wave propagating with speed c_{p+} in the half-space $x > 0$ in the positive x-direction, and the reflected wave propagating with speed c_{p-} in the half-space $x < 0$ in the negative x-direction. We know that $\cos[\omega(t - x/c_{p-})]$ and $\cos[\omega(t + x/c_{p-})]$ satisfy the equation for longitudinal waves. Since this equation is linear, a linear combination of these functions also satisfies it. Hence, we can look for a solution in the form

$$u = \begin{cases} a\cos[\omega(t - x/c_{p-})] + b\cos[\omega(t + x/c_{p-})], & x < 0, \\ c\cos[\omega(t - x/c_{p+})], & x > 0. \end{cases}$$

Substituting this expression into the boundary conditions ($*$) and ($\dagger$), we obtain

$$a\cos(\omega t) + b\cos(\omega t) = c\cos(\omega t),$$

$$\frac{(\lambda_- + 2\mu_-)(a - b)}{c_{p-}}\sin(\omega t) = \frac{(\lambda_+ + 2\mu_+)c}{c_{p+}}\sin(\omega t).$$

This system can be rewritten as

$$a + b = c, \quad \frac{(\lambda_- + 2\mu_-)(a - b)}{c_{p-}} = \frac{(\lambda_+ + 2\mu_+)c}{c_{p+}}.$$

From this system we easily find

$$\frac{b}{a} = \frac{c_{p+}(\lambda_- + 2\mu_-) - c_{p-}(\lambda_+ + 2\mu_+)}{c_{p+}(\lambda_- + 2\mu_-) + c_{p-}(\lambda_+ + 2\mu_+)}, \quad \frac{c}{a} = \frac{2c_{p+}(\lambda_- + 2\mu_-)}{c_{p+}(\lambda_- + 2\mu_-) + c_{p-}(\lambda_+ + 2\mu_+)}.$$

Then

$$R = \frac{b^2}{a^2} = \left(\frac{c_{p+}(\lambda_- + 2\mu_-) - c_{p-}(\lambda_+ + 2\mu_+)}{c_{p+}(\lambda_- + 2\mu_-) + c_{p-}(\lambda_+ + 2\mu_+)}\right)^2.$$

6.9: The rotational motion of the disc is described by

$$I\frac{d^2\alpha}{dt^2} = M, \tag{$*$}$$

where α is the rotation angle of the disc, I its moment of inertia, and M the moment of force applied to the disc. In the solution to problem **6.4** the following relation is obtained:

$$\alpha = \frac{2LM}{\pi\mu R^4}.$$

Substituting this expression into equation ($*$) we obtain

$$\frac{d^2\alpha}{dt^2} = \frac{\pi\mu R^4}{2LI}\alpha.$$

This equation describes oscillations with the frequency

$$\omega = R^2\sqrt{\frac{\pi\mu}{2LI}}.$$

6.10: Since the problem is one-dimensional we look for a solution to Eq. (6.6.2) in the form $\boldsymbol{u} = u(t, x)\boldsymbol{e}_x$. Then we obtain

$$\rho_0\frac{\partial^2 u}{\partial t^2} = (\lambda + 2\mu)\frac{\partial^2}{\partial x^2}\left(u + \varsigma\frac{\partial u}{\partial t}\right),$$

where $\varsigma = (\lambda + 2\mu)^{-1}[(\lambda + \mu)\theta_\lambda + \mu\theta_\mu]$. We look for a solution to this equation in the form $u = \Re\left(w(x)e^{-i\omega t}\right)$, where $\Re$ indicates the real part of a quantity. Substituting this expression into the equation for u yields

$$c_p^2(1 - i\varsigma\omega)\frac{d^2 w}{dx^2} + \omega^2 w = 0, \tag{$*$}$$

where $c_p^2 = (\lambda + 2\mu)/\rho_0$. The function w must also satisfy the boundary condition

$$w = ia \quad \text{at} \quad x = 0.$$

The general solution to equation ($*$) is

$$w = C_1 e^{\kappa x} + C_2 e^{-\kappa x},$$

where C_1 and C_2 are constants, and

$$\kappa = \frac{\omega\left(\sqrt{\sqrt{1+\varsigma^2\omega^2}-1} - i\sqrt{\sqrt{1+\varsigma^2\omega^2}+1}\right)}{c_p\sqrt{2(1+\varsigma^2\omega^2)}}.$$

The first term on the right-hand side of the expression for w grows exponentially as x increases. To eliminate this unphysical behaviour we must take $C_1 = 0$. Then, using the boundary condition at $x = 0$ we obtain $w = iae^{-\kappa x}$. Hence, we eventually arrive at

$$u = \Re\big(ia\exp(-\kappa x - i\omega t)\big) = A(x)\sin(\omega t - kx),$$

where the wave amplitude $A(x)$ and the wavenumber k are given by

$$A(x) = a\exp\left(-\frac{\omega x\sqrt{\sqrt{1+\varsigma^2\omega^2}-1}}{c_p\sqrt{2(1+\varsigma^2\omega^2)}}\right), \quad k = \frac{\omega\sqrt{\sqrt{1+\varsigma^2\omega^2}+1}}{c_p\sqrt{2(1+\varsigma^2\omega^2)}}.$$

6.11: We look for the solution to Eq. (6.6.2) in the form $\boldsymbol{u} = v(t,x)\boldsymbol{e}_y$. Then we obtain

$$\rho_0\frac{\partial^2 v}{\partial t^2} = \mu\frac{\partial^2}{\partial x^2}\left(v + \theta_\mu\frac{\partial v}{\partial t}\right).$$

Now we look for the solution to this equation in the form $v = \Re\left(w(x)e^{-i\omega t}\right)$. As a result we obtain

$$c_s^2(1 - i\theta_\mu\omega)\frac{d^2 w}{dx^2} + \omega^2 w = 0,$$

and the boundary condition

$$w = ia \quad \text{at} \quad x = 0,$$

where $c_s^2 = \mu/\rho_0$. Then the solution is similar to that of the previous problem. Hence, we skip the details and only give the final answer:

$$v = A(x)\sin(\omega t - kx),$$

where the wave amplitude $A(x)$ and the wavenumber k are given by

$$A(x) = a\exp\left(-\frac{\omega x\sqrt{(1+\theta_\mu^2\omega^2)^{1/2}-1}}{c_s\sqrt{2}(1+\theta_\mu^2\omega^2)^{1/2}}\right), \quad k = \frac{\omega\sqrt{(1+\theta_\mu^2\omega^2)^{1/2}+1}}{c_s\sqrt{2}(1+\theta_\mu^2\omega^2)^{1/2}}.$$

6.12: (i) It is shown in the solution to problem **6.4** that in cylindrical coordinates with the z-axis coinciding with the cylinder axis and the origin at the lower plane the displacement is given by $\boldsymbol{u} = u\boldsymbol{e}_\phi$ with u defined by

$$u = \frac{\alpha r z}{L}.$$

It is also shown that the stress tensor is

$$\mathsf{T} = \mu\left(\frac{\partial u}{\partial r} - \frac{u}{r}\right)(\boldsymbol{e}_r\boldsymbol{e}_\phi + \boldsymbol{e}_\phi\boldsymbol{e}_r) + \mu\frac{\partial u}{\partial z}(\boldsymbol{e}_\phi\boldsymbol{e}_z + \boldsymbol{e}_z\boldsymbol{e}_\phi).$$

Substituting the expression for u into this formula yields

$$\mathsf{T} = \frac{\mu\alpha r}{L}(\boldsymbol{e}_\phi\boldsymbol{e}_z + \boldsymbol{e}_z\boldsymbol{e}_\phi).$$

It is straightforward to obtain that the principal stresses are $T_1 = -\mu\alpha r/L$, $T_2 = 0$, and $T_3 = \mu\alpha r/L$. Substituting these expressions into Eq. (6.7.2) we obtain that the yield condition is

$$\alpha = \frac{LY}{\sqrt{3}\mu r}. \qquad (*)$$

It is obvious that this condition is first satisfied at $r = R$ and we obtain

$$\alpha_Y = \frac{LY}{\sqrt{3}\mu R}.$$

Using the relation between the momentum M applied to the upper plane and α derived in the solution to problem **6.4** we obtain that the corresponding momentum is

$$M_Y = \frac{\pi Y R^3}{2\sqrt{3}}.$$

(ii) When $\alpha > \alpha_Y$ the yield spreads inward from $r = R$. We make the assumption that in the region of plastic behaviour the displacement is still given by $\boldsymbol{u} = rf(z)\boldsymbol{e}_\phi$. Then it follows from the expressions for $\nabla\boldsymbol{u}$ and $(\nabla\boldsymbol{u})^T$ obtained in the solution to problem **6.4** that the strain tensor is given by

$$\mathsf{E} = \frac{r}{2}f'(z)(\boldsymbol{e}_\phi\boldsymbol{e}_z + \boldsymbol{e}_z\boldsymbol{e}_\phi),$$

where the prime indicates the derivative. Then it follows from Eq. (6.7.5) that

$$\mathsf{T} = G(r, z)(\boldsymbol{e}_\phi\boldsymbol{e}_z + \boldsymbol{e}_z\boldsymbol{e}_\phi).$$

The principal stresses are $-G(r, z)$, 0, and $G(r, z)$. Substituting these expressions into Eq. (6.7.2) we obtain $G = Y/\sqrt{3} = \text{const}$. It follows from Eq. (6.7.5) that

$$d(rf') = 2G\,d\Lambda,$$

however, this relation is not used in what follows. Equation $(*)$ implies that the yield condition is satisfied for $r \geq r_0$, while the material behaviour is described by the linear elasticity for $r < r_0$, where

$$r_0 = \frac{LY}{\sqrt{3}\mu\alpha} = R\frac{\alpha_Y}{\alpha}.$$

The surface traction at the upper edge of the rod is given by

$$\boldsymbol{t} = \mathbf{T} \cdot \boldsymbol{e}_z = \boldsymbol{e}_\phi \begin{cases} \mu\alpha r/L, \ r < r_0, \\ Y/\sqrt{3}, \ \ r \geq r_0. \end{cases}$$

Then the momentum of the force needed to rotate the upper surface by the angle α is

$$M = \int_0^{2\pi} d\phi \int_0^R (r\|\boldsymbol{t}\|)r\,dr = 2\pi\left(\frac{\mu\alpha}{L}\int_0^{r_0} r^3 dr + \frac{Y}{\sqrt{3}}\int_{r_0}^R r^2 dr\right)$$
$$= \frac{2\pi Y R^3}{3\sqrt{3}}\left[1 - \frac{1}{4}\left(\frac{\alpha_Y}{\alpha}\right)^3\right].$$

This expression is only valid for $\alpha \geq \alpha_Y$.

A.7 Solutions for Chap. 7

7.1: We introduce cylindrical coordinates $r,\ \phi,\ z$ with the z-axis coinciding with the cylinder axis. Since $\boldsymbol{v} = v(r)\boldsymbol{e}_\phi$ it follows from Eq. (2.13.10) that

$$(\boldsymbol{v} \cdot \nabla)\boldsymbol{v} = -\frac{v^2}{r}\boldsymbol{e}_r. \qquad (*)$$

Using Eqs. (2.13.1), (2.13.8), (2.13.9) and (6.3.12) we obtain

$$\nabla^2 \boldsymbol{v} = -\nabla \times (\nabla \times \boldsymbol{v}) = \frac{d}{dr}\left[\frac{1}{r}\frac{d(rv)}{dr}\right]\boldsymbol{e}_\phi. \qquad (\dagger)$$

Then, using Eqs. (2.13.1), (*), and (†) we reduce Eq. (7.1.6) to

$$\frac{dp}{dr} = \frac{\rho v^2}{r}, \quad \frac{d}{dr}\left[\frac{1}{r}\frac{d(rv)}{dr}\right] = 0. \qquad (\ddagger)$$

Below we only use the second of these two equations. Integrating this equation we obtain

$$v = Ar + \frac{B}{r},$$

where A and B are arbitrary constants. The velocity at the cylinder surfaces must be equal to the velocity of these surfaces. As a result we have the boundary conditions

$$v = R_1\Omega_1 \quad \text{at} \quad r = R_1, \qquad v = R_2\Omega_2 \quad \text{at} \quad r = R_2.$$

Substituting the expression for v in these boundary conditions yields

$$AR_1^2 + B = R_1^2\Omega_1, \quad AR_2^2 + B = R_2^2\Omega_2.$$

Solving this system of two linear algebraic equations we obtain

$$A = \frac{R_2^2\Omega_2 - R_1^2\Omega_1}{R_2^2 - R_1^2}, \quad B = \frac{R_1^2 R_2^2(\Omega_1 - \Omega_2)}{R_2^2 - R_1^2}.$$

Substituting this expressions into the equation for v results in

$$v = \frac{r^2(R_2^2\Omega_2 - R_1^2\Omega_1) + R_1^2 R_2^2(\Omega_1 - \Omega_2)}{r(R_2^2 - R_1^2)}.$$

We now calculate the tangential component of the surface traction at $r = R_1$ and $r = R_2$. Since the surface traction component related to the pressure is orthogonal to any surface, we only need to find the expression for T'. It follows from Eq. (7.1.3) that

$$\mathsf{T}' = \eta\big[\nabla\boldsymbol{v} + (\nabla\boldsymbol{v})^T\big].$$

Using Eq. (2.13.7) we obtain

$$\nabla\boldsymbol{v} = \frac{dv}{dr}\boldsymbol{e}_r\boldsymbol{e}_\phi - \frac{v}{r}\boldsymbol{e}_\phi\boldsymbol{e}_r, \quad (\nabla\boldsymbol{v})^T = \frac{dv}{dr}\boldsymbol{e}_\phi\boldsymbol{e}_r - \frac{v}{r}\boldsymbol{e}_r\boldsymbol{e}_\phi.$$

Substituting this result into the expression for T' yields

$$\mathsf{T}' = \eta\left(\frac{dv}{dr} - \frac{v}{r}\right)(\boldsymbol{e}_r\boldsymbol{e}_\phi + \boldsymbol{e}_\phi\boldsymbol{e}_r).$$

Then the part of the surface traction caused by viscosity is

$$\mathsf{T}' \cdot \boldsymbol{n} = \eta\left(\frac{dv}{dr} - \frac{v}{r}\right)(\boldsymbol{e}_r \cdot \boldsymbol{n})\boldsymbol{e}_\phi,$$

where $\boldsymbol{n} = \boldsymbol{e}_r$ at the inner cylinder and $\boldsymbol{n} = -\boldsymbol{e}_r$ at the outer cylinder. We see that $\mathsf{T}' \cdot \boldsymbol{n}$ is in the ϕ-direction, that is, it is tangential to the cylindrical surfaces. Substituting the expression for v into this formula we obtain

$$\boldsymbol{t}_\tau = \mathsf{T}' \cdot \boldsymbol{n} = \frac{2\eta R_1^2 R_2^2(\Omega_2 - \Omega_1)}{r^2(R_2^2 - R_1^2)}(\boldsymbol{e}_r \cdot \boldsymbol{n})\boldsymbol{e}_\phi.$$

Then it follows that the tangential components of surface traction at the inner and outer cylinder are given by

$$\boldsymbol{t}_{\tau 1} = \frac{2\eta R_2^2(\Omega_2 - \Omega_1)}{R_2^2 - R_1^2}\boldsymbol{e}_\phi, \quad \boldsymbol{t}_{\tau 2} = -\frac{2\eta R_1^2(\Omega_2 - \Omega_1)}{R_2^2 - R_1^2}\boldsymbol{e}_\phi.$$

7.2: We introduce cylindrical coordinates r, ϕ, z with the z-axis coinciding with the cylinder axis. Then we look for the velocity in the form $\boldsymbol{v} = v(r)\boldsymbol{e}_z$. It follows from Eq. (2.13.10) that $(\boldsymbol{v} \cdot \nabla)\boldsymbol{v} = 0$. Using Eq. (2.13.8) we obtain $\nabla \cdot \boldsymbol{v} = 0$. Then it follows from Eqs. (2.13.9) and (6.3.12) that

$$\nabla^2 \boldsymbol{v} = -\nabla \times (\nabla \times \boldsymbol{v}) = \frac{1}{r}\frac{d}{dr}r\frac{dv}{dr}\boldsymbol{e}_z. \qquad (*)$$

Now, assuming that the pressure is independent of z, we obtain from Eq. (7.1.6) that

$$\frac{d}{dr}r\frac{dv}{dr} = 0.$$

In addition, it also follows from Eq. (7.1.6) that $p = \text{const}$. Integrating the equation for v yields

$$v = A\ln r + B,$$

where A and B are constants. Since the fluid velocity at the cylinders must coincide with the velocity of cylindrical surfaces, we have the boundary conditions

$$v = V \quad \text{at} \quad r = R_1, \qquad v = 0 \quad \text{at} \quad r = R_2.$$

Substituting the expression for v into these boundary conditions we obtain

$$A\ln R_1 + B = V, \quad A\ln R_2 + B = 0.$$

We found from these equations that

$$A = -\frac{V}{\ln(R_2/R_1)}, \quad B = -A\ln R_2.$$

Substituting this result into the expression for v yields

$$v = -\frac{V\ln(r/R_2)}{\ln(R_2/R_1)}.$$

Using Eq. (2.13.7) and the expression for v we obtain

$$\nabla\boldsymbol{v} = \frac{dv}{dr}\boldsymbol{e}_r\boldsymbol{e}_z, \quad (\nabla\boldsymbol{v})^T = \frac{dv}{dr}\boldsymbol{e}_z\boldsymbol{e}_r.$$

Then it follows from Eq. (7.1.3) that

$$\mathbf{T}' = \eta\big[\nabla \boldsymbol{v} + (\nabla \boldsymbol{v})^T\big] = \eta \frac{dv}{dr}(\boldsymbol{e}_r\boldsymbol{e}_z + \boldsymbol{e}_z\boldsymbol{e}_r).$$

Since the pressure exerts a force orthogonal to any surface, the tangential component of the surface traction at the cylindrical surfaces is $\boldsymbol{t}_\tau = \mathbf{T}' \cdot \boldsymbol{n}$, where $\boldsymbol{n} = \boldsymbol{e}_r$ at the internal cylinder and $\boldsymbol{n} = -\boldsymbol{e}_r$ at the external cylinder. Then it follows that

$$\boldsymbol{t}_\tau = \eta \frac{dv}{dr}(\boldsymbol{e}_r \cdot \boldsymbol{n})\boldsymbol{e}_z.$$

Substituting the expression for $\boldsymbol{v}$ into this formula we obtain

$$\boldsymbol{t}_{\tau 1} = -\frac{\eta V}{R_1 \ln(R_2/R_1)}\boldsymbol{e}_z, \quad \boldsymbol{t}_{\tau 2} = \frac{\eta V}{R_2 \ln(R_2/R_1)}\boldsymbol{e}_z.$$

The tangential forces acting on the surfaces of the internal and external cylinders are in the anti-z and z-direction, respectively, and are equal to

$$f_1 = -2\pi R_1 \boldsymbol{t}_{\tau 1} \cdot \boldsymbol{e}_z = \frac{2\pi\eta V}{\ln(R_2/R_1)}, \quad f_2 = 2\pi R_2 \boldsymbol{t}_{\tau 2} \cdot \boldsymbol{e}_z = \frac{2\pi\eta V}{\ln(R_2/R_1)}.$$

These forces are calculated per unit length in the axial direction of each of the two cylinders.

7.3: We introduce cylindrical coordinates $r,\ \phi,\ z$ with the z-axis coinciding with the cylinder axis. We look for the velocity in the form $\boldsymbol{v} = v(r)\boldsymbol{e}_z$. Then it follows from Eqs. (2.13.8) and (2.13.10) that $\nabla \cdot \boldsymbol{v} = 0$ and $(\boldsymbol{v} \cdot \nabla)\boldsymbol{v} = 0$, while Eq. (6.3.12) gives

$$\nabla^2 \boldsymbol{v} = -\nabla \times (\nabla \times \boldsymbol{v}) = \frac{1}{r}\frac{d}{dr} r \frac{dv}{dr}\boldsymbol{e}_z.$$

Using these results we obtain from Eq. (7.1.6)

$$\frac{\partial p}{\partial r} = 0, \quad \frac{\partial p}{\partial \phi} = 0, \quad \frac{\partial p}{\partial z} = \frac{\eta}{r}\frac{d}{dr} r \frac{dv}{dr}. \qquad (*)$$

It follows from the first two equations that p only depends on z. Since the right-hand side of the third equation only depends on r we conclude that $dp/dz = \text{const}$. We introduce the notation $dp/dz = -\Delta p$. Then integrating the third equation in $(*)$ yields

$$v = -\frac{r^2(\Delta p)}{4\eta} + A \ln r + B. \qquad (\dagger)$$

Since the velocity must be regular at $r = 0$ we conclude that $A = 0$. Since the fluid cannot slide along the cylinder surface, the velocity must satisfy the boundary condition

$$v = 0 \quad \text{at} \quad r = R.$$

It follows from this condition that $B = R^2(\Delta p)/4\eta$ and we obtain for the velocity

$$v = \frac{(\Delta p)(R^2 - r^2)}{4\eta}.$$

The pressure only imposes a normal force on the cylinder surface, so the tangential component of the force is determined by $\mathbf{T}'$. It follows from Eqs. (2.13.7) and (7.1.3) that

$$\mathbf{T}' = \eta\big[\nabla \boldsymbol{v} + (\nabla \boldsymbol{v})^T\big] = \eta \frac{dv}{dr}(\boldsymbol{e}_r \boldsymbol{e}_z + \boldsymbol{e}_z \boldsymbol{e}_r).$$

Then, taking into account that the unit normal at the cylinder surface is $-\boldsymbol{e}_r$ we obtain

$$\boldsymbol{t}_\tau = -\eta \frac{dv}{dr} \boldsymbol{e}_z = \frac{R(\Delta p)}{2} \boldsymbol{e}_z.$$

The tangential force acting on the tube surface per unit length along the cylinder axis is

$$f = 2\pi \boldsymbol{t}_\tau \cdot \boldsymbol{e}_z = \pi R(\Delta p).$$

7.4: The solution to this problem is the same as the one to problem **7.3** up to equation (†). The boundary conditions are now

$$v = 0 \quad \text{at} \quad r = R_1,\ R_2.$$

Using equation (†) from the solution to problem **7.3** we obtain

$$-\frac{R_1^2(\Delta p)}{4\eta} + A \ln R_1 + B = 0, \qquad -\frac{R_2^2(\Delta p)}{4\eta} + A \ln R_2 + B = 0.$$

Solving this system of two algebraic equations we obtain

$$A = \frac{(\Delta p)(R_2^2 - R_1^2)}{4\eta \ln(R_2/R_1)}, \qquad B = \frac{(\Delta p)(R_1^2 \ln R_2 - R_2^2 \ln R_1)}{4\eta \ln(R_2/R_1)}.$$

Substituting these expressions into equation (†) from the solution to problem **7.3** yields

$$v = \frac{\Delta p}{4\eta}\left(\frac{R_2^2 \ln(r/R_1) - R_1^2 \ln(r/R_2)}{\ln(R_2/R_1)} - r^2\right).$$

Similar to the solution to problem **7.3** we obtain

$$\mathbf{T}' = \eta\big[\nabla \boldsymbol{v} + (\nabla \boldsymbol{v})^T\big] = \eta \frac{dv}{dr}(\boldsymbol{e}_r \boldsymbol{e}_z + \boldsymbol{e}_z \boldsymbol{e}_r).$$

Then the tangential component of the surface traction at the cylinder surfaces is

$$\boldsymbol{t}_\tau = \eta \frac{dv}{dr}(\boldsymbol{e}_r \cdot \boldsymbol{n})\boldsymbol{e}_z,$$

where $\boldsymbol{n} = \boldsymbol{e}_r$ at the internal cylinder and $\boldsymbol{n} = -\boldsymbol{e}_r$ at the external cylinder. Now the tangential forces acting on the surfaces of each of the cylinders per unit length along the cylinder axis are $f = 2\pi \boldsymbol{t}_\tau \cdot \boldsymbol{e}_z$, and we obtain

$$f_1 = -\frac{\pi(\Delta p)}{2}\left(2R_1 - \frac{(R_2^2 - R_1^2)}{R_1 \ln(R_2/R_1)}\right), \quad f_2 = \frac{\pi(\Delta p)}{2}\left(2R_2 - \frac{(R_2^2 - R_1^2)}{R_2 \ln(R_2/R_1)}\right).$$

7.5: We assume that $\boldsymbol{v} = v(r, \theta)\boldsymbol{e}_\phi$. Then it follows from Eq. (2.13.17) that $\nabla \cdot \boldsymbol{v} = 0$, while Eqs. (2.13.18) and (6.3.12) give

$$\nabla^2 \boldsymbol{v} = \frac{\boldsymbol{e}_\phi}{r}\left[\frac{\partial^2(rv)}{\partial r^2} + \frac{1}{r}\frac{\partial}{\partial \theta}\left(\frac{1}{\sin\theta}\frac{\partial(v \sin\theta)}{\partial \theta}\right)\right].$$

Using this result and Eq. (2.13.11) we obtain from Eq. (7.4.3)

$$\frac{\partial p}{\partial r} = 0, \quad \frac{\partial p}{\partial \theta} = 0, \quad \frac{\partial p}{\partial \phi} = \eta \sin\theta \left[\frac{\partial^2(rv)}{\partial r^2} + \frac{1}{r}\frac{\partial}{\partial \theta}\left(\frac{1}{\sin\theta}\frac{\partial(v \sin\theta)}{\partial \theta}\right)\right].$$

It follows from the first two equations that p is independent of r and θ. Since the right-hand side of the third equation is independent of ϕ it follows from this equation that p is a linear function of ϕ unless the right-hand side of the third equation is zero. However p must be a periodic function of ϕ, which implies that the right-hand side must be zero,

$$\frac{\partial^2(rv)}{\partial r^2} + \frac{1}{r}\frac{\partial}{\partial \theta}\left(\frac{1}{\sin\theta}\frac{\partial(v \sin\theta)}{\partial \theta}\right) = 0. \qquad (*)$$

We look for the solution to this equation in the form $v = f(r)\sin\theta$. Substituting this expression into equation $(*)$ we obtain

$$r\frac{d^2(rf)}{dr^2} - 2f = 0. \qquad (\dagger)$$

We look for a solution to this equation in the form $f = r^\alpha$. Substituting it into this equation yields $\alpha(\alpha + 1) = 2$. This equation has two solutions, $\alpha = 1$ and $\alpha = -2$. Then the general solution to equation $(\dagger)$ is $f(r) = Ar + Br^{-2}$, where A and B are arbitrary constants. Hence, we obtain

$$v = (Ar + Br^{-2})\sin\theta.$$

The velocity of the point with coordinates (R_2, θ, ϕ) is $(R_2\Omega\sin\theta)\boldsymbol{e}_\phi$. Then v must satisfy the boundary conditions

$$v = 0 \quad \text{at} \quad r = R_1, \qquad v = R_2\Omega\sin\theta \quad \text{at} \quad r = R_2.$$

Substituting the expression for v into these boundary conditions yields

$$AR_1 + BR_1^{-2} = 0, \quad AR_2 + BR_2^{-2} = R_2\Omega.$$

It follows from these equations that

$$A = \frac{\Omega R_2^3}{R_2^3 - R_1^3}, \quad B = -\frac{\Omega R_1^3 R_2^3}{R_2^3 - R_1^3}.$$

Hence, eventually we obtain

$$v = \frac{\Omega R_2^3(r - R_1^3 r^{-2})\sin\theta}{R_2^3 - R_1^3}.$$

Using Eqs. (2.13.15) and (2.13.16) we obtain

$$\nabla\boldsymbol{v} = \frac{\partial v}{\partial r}\,\boldsymbol{e}_r\boldsymbol{e}_\phi + \frac{1}{r}\frac{\partial v}{\partial\theta}\,\boldsymbol{e}_\theta\boldsymbol{e}_\phi - \frac{v}{r}\,\boldsymbol{e}_\phi\boldsymbol{e}_r - \frac{v\cot\theta}{r}\,\boldsymbol{e}_\phi\boldsymbol{e}_\theta.$$

Substituting the expression for v into this equation yields

$$\nabla\boldsymbol{v} = \frac{\Omega R_2^3}{R_2^3 - R_1^3}\big[(1 + 2R_1^3 r^{-3})\sin\theta\,\boldsymbol{e}_r\boldsymbol{e}_\phi \\ + (1 - R_1^3 r^{-3})\cos\theta(\boldsymbol{e}_\theta\boldsymbol{e}_\phi - \boldsymbol{e}_\phi\boldsymbol{e}_\theta) - (1 - R_1^3 r^{-3})\sin\theta\,\boldsymbol{e}_\phi\boldsymbol{e}_r\big].$$

The unit normal vector at the external sphere is $-\boldsymbol{e}_r$. Then the tangent component of the surface traction at the external sphere is

$$\boldsymbol{t}_\tau = -\mathbf{T}'\cdot\boldsymbol{e}_r\big|_{r=R_2} = \eta\big[\nabla\boldsymbol{v} + (\nabla\boldsymbol{v})^T\big]\cdot\boldsymbol{e}_r\big|_{r=R_2} = -\frac{3\eta\Omega R_1^3\sin\theta}{R_2^3 - R_1^3}\,\boldsymbol{e}_\phi.$$

The magnitude of the tangential force applied to an elementary surface of the external sphere dS at the point with coordinates (R_2, θ, ϕ) is $\|\boldsymbol{t}_\tau\|\,dS$. The momentum created by this force is $\|\boldsymbol{t}_\tau\|R_2\sin\theta\,dS$ and it is antiparallel to the z-axis. Then to keep the external sphere rotating with constant angular speed one needs to apply a momentum with magnitude

$$M = \int_S \|\boldsymbol{t}_\tau\| R_2 \sin\theta \, dS = \frac{6\pi\eta\Omega R_1^3 R_2^3}{R_2^3 - R_1^3} \int_0^\pi \sin^2\theta \, d\theta = \frac{3\pi^2\eta\Omega R_1^3 R_2^3}{R_2^3 - R_1^3}.$$

This momentum is parallel to the z-axis.

Since the characteristic speed in this problem is ΩR_2 we can define the Reynolds number as $\mathrm{Re} = \eta\Omega R_2^2/\rho$. Hence, formally the solution using Stokes' approximation is valid when $\mathrm{Re} = \eta\Omega R_2^2/\rho \ll 1$. However, the comparison with the numerical solution obtained using the full Navier–Stokes equation revealed that the solution based on Stokes' approximation provides very good description of the flow between the spheres even when Re is of the order of unity.

7.6: The ball volume is $\frac{4}{3}\pi R^3$. Hence, according to Archimedes' law there is a force acting on the ball that is directed upward and equal to $\frac{4}{3}\pi g\rho_f R^3$, where g is the acceleration due to gravity. The force of gravity acting on the ball is $\frac{4}{3}\pi g\rho_b R^3$. On the other hand, there is a resistance force acting on the ball. In accordance with Eq. (7.4.22) it is equal to $6\pi\eta RU$. Since the ball moves with a constant speed, the three forces are in equilibrium. Hence, we have

$$\frac{4}{3}\pi g(\rho_b - \rho_f)R^3 = 6\pi\eta RU,$$

which gives

$$\eta = \frac{2g(\rho_b - \rho_f)R^2}{3U} = \frac{5}{9}\,\mathrm{kg\,m^{-1}\,s^{-1}}.$$

The Reynolds number is $\mathrm{Re} = \rho_f RU/\eta = 0.9$, which is smaller than unity. Hence, we expect that the expression for the viscous force given by Eq. (7.4.22) works very well.

7.7: We introduce Cartesian coordinates x, y, z with the x-axis parallel to the flow velocity and the z-axis perpendicular to the plate. The tangential force acting on the strip of length dx and with unit length in the y-direction is $t_x\,dx$. The tangential component of the surface traction is given by Eq. (7.5.23). Then the tangential force acting on the plate per unit length in the direction perpendicular to the fluid velocity is

$$f = \int_0^l t_x \, dx = 0.332\sqrt{\rho\eta U^3} \int_0^l \frac{dx}{\sqrt{x}} = 0.664\sqrt{\rho\eta l U^3}.$$

A.8 Solutions for Chap. 8

8.1: Substituting Eq. (8.3.6) into Eq. (8.3.5) we obtain

$$\rho T \frac{DS}{Dt} = \mathbf{D} : \mathbf{T}' - \nabla \cdot \boldsymbol{q}.$$

8.2: In a sound wave the pressure and density perturbation are related by $\delta p = a_s^2 \delta\rho$, where $a_s^2 = a_s^2(\rho_0) = \gamma p_0/\rho_0$. Then it follows from Eq. (8.4.9) that

$$\delta p = f(x - a_s t) + g(x + a_s t).$$

Since the wave propagating in the positive x-direction is harmonic and the density oscillation amplitude in this wave is $\epsilon\rho_0$, it follows that

$$f(x - a_s t) = \epsilon\rho_0 a_s^2 \sin[\omega(x/a_s - t)].$$

Hence we transform the expression for δp into

$$\delta p = \epsilon\rho_0 a_s^2 \sin[\omega(x/a_s - t)] + g(x + a_s t).$$

It follows from Eq. (8.4.6) that

$$\frac{\partial v}{\partial t} = -\frac{1}{\rho_0}\frac{\partial \delta p}{\partial x} = -\epsilon a_s \omega \cos[\omega(x/a_s - t)] - \frac{1}{\rho_0} g'(x + a_s t),$$

where v is the velocity. Integrating this equation we obtain

$$v = \epsilon a_s \sin[\omega(x/a_s - t)] - \frac{1}{\rho_0 a_s} g(x + a_s t) + \Phi(x),$$

where $\Phi(x)$ is an arbitrary function. It follows from Eq. (8.4.5) that

$$\frac{\partial v}{\partial x} = -\frac{1}{\rho_0 a_s^2}\frac{\partial \delta p}{\partial t}.$$

Substituting the expressions for v and δp into this equation yields $\Phi'(x) = 0$, which implies that $\Phi(x) = \text{const}$. Then we can include Φ in the function g and take $\Phi = 0$ in the expression for v. Since $v = 0$ at the rigid wall, that is, at $x = 0$, it follows that

$$g(a_s t) = -\epsilon\rho_0 a_s^2 \sin(\omega t).$$

Using this result we transform the expression for v and δp to

$$v = \epsilon a_s \{\sin[\omega(x/a_s - t)] + \sin[\omega(x/a_s + t)]\},$$

$$\delta p = \epsilon\rho_0 a_s^2 \{\sin[\omega(x/a_s - t)] - \sin[\omega(x/a_s + t)]\}.$$

The pressure imposed by the wave on the rigid wall is

$$\delta p = -2\epsilon\rho_0 a_s^2 \sin(\omega t).$$

8.3: The speed of sound in the left half-space is $a_{s1} = (\gamma p_0/\rho_1)^{1/2}$, and it is $a_{s2} = (\gamma p_0/\rho_2)^{1/2}$ in the right half-space. We look for the solution to the problem in the form

$$\delta\rho_1 = \epsilon\rho_1 \sin[\omega(x/a_{s1} - t)] + A\sin[\omega(x/a_{s1} + t) + \varphi_1],$$

in the left half-space, and in the form

$$\delta\rho_2 = B\sin[\omega(x/a_{s2} - t) + \varphi_2],$$

in the right half-space, where A, B, φ_1, and φ_2 are constant, $|A|$ and $|B|$ being the amplitudes of the density perturbation in the reflected and transmitted waves, respectively. The pressure perturbation is given by

$$\delta p = \begin{cases} a_{s1}^2\delta\rho_1, & x < 0, \\ a_{s2}^2\delta\rho_2, & x > 0. \end{cases}$$

It follows from Eq. (8.4.6) that

$$\rho_0 \frac{\partial v}{\partial t} = -a_s^2 \frac{\partial \delta\rho}{\partial x}.$$

Then, imposing the condition that the average of the velocity over the wave period is zero we obtain

$$v = \begin{cases} a_{s1}\{\epsilon \sin[\omega(x/a_{s1} - t)] - (A/\rho_1)\sin[\omega(x/a_{s1} + t) + \varphi_1]\}, & x < 0, \\ (a_{s2}B/\rho_2)\sin[\omega(x/a_{s2} - t) + \varphi_2], & x > 0. \end{cases}$$

Now we use the condition that v and p must be continuous at $x = 0$. As a result we obtain

$$a_{s1}\epsilon \sin(\omega t) + \frac{a_{s1}A}{\rho_1}\sin(\omega t + \varphi_1) = \frac{a_{s2}B}{\rho_2}\sin(\omega t - \varphi_2),$$

$$a_{s1}^2[\epsilon\rho_1 \sin(\omega t) - A\sin(\omega t + \varphi_1)] = a_{s2}^2 B\sin(\omega t - \varphi_2).$$

It follows from these equations that

$$A\sin(\omega t + \varphi_1) = \epsilon\rho_1 \frac{\rho_1 a_{s1} - \rho_2 a_{s2}}{\rho_1 a_{s1} + \rho_2 a_{s2}}\sin(\omega t), \quad B\sin(\omega t - \varphi_2) = \frac{2\epsilon\rho_2^2 a_{s2}\sin(\omega t)}{\rho_1 a_{s1} + \rho_2 a_{s2}}.$$

When deriving the expression for B we used the relation $\rho_1 a_{s1}^2 = \rho_2 a_{s2}^2$. It is obvious that these equations can only be satisfied if $\varphi_1 = \pi n$ and $\varphi_2 = \pi m$, where n and m are integers. Since 2π is the period of sine it is enough to take either $n = 0$ or $n = 1$, and $m = 0$ or $m = 1$. If $n = 1$ then it is equivalent to changing the sign of A, and the same with m and B. Since we are only interested in $|A|$ and $|B|$ we can take without loss of generality $n = m = 0$. Hence, the amplitudes of the density perturbation in the reflected and transmitted waves are

$$A = \epsilon\rho_1 \frac{|\rho_1 a_{s1} - \rho_2 a_{s2}|}{\rho_1 a_{s1} + \rho_2 a_{s2}}, \quad B = \frac{2\epsilon\rho_2^2 a_{s2}}{\rho_1 a_{s1} + \rho_2 a_{s2}}.$$

8.4: Using Eq. (8.5.9) we obtain that the Mach number at the nozzle exit is

$$M_1 = \sqrt{\left(M_0^2 + \frac{2}{\gamma - 1}\right)\left(\frac{p_0}{p_1}\right)^{1-1/\gamma} - \frac{2}{\gamma - 1}} = \sqrt{9 \times 2^{2/7} - 5} \approx 2.44.$$

Then using Eqs (8.5.8) and (8.5.11) we obtain

$$\rho_1 = \rho_0 \left[\frac{2 + (\gamma - 1)M_0^2}{2 + (\gamma - 1)M_1^2}\right]^{\frac{1}{\gamma-1}} \approx 0.61\rho_0,$$

$$A = \frac{A_0 M_0}{M_1}\left[\frac{2 + (\gamma - 1)M_1^2}{2 + (\gamma - 1)M_0^2}\right]^{\frac{\gamma+1}{2(\gamma-1)}} \approx 1.48 A_0.$$

8.5: We write $M_1 = 1 + \epsilon$. Then using Eq. (8.6.17) we obtain

$$\begin{aligned}
F &= \frac{2\gamma(1+\epsilon)^2 - (\gamma - 1)}{\gamma + 1}\left[\frac{2 + (\gamma - 1)(1+\epsilon)^2}{(\gamma + 1)(1+\epsilon)^2}\right]^\gamma = \left(1 + \frac{4\gamma\epsilon}{\gamma + 1} + \frac{2\gamma\epsilon^2}{\gamma + 1}\right) \\
&\quad \times \left[\left(1 + \frac{2(\gamma - 1)\epsilon}{\gamma + 1} + \frac{(\gamma - 1)\epsilon^2}{\gamma + 1}\right)\left(1 - 2\epsilon + 3\epsilon^2 - 4\epsilon^3 + \mathcal{O}(\epsilon^4)\right)\right]^\gamma \\
&= \left(1 + \frac{4\epsilon}{\gamma + 1} + \frac{2\gamma\epsilon^2}{\gamma + 1}\right)\left(1 - \frac{4\epsilon}{\gamma + 1} + \frac{6\epsilon^2}{\gamma + 1} - \frac{8\epsilon^3}{\gamma + 1} + \mathcal{O}(\epsilon^4)\right)^\gamma \\
&= \left(1 + \frac{4\gamma\epsilon}{\gamma + 1} + \frac{2\gamma\epsilon^2}{\gamma + 1}\right)\left(1 - \frac{4\gamma\epsilon}{\gamma + 1} + \frac{2\gamma(7\gamma - 1)\epsilon^2}{(\gamma + 1)^2}\right. \\
&\quad \left. - \frac{16\gamma(8\gamma^2 - 3\gamma + 1)\epsilon^3}{3(\gamma + 1)^3} + \mathcal{O}(\epsilon^4)\right) = 1 + \frac{16\gamma(\gamma - 1)\epsilon^3}{3(\gamma + 1)^2} + \mathcal{O}(\epsilon^4).
\end{aligned}$$

Substituting this result into Eq. (8.6.16) yields

$$S_2 - S_1 = \frac{16\gamma C_v(\gamma - 1)\epsilon^3}{3(\gamma + 1)^2} + \mathcal{O}(\epsilon^4).$$

8.6: We introduce the x-axis in the direction of the shock motion. The coordinate of the wall is $x = 0$. The gas occupies the region $x < 0$. Using Eqs. (8.6.7) and (8.6.10) we obtain that the density, pressure, and Mach number behind the shock are

$$\rho_2 = \frac{(\gamma+1)\rho_1 M_1^2}{2+(\gamma-1)M_1^2} = \frac{27\rho_1}{7} \approx 3.86\rho_1,$$

$$p_2 = p_1\frac{2\gamma M_1^2 - \gamma + 1}{\gamma+1} = \frac{31p_1}{3} \approx 10.33p_1,$$

$$M_2^2 = \frac{(\gamma-1)M_1^2+2}{2\gamma M_1^2 - \gamma + 1} = \frac{7}{31} \approx 0.226, \quad M_2 \approx 0.475.$$

We now introduce three reference frames: $\mathcal{S}_1$ attached to the wall, $\mathcal{S}_2$ attached to the shock propagating toward the wall, and $\mathcal{S}_3$ attached to the reflected shock. The direction of the x-axis is the same in all three reference frames. The sound speed behind the shock is $a_2 = (\gamma p_2/\rho_2)^{1/2} = (a_1/9)\sqrt{217} \approx 1.64a_1$, where $a_1 = (\gamma p_1/\rho_1)^{1/2}$. Then in $\mathcal{S}_2$ the velocity behind the shock is $\tilde{v}_2 = -a_2M_2 \approx -0.777a_1$, where the minus sign indicates that the velocity is in the negative x-direction. Then in $\mathcal{S}_1$ the velocity of the gas behind the shock is $v_2 = M_1a_1 + \tilde{v}_2 \approx 2.22a_1$.

After the shock hits the wall it is reflected and the reflected shock propagates in the negative x-direction in the gas with the density, pressure, and velocity equal to ρ_2, p_2, and v_2. Let the reflected shock Mach number be M_3. Then it moves with velocity $-M_3a_2$ with respect to the gas ahead of it, and with velocity $v_2 - M_3a_2 \approx 2.22a_1 - M_3a_2$ in $\mathcal{S}_1$. The density, pressure, and Mach number behind the reflected shock are

$$\rho_3 = \frac{(\gamma+1)\rho_2 M_3^2}{2+(\gamma-1)M_3^2} \approx \frac{4.63\rho_1 M_3^2}{1+0.2M_3^2},$$

$$p_3 = p_2\frac{2\gamma M_3^2 - \gamma + 1}{\gamma+1} \approx (12.05M_3^2 - 1.72)p_1,$$

$$M_4^2 = \frac{(\gamma-1)M_3^2+2}{2\gamma M_3^2 - \gamma + 1} = \frac{0.2M_3^2+1}{1.4M_3^2 - 0.2}.$$

Then the speed of sound behind the reflected shock is

$$a_3 = \sqrt{\frac{\gamma p_3}{\rho_3}} \approx \frac{a_1}{M_3}\sqrt{0.52M_3^4 + 2.53M_3^2 - 0.37}.$$

In $\mathcal{S}_3$ the flow velocity behind the reflected show is M_4a_3. The shock velocity in $\mathcal{S}_1$ is $v_2 - M_3a_2$, and the flow velocity is $M_4a_3 + v_2 - M_3a_2$ in $\mathcal{S}_1$. Since the flow velocity at the wall must be zero, we obtain the equation determining M_3. It is $M_4a_3 + v_2 - M_3a_2 = 0$. With the aid of the previously obtained results this gives

$$\sqrt{\frac{(0.2M_3^2+1)(0.52M_3^4+2.53M_3^2-0.37)}{1.4M_3^2-0.2}} + 2.22M_3 - 1.64M_3^2 = 0.$$

It is straightforward to show that the solution to this equation must satisfy the condition for $M_3 > 1.35$. If this condition is violated then the right-hand side of the equation M_3 is positive. The equation for M_3 can be reduced to

$$F(M_3) \equiv M_3^6 - 2.78M_3^5 + 1.46M_3^4 + 0.40M_3^3 - 0.94M_3^2 + 0.10 = 0.$$

The graph of the function $F(M_3)$ is shown in Fig. A.9. We see that there are two positive zeros of function $F(M_3)$. One of them is less than 1.35. Hence we reject it. The second zero is $M_3 \approx 2.09$. Substituting this value into the expression for p_3 we obtain $p_3 \approx 50.9p_1 > 40p_1$. Hence, we conclude that the wall will be destroyed.

8.7: We look for the solution to this problem in the form of a simple wave described by Eq. (8.7.12). The position of the piston is defined by $x = X(t)$, where $X(t) = \int_0^t U(t')\,dt' = TU_0 \ln\cosh(t/T)$. The solution to Eq. (8.7.12) is defined in the region

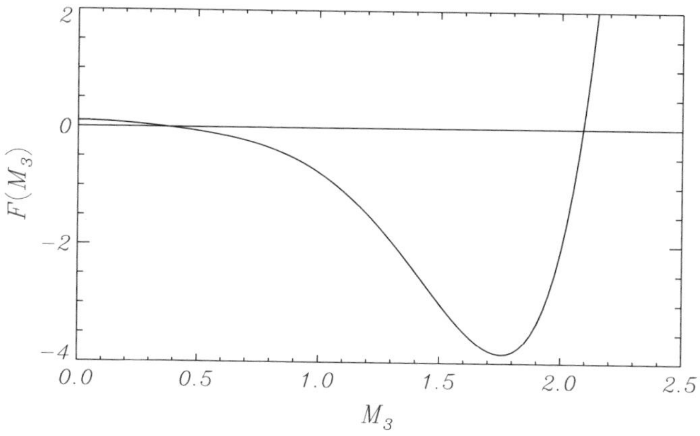

Fig. A.9 Graph of the function $F(M_3)$

Fig. A.10 The thick solid curve is defined by the equation $x = X(t)$. The straight lines are the characteristics

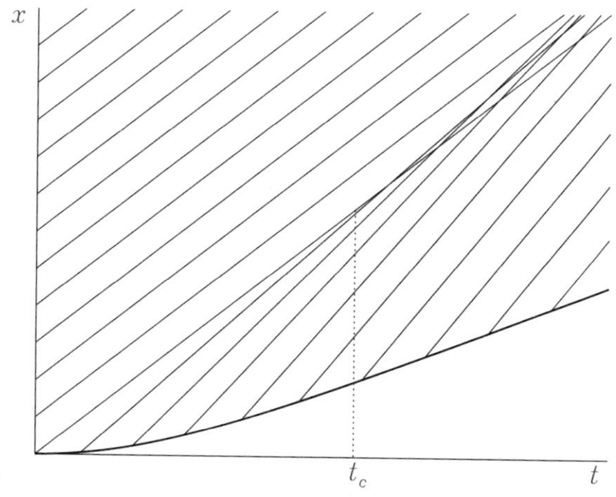

$t > 0$, $x > X(t)$ (see Fig. A.10). The characteristics of Eq. (8.7.12) are defined by Eq. (8.7.13). The characteristics are shown in Fig. A.10. The gas in the region $x > a_0 t$ remains at rest. The gradient catastrophe occurs at the time t_c when two characteristics starting at two infinitesimally close points on the curve $x = X(t)$ first intersect. The equations of characteristics starting from points t_1 and $t_1 + \delta t$ are

$$x = \left(a_0 + \frac{\gamma+1}{2}U(t_1)\right)(t - t_1) + X(t_1),$$

$$\begin{aligned} x &= \left(a_0 + \frac{\gamma+1}{2}U(t_1+\delta t)\right)(t - t_1 - \delta t) + X(t_1+\delta t) \\ &= \left(a_0 + \frac{\gamma+1}{2}U(t_1)\right)(t - t_1) + X(t_1) \\ &\quad + \left(\frac{\gamma+1}{2}(t-t_1)U'(t_1) - a_0 - \frac{\gamma-1}{2}U(t_1)\right)\delta t + \mathcal{O}(\delta t^2). \end{aligned}$$

These two infinitesimally close characteristics intersect at

$$t = t_1 + \frac{(\gamma-1)U(t_1) + 2a_0}{(\gamma+1)U'(t_1)} = t_1 + \frac{T[(\gamma-1)\tanh(t_1/T) + 2/M]\cosh^2(t_1/T)}{\gamma+1},$$

where $M = U_0/a_0$. Then $t_c = \min_{t_1} t$. To calculate this minimum we differentiate t with respect to t_1 to obtain

$$\frac{dt}{dt_1} = \frac{2\gamma}{\gamma+1} + 2\sinh(t_1/T)\frac{(\gamma-1)\sinh(t_1/T)] + (2/M)\cosh(t_1/T)}{\gamma-1} > 0.$$

It follow from this result that t is a monotonically increasing function of t_1. Hence, it takes its minimum value at $t_1 = 0$ and we obtain that

$$t_c = \frac{2a_0 T}{U_0(\gamma+1)}.$$

8.8: A picture of the characteristics is shown in Fig. A.11. We consider the reference frame where the gas before the rarefaction wave is at rest. The density and pressure before the wave are ρ_0 and p_0 and, consequently, the speed of sound is $a_0 = (\gamma p_0/\rho_0)^{1/2}$. The velocity behind the wave is $v_1 < 0$. Without loss of generality we can assume that the origin of the fan of characteristics coincides with the coordinate origin. The general equation of the characteristics is given by Eq. (8.7.13) with $v = \text{const}$ on a characteristic. Then the equation of the characteristics above the fan is

$$x = x_0 + ta_0, \quad x_0 > 0,$$

while below the fan they are

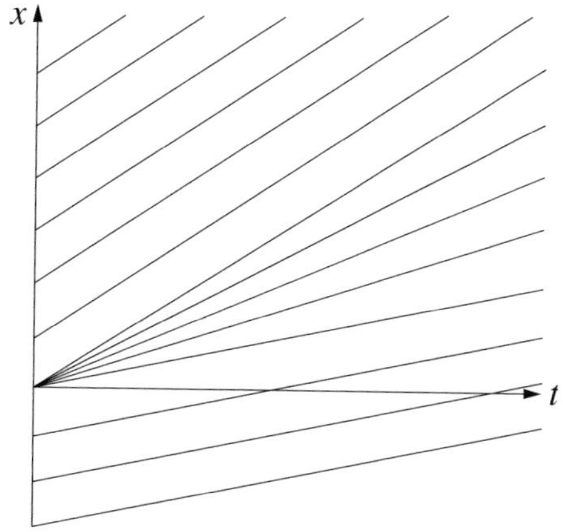

Fig. A.11 Picture of characteristics for a rarefaction wave

$$x = x_0 + t\left(a_0 + \frac{\gamma + 1}{2}v_1\right), \quad x_0 < 0.$$

The equation of a characteristic in the fan is

$$x = t\left(a_0 + \frac{\gamma + 1}{2}v\right), \quad v_1 < v < 0. \tag{A8.1}$$

Then the velocity in the rarefaction wave is given by

$$v = \frac{2}{\gamma + 1}\left(\frac{x}{t} - a_0\right). \tag{A8.2}$$

Since $v < 0$ it follows that $x < a_0 t$. Now it follows from Eq. (8.7.11) that

$$a = a_0 + \frac{\gamma - 1}{2}v = \frac{2}{\gamma + 1}\left(a_0 + \frac{(\gamma - 1)x}{2t}\right). \tag{A8.3}$$

In accordance with Eq. (A8.1) the rarefaction wave occupies the region

$$t\left(a_0 + \frac{\gamma + 1}{2}v_1\right) < x < t a_0. \tag{A8.4}$$

Substituting Eq. (A8.3) into Eq. (8.7.3) we obtain

$$\rho = \rho_0\left[\frac{2}{\gamma + 1}\left(1 + \frac{(\gamma - 1)x}{2a_0 t}\right)\right]^{\frac{2}{\gamma - 1}}, \quad p = p_0\left[\frac{2}{\gamma + 1}\left(1 + \frac{(\gamma - 1)x}{2a_0 t}\right)\right]^{\frac{2\gamma}{\gamma - 1}}. \tag{A8.5}$$

It follows from Eq. (A8.4) that in the rarefaction wave $\rho < \rho_0$, thus justifying the name of the wave. The coordinate of the trailing edge of the wave is $x = t\left(a_0 + \frac{\gamma+1}{2}v_1\right)$. Substituting this result into Eq. (A8.5) we obtain

$$\rho_1 = \rho_0\left(1 + \frac{v_1(\gamma-1)}{2a_0}\right)^{\frac{2}{\gamma-1}}, \quad p_1 = p_0\left(1 + \frac{v_1(\gamma-1)}{2a_0}\right)^{\frac{2\gamma}{\gamma-1}}. \tag{A8.6}$$

8.9: The solution of any problem must only depend on dimensionless variables and parameters. In this problem there are quantities with the dimension of density and pressure. We can also construct two quantities with the dimension of velocity, the sound speeds $a_1 = \sqrt{\gamma p_1/\rho_1}$ and $a_2 = \sqrt{\gamma p_2/\rho_2}$. However there are no quantities with the dimension of time and length. This implies that the solution must only depend on the combination x/t which has the dimension of velocity. The only solutions of the one-dimensional equations describing the motion of a compressible fluid that depend on x/t are a shock and a self-similar rarefaction wave.

After the membrane is removed the two gases will be separated by a contact discontinuity where there is a jump of the density, but the pressure must be the same at its two sides. Since $p_1 < p_2$ we need to reduce the pressure at the right of the membrane and increase it at the left. This implies that there is a shock propagating in the left gas and a self-similar rarefaction wave in the right gas.

Let the Mach number of the shock be M. Then the gas velocity before the shock in the reference frame moving together with the shock is $a_1 M$. It follows from Eqs. (8.6.6) and (8.6.7) that the flow velocity and pressure behind the shock in the reference frame moving together with the shock are

$$v_1 = a_1 M\left(1 - \frac{2(M^2-1)}{(\gamma+1)M^2}\right), \quad p_1' = p_1\left(1 + \frac{2\gamma(M^2-1)}{\gamma+1}\right).$$

The velocity of the left gas behind the shock in the laboratory reference frame is

$$v_1' = v_1 - a_1 M = -\frac{2a_1(M^2-1)}{(\gamma+1)M}.$$

Now we use the results obtained in the solution to the previous problem. Let in the rarefaction wave x/t vary from w to a_2. Then it follows from Eqs. (A8.2) and (A8.5) that the velocity and pressure in this wave drop from zero and p_2 at its leading edge to

$$v_2' = \frac{w - a_2}{\gamma+1}, \quad p_2' = p_2\left[\frac{2}{\gamma+1}\left(1 + \frac{(\gamma-1)w}{2a_2}\right)\right]^{\frac{2\gamma}{\gamma-1}}.$$

Since the velocity and pressure must be the same at the two sides of the contact discontinuity we have $v_1' = v_2'$ and $p_1' = p_2'$. These relations give the system of two equations determining M and w,

$$\frac{2a_1(M^2-1)}{M} = a_2 - w,$$

$$p_1\left(1 + \frac{2\gamma(M^2-1)}{\gamma+1}\right) = p_2\left[\frac{2}{\gamma+1}\left(1 + \frac{(\gamma-1)w}{2a_2}\right)\right]^{\frac{2\gamma}{\gamma-1}}.$$

Eliminating w and recalling that $p_2/p_1 = 1+\epsilon$ yields

$$1 + \frac{2\gamma(M^2-1)}{\gamma+1} = (1+\epsilon)\left(1 - \frac{2a_1(\gamma-1)(M^2-1)}{a_2 M(\gamma+1)}\right)^{\frac{2\gamma}{\gamma-1}}. \tag{A8.7}$$

When $\epsilon = 0$ we have $M = 1$. Hence, $M - 1$ must be small when $\epsilon \ll 1$. Then we can approximate the right-hand side of Eq. (A8.7) by the first two terms of its Taylor expansion with respect to $M^2 - 1$ to obtain

$$\frac{2\gamma(M^2-1)}{\gamma+1} \approx \epsilon - \frac{4\gamma a_1(M^2-1)}{a_2(\gamma+1)}.$$

When deriving this equation we neglected the term proportional to $\epsilon(M^2-1)$. Now we easily find

$$M \approx 1 + \frac{\epsilon a_2(\gamma+1)}{4\gamma(2a_1+a_2)}.$$

The velocity of the contact discontinuity is

$$v_1' = v_2' = -\frac{2a_1(M^2-1)}{(\gamma+1)M} \approx -\frac{\epsilon a_1 a_2}{\gamma(2a_1+a_2)}.$$

Finally, using Eqs. (8.6.7) and (A8.6) with $v_1 = v_1'$ we obtain that the gas densities at the left and the right of the contact discontinuity are given by

$$\rho_1' \approx \rho_1\left(1 + \frac{\epsilon a_2}{\gamma(2a_1+a_2)}\right), \quad \rho_2' \approx \rho_2\left(1 - \frac{\epsilon a_1}{\gamma(2a_1+a_2)}\right).$$

8.10: The velocity and time are related by Eq. (8.8.5). It is obvious from this equation that v is an odd function of τ. Hence, the expansion of v in the Fourier series is

$$\frac{v}{U} = \sum_{n=1}^{\infty} b_n \sin(n\omega\tau). \tag{$*$}$$

Using Eqs. (8.8.5) and (8.8.6) we obtain

$$b_n = \frac{2\omega}{\pi}\int_0^{\pi/\omega} \sin\left(\omega\tau + \frac{xv}{x_c U}\right)\sin(n\omega\tau)\,d\tau.$$

Now we introduce $\sigma = \omega\tau + xv/(x_c U)$. Then it follows from Eq. (8.8.5) that $v = U \sin\sigma$ and, consequently, $\omega\tau = \sigma - (x/x_c)\sin\sigma$. It follows from equation (∗) that $v = 0$ at $\tau = 0$ and $\omega\tau = \pi$. Then the expression for b_n reduces to

$$b_n = \frac{2}{\pi}\int_0^\pi \sin\sigma \sin(n\sigma - (nx/x_c)\sin\sigma)[1 - (x/x_c)\cos\sigma]\, d\sigma.$$

We can further reduce this expression to

$$b_n = \frac{1}{\pi}\int_0^\pi [\cos((n-1)\sigma - (nx/x_c)\sin\sigma) - \cos((n+1)\sigma - (nx/x_c)\sin\sigma)]\, d\sigma$$
$$- \frac{x}{2\pi x_c}\int_0^\pi [\cos((n-2)\sigma - (nx/x_c)\sin\sigma) - \cos((n+2)\sigma - (nx/x_c)\sin\sigma)]\, d\sigma.$$

Now, using the expression

$$J_m(y) = \frac{1}{\pi}\int_0^\pi \cos(m\sigma - y\sin\sigma)\, d\sigma,$$

where $J_m(y)$ is the Bessel function of the first kind and order m, and the identity

$$J_{m-1}(y) + J_{m+1}(y) = \frac{2m}{y} J_m(y),$$

we eventually obtain

$$b_n = \frac{2x_c J_n(nx/x_c)}{nx}.$$

Hence, the Fourier series for v is

$$\frac{v}{U} = \frac{2x_c}{x}\sum_{n=1}^{\infty} \frac{J_n(nx/x_c)}{n}\sin(n\omega\tau).$$

The dependences of b_1, b_2, and b_3 on x are shown in Fig. A.12.

A.9 Solutions for Chap. 9

9.1: Since the walls are rigid the velocity must be equal to the boundary velocity at the boundaries. Since the walls are thermally insulated it follows from Eq. (9.1.1) that the temperature gradient is zero at the boundaries. Hence, we have the boundary conditions

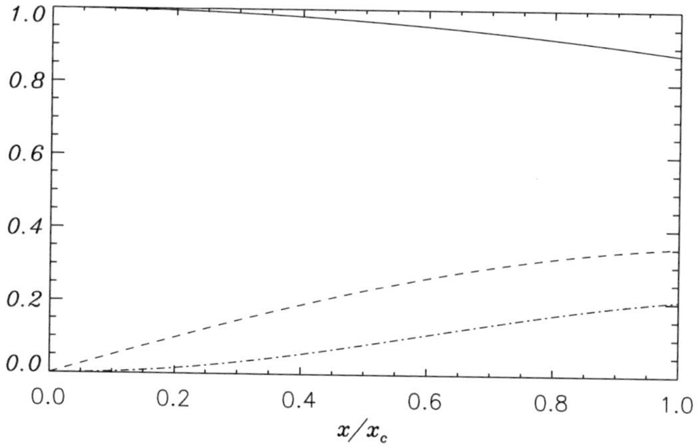

Fig. A.12 Dependences of b_1 (solid curve), b_2 (dashed curve), and b_3 (dash-dotted curve) on x

$$v = 0 \text{ at } x = 0, \quad v = l\omega \sin(\omega t) \text{ at } x = L, \quad \frac{\partial T}{\partial x} = 0 \text{ at } x = 0, L. \tag{A9.1}$$

Now we use Eqs. (9.2.4)–(9.2.7). Eliminating $\delta\rho$ and δp from these equations we obtain

$$\frac{\partial^2 v}{\partial t^2} - \frac{a_s^2}{\gamma}\frac{\partial^2 v}{\partial x^2} + \frac{a_s^2}{\gamma T_0}\frac{\partial(\delta T)}{\partial t \partial x} = 0, \tag{A9.2}$$

$$\rho_0\left(C_v\frac{\partial(\delta T)}{\partial t} + \frac{a_s^2}{\gamma}\frac{\partial v}{\partial x}\right) = \kappa\frac{\partial^2(\delta T)}{\partial x^2}. \tag{A9.3}$$

When deriving Eqs. (A9.2) and (A9.3) we used the relations $a_s^2 = \gamma \widetilde{R} T_0/\tilde{\mu}$ and $\gamma(\gamma - 1)C_v T_0 = a_s^2$ that follow from Eqs. (8.3.10), (8.3.14), and (8.4.4). The set of Eqs. (A9.2) and (A9.3) is of third order with respect to time, which implies that we need three initial conditions. Two of them are the values of v and T at $t = 0$. It also follows from Eq. (9.2.6) that the time derivative of the velocity is zero at the initial time. Hence, we have the initial conditions

$$v = 0, \quad \frac{\partial v}{\partial t} = 0, \quad \delta T = 0 \quad \text{at} \quad t = 0. \tag{A9.4}$$

Now we make the variable substitution

$$v = \tilde{v} + \frac{l\omega x}{L}\sin(\omega t). \tag{A9.5}$$

Then the boundary conditions for v reduce to

$$\tilde{v} = 0 \quad \text{at} \quad x = 0, L, \tag{A9.6}$$

while Eqs. (A9.2) and (A9.3) reduce to

$$\frac{\partial^2 \tilde{v}}{\partial t^2} - \frac{a_s^2}{\gamma}\frac{\partial^2 \tilde{v}}{\partial x^2} + \frac{a_s^2}{\gamma T_0}\frac{\partial(\delta T)}{\partial t \partial x} = \frac{l\omega^3 x}{L}\sin(\omega t), \tag{A9.7}$$

$$C_v \frac{\partial(\delta T)}{\partial t} + \frac{a_s^2}{\gamma}\frac{\partial \tilde{v}}{\partial x} - \frac{\kappa}{\rho_0}\frac{\partial^2(\delta T)}{\partial x^2} = -\frac{l\omega a_s^2}{\gamma L}\sin(\omega t). \tag{A9.8}$$

We expand $\tilde{v}$ and δT in Fourier series with respect to x. In accordance with the boundary Eqs. (A9.1) and (A9.6) these expansions are

$$\tilde{v} = \sum_{n=1}^{\infty} v_n(t) \sin\frac{\pi n x}{L}, \quad \delta T = \sum_{n=0}^{\infty} \delta T_n(t) \cos\frac{\pi n x}{L}. \tag{A9.9}$$

Substituting these expansions into Eqs. (A9.7) and (A9.8) and using the identity

$$\sum_{n=1}^{\infty} \frac{(-1)^n}{n} \sin\frac{\pi n x}{L} = -\frac{\pi x}{2L},$$

we obtain

$$\frac{d^2 v_n}{dt^2} + \frac{\omega_n^2 v_n}{\gamma} - \frac{\omega_n a_s}{\gamma T_0}\frac{d(\delta T_n)}{dt} = -\frac{2l\omega^3(-1)^n}{\pi n}\sin(\omega t), \tag{A9.10}$$

$$\frac{d(\delta T_n)}{dt} + \frac{\omega_n a_s v_n}{\gamma C_v} + n^2\chi\,\delta T_n = 0, \tag{A9.11}$$

$$\frac{d(\delta T_0)}{dt} = -\frac{l\omega a_s^2}{\gamma L C_v}\sin(\omega t), \tag{A9.12}$$

where n varies from 1 to ∞ and

$$\omega_n = \frac{\pi n a_s}{L}, \quad \chi = \frac{\pi^2 \kappa}{\rho_0 C_v L^2} = \frac{\pi^2 \gamma a_s}{L}\mathrm{Pe}^{-1}. \tag{A9.13}$$

Eliminating δT_n from Eqs. (A9.10) and (A9.11), and using the relation $\gamma(\gamma-1)C_v T_0 = a_s^2$ we obtain

$$\frac{d^3 v_n}{dt^3} + \omega_n^2\frac{dv_n}{dt} + n^2\chi\left(\frac{d^2 v_n}{dt^2} + \frac{\omega_n^2 v_n}{\gamma}\right) = -\frac{2l\omega^3(-1)^n}{\pi n}\left[\omega\cos(\omega t) + n^2\chi\sin(\omega t)\right]. \tag{A9.14}$$

It follows from Eqs. (A9.4) and (A9.8) that $\partial(\delta T)/\partial t = 0$ at $t = 0$. Then we obtain from Eq. (A9.7) that $\partial^2\tilde{v}/\partial t^2 = 0$ at $t = 0$. Hence, the solution to Eq. (A9.14) must satisfy the initial conditions

$$v_n = 0, \quad \frac{dv_n}{dt} = 0, \quad \frac{d^2 v_n}{dt^2} = 0 \quad \text{at} \quad t = 0. \tag{A9.15}$$

The characteristic equation of the differential Eq. (A9.14) is

$$\lambda^3 + n^2\chi\lambda^2 + \omega_n^2\lambda + \frac{n^2\omega_n^2\chi}{\gamma} = 0.$$

Since the dissipation is weak it is enough only to keep the terms proportional to χ and neglect terms of higher order when calculating λ. As a result we have

$$\lambda_{1n} = i\omega_n - \frac{n^2\chi(\gamma-1)}{2\gamma}, \quad \lambda_{2n} = -i\omega_n - \frac{n^2\chi(\gamma-1)}{2\gamma}, \quad \lambda_{3n} = -\frac{n^2\chi}{\gamma}.$$

The complimentary function is

$$v_{nc} = C_1 e^{\lambda_{1n}t} + C_2 e^{\lambda_{2n}t} + C_3 e^{\lambda_{3n}t},$$

where C_1, C_2, and C_3 are arbitrary constants. A particular integral to Eq. (A9.14) is

$$v_{np} = H_n\left[n^2\chi\omega\omega_n^2(\gamma-1)\cos(\omega t) + A_n\sin(\omega t)\right], \tag{A9.16}$$

where

$$H_n = \frac{2\gamma l\omega^3(-1)^n}{\pi n[\gamma^2\omega^2(\omega^2-\omega_n^2)^2 + n^4\chi^2(\gamma\omega^2-\omega_n^2)^2]}, \tag{A9.17}$$

$$A_n = \gamma\omega^2(\omega^2-\omega_n^2) + n^4\chi^2(\gamma\omega^2-\omega_n^2). \tag{A9.18}$$

The general solution to Eq. (A9.14) is $v_n = v_{nc} + v_{np}$. Substituting this solution into the initial conditions (A9.15) we obtain the system of equations for C_1, C_2, and C_3,

$$\begin{aligned} C_1 + C_2 + C_3 + n^2\chi\omega\omega_n^2(\gamma-1)H_n &= 0, \\ \lambda_{1n}C_1 + \lambda_{2n}C_2 + \lambda_{3n}C_3 + \omega A_n H_n &= 0, \\ \lambda_{1n}^2 C_1 + \lambda_{2n}^2 C_2 + \lambda_{3n}^2 C_3 - n^2\chi\omega^3\omega_n^2(\gamma-1)H_n &= 0. \end{aligned} \tag{A9.19}$$

The solution to this system of equations is

$$C_1 = -C_2 = \frac{i\gamma\omega^3 H_n(\omega^2-\omega_n^2)}{2\omega_n} + \mathcal{O}(\mathrm{Pe}^{-1}), \quad C_3 = \mathcal{O}(\mathrm{Pe}^{-1}). \tag{A9.20}$$

Using Eqs. (A9.16), (A9.18) and (A9.20) we obtain

$$v_n = \gamma\omega^2 H_n(\omega^2-\omega_n^2)\left\{\sin(\omega t) - \frac{\omega}{\omega_n}e^{-\frac{n^2\chi(\gamma-1)t}{2\gamma}}\sin(\omega_n t)\right\} + \mathcal{O}(\mathrm{Pe}^{-1}). \tag{A9.21}$$

The expression for v is defined by Eqs. (A9.5), (A9.9), (A9.17), and (A9.20).

It follows from Eq. (A9.4) that $\delta T_n = 0$ at $t = 0$. Then, using Eq. (A9.11) we obtain for $n > 0$

$$\delta T_n = \frac{\omega a_s H_n(\omega^2 - \omega_n^2)}{\omega_n C_v}\Big\{\omega_n^2 \cos(\omega t) + \big(\omega^2 - \omega_n^2\big)e^{-n^2\chi t}$$
$$-\omega^2 e^{-\frac{n^2\chi(\gamma-1)t}{2\gamma}} \cos(\omega_n t)\Big\} + \mathcal{O}(\mathrm{Pe}^{-1}). \quad \text{(A9.22)}$$

It follows from Eq. (A9.12) that

$$\delta T_0 = -\frac{l(\gamma - 1)T_0}{L}[1 - \cos(\omega t)] + \mathcal{O}(\mathrm{Pe}^{-1}). \quad \text{(A9.23)}$$

The expression for δT is given by Eqs. (A9.9), (A9.17), (A9.22), and (A9.23). The expressions for $\delta\rho$ and δp are similar to that for δT and we do not give them.

The term proportional to $\sin(\omega_n t)$ in Eq. (A9.21) and the terms proportional to $e^{-n^2\chi t}$ and $e^{-\frac{n^2\chi(\gamma-1)t}{2\gamma}}$ in Eq. (A9.22) describe the effect of the initial conditions. We see that this effect disappears and the system arrives at the steady-state oscillation after a transitional time of the order of χ^{-1} which, in accordance with Eq. (A9.13), is proportional to Pe. To formally obtain the expression for v_n and δT_n describing the steady-state of oscillations, we take $t \gg \chi^{-1}$ in Eqs. (A9.21) and (A9.22). This results in

$$v_n = \gamma\omega^2 H_n(\omega^2 - \omega_n^2)\sin(\omega t) + \mathcal{O}(\mathrm{Pe}^{-1}), \quad \text{(A9.24)}$$

$$\delta T_n = \frac{\omega\omega_n a_s}{C_v} H_n(\omega^2 - \omega_n^2)\cos(\omega t) + \mathcal{O}(\mathrm{Pe}^{-1}). \quad \text{(A9.25)}$$

Hence, in the steady-state the perturbations of all variables oscillate with the same frequency ω.

An interesting phenomenon arises when $\omega = \omega_n + \alpha\chi$, where α is a constant of the order of unity, and we assume that n is moderate, $n \ll \mathrm{Pe}$. In this case we say that the driver is in quasi-resonance with the nth harmonic. It follows from Eqs. (A9.17) and (A9.21) that $v_n = \mathcal{O}(\mathrm{Pe})$, while $v_m = \mathcal{O}(1)$, where $m \neq n$, and the same estimates are valid for δT_n and δT_m. As a result, in the steady-state of oscillation the expressions for v and δT take an especially simple form:

$$v = \frac{4n\alpha\gamma l a_s(-1)^n \mathrm{Pe}}{\pi L[4\alpha^2 + n^2(\gamma - 1)^2]}\sin(\omega t)\sin\frac{\pi n x}{L} + \mathcal{O}(l/L), \quad \text{(A9.26)}$$

$$\delta T = \frac{4n\alpha a_s^2(-1)^n \mathrm{Pe}}{\pi L C_v[4\alpha^2 + n^2(\gamma - 1)^2]}\cos(\omega t)\cos\frac{\pi n x}{L} + \mathcal{O}(l/L). \quad \text{(A9.27)}$$

We assumed that $\omega \neq \omega_n$ when deriving Eq. (A9.16), so formally we cannot consider the case of exact resonance, $\omega = \omega_n$. However, in this case we can obtain the solution just by taking $\alpha \to 0$.

Another interesting property of the obtained solution is that the mean value of the temperature perturbation over the period is not zero. In accordance with Eq. (A9.23) it is equal to $l(\gamma - 1)T_0/L$.

9.2: Using Eqs. (8.3.10), (8.3.14), and (8.4.2) we obtain from Eq. (9.3.8) with $\mu = 0$

$$v^2 - vv_1\left(1 + \frac{1}{\gamma M_1^2}\right) + \frac{Tv_1^2}{\gamma T_1 M_1^2} = 0. \tag{A9.28}$$

In the shock structure T is a monotonically increasing function. It is obvious that the discriminant of Eq. (A9.28) is a monotonically decreasing function of T. Hence, it takes its minimum value at $T = T_2$. Using Eq. (8.6.8) to express T_2/T_1 in terms of M_1 it is easy to verify that the discriminant is positive when $T = T_2$. Hence, we conclude that Eq. (A9.28) has two real positive roots for any $T \in [T_1, T_2]$. When $T = T_1$ we must have $v = v_1$. This expression is given by the larger root of Eq. (A9.28) with $T = T_1$. Since v must be a continuous function of T it follows from this result that we must choose the larger root of this equation for any $T \in [T_1, T_2]$. Hence,

$$v = \frac{v_1}{2}\left(1 + \frac{1}{\gamma M_1^2} + \sqrt{\left(1 + \frac{1}{\gamma M_1^2}\right)^2 - \frac{4T}{\gamma T_1 M_1^2}}\right). \tag{A9.29}$$

We introduce the dimensionless variables

$$w = \frac{v}{v_1} - 1, \quad \Theta = \frac{T}{T_1} - 1, \quad y = \frac{(\gamma + 1)C_v\rho_1 v_1}{4\kappa}x. \tag{A9.30}$$

Using these variables and Eq. (A9.28) we transform Eq. (9.3.7) into

$$\frac{d\Theta}{dy} = \frac{2(\gamma - 1)(\gamma M_1^2 + 1)w}{\gamma + 1} + 2\Theta. \tag{A9.31}$$

Now we make the variable substitution

$$s = \sqrt{\left(1 - \frac{1}{\gamma M_1^2}\right)^2 - \frac{4\Theta}{\gamma M_1^2}}. \tag{A9.32}$$

Then, using Eqs. (8.3.10), (8.3.14), (8.4.2), and (A9.29) we reduce Eq. (A9.31) to

$$s\frac{ds}{dy} = (s - s_1)(s - s_2), \tag{A9.33}$$

where

$$s_1 = \frac{\gamma M_1^2 - 1}{\gamma M_1^2}, \quad s_2 = \frac{3\gamma - 1 - \gamma(3 - \gamma)M_1^2}{\gamma(\gamma + 1)M_1^2}. \tag{A9.34}$$

It is easy to show that $s_1 > s_2$ when $M_1 > 1$. It follows from Eq. (A9.34) that in the shock structure s decreases from s_1 to s_2 when y varies from $-\infty$ to ∞. Since $s \geq 0$ we conclude that the solution describing the shock structure only exists when $s_2 > 0$. This condition reduces to

$$M_1 < M_c(\gamma) = \sqrt{\frac{3\gamma - 1}{\gamma(3-\gamma)}}. \tag{A9.35}$$

Using separation of variables we obtain from Eq. (A9.33)

$$\frac{(s_1 - s)^{s_1}}{(s - s_2)^{s_2}} = e^y. \tag{A9.36}$$

It follows from Eq. (A9.32) that

$$\frac{T}{T_1} = \frac{\gamma M_1^2 + 1}{4\gamma M_1^2} - \frac{\gamma M_1^2 s^2}{4}. \tag{A9.37}$$

Equations (A9.36) and Eq. (A9.37) define $T(x)$ in the shock structure as an implicit function. The velocity v is then given by Eq. (A9.29). Then we can also find the dependence of ρ and p on x using Eqs. (9.3.4) and (9.3.5) with $\mu = 0$. It is easy to verify that T, v, ρ, and p tend to T_2, v_2, ρ_2, and p_2 given by Eqs. (8.6.6)–(8.6.8) as $x \to \infty$.

It is easy to show that the function $M_c(\gamma)$ is monotonically increasing from 1 to $3/\sqrt{5} \approx 1.342$ in the interval $[1, 5/3]$. The condition given by Eq. (A9.35) imposes a restriction on the value of T_2/T_1. Using Eq. (8.6.8) we obtain that

$$\frac{T_2}{T_1} < R(\gamma) = \frac{4\gamma^2 - 5\gamma + 3}{3\gamma - 1}. \tag{A9.38}$$

The function $R(\gamma)$ monotonically increases from 1 to 13/9 in the interval $[1, 5/3]$. We see that the solution describing the shock structure only exists for shocks with very moderate intensity with the Mach number bounded by the inequality (A9.35). It is worth noting that in the opposite case where $\kappa = 0$ and $\mu \neq 0$ the solution describing the shock structure exists for any value of M_1.

9.3: Equation (9.4.27) describes perturbations propagating with a speed close to a_0 in the positive x-direction, while Eq. (9.3.26) describes a weak shock-like perturbation propagating in the negative x-direction. To reconcile Eqs. (9.3.26) and (9.4.27) we make the substitution $x' = -x$ and $\tilde{v} = -v$. As a result, dropping the prime and tilde, we obtain

$$\frac{\partial v}{\partial t} - a_0\left(1 - \frac{\gamma + 1}{2a_0} v\right)\frac{\partial v}{\partial x} - a_0^3 \Gamma \frac{\partial^2 v}{\partial x^2} = 0. \tag{A9.39}$$

When deriving Eq. (9.4.27) we assumed that the perturbations tend to zero at least at one infinity, either plus or minus. We assume it is at plus infinity. Then Eq. (A9.39) describes perturbations tending to zero as $x \to -\infty$. However, we need to obtain a solution where the velocity tends to v_1 as $x \to -\infty$. In accordance with this we introduce a new reference frame moving with velocity v_1 in the negative x-direction.

The variable transformation to this new reference frame is $y = x + v_1 t$, $v' = v + v_1$. In this reference frame we look for a solution that is stationary, that is, independent of t. Then, dropping the prime, we obtain from Eq. (A9.39)

$$\frac{\gamma - 1}{2} v_1 \frac{dv}{dy} + a_0 \left(1 - \frac{\gamma + 1}{2a_0} v\right) \frac{dv}{dy} + a_0^3 \Gamma \frac{d^2 v}{dy^2} = 0. \tag{A9.40}$$

Integrating this equation and taking into account that $v \to v_1$ as $x \to -\infty$ yields

$$a_0^3 \Gamma \frac{dv}{dy} = \frac{\gamma + 1}{4}(v - v_1)(v - v_2), \quad v_2 = a_0 \frac{4 - (3 - \gamma) M_1}{\gamma + 1}, \quad M_1 = \frac{v_1}{a_0}. \tag{A9.41}$$

It is obvious that in a non-singular solution to Eq. (A9.41) v must be between v_1 and v_2. Then, integrating Eq. (A9.41), we obtain

$$v = \frac{v_1 + v_2}{2} - \frac{v_1 - v_2}{2} \tanh \frac{\epsilon y}{\ell}, \tag{A9.42}$$

where

$$\epsilon = M_1^2 - 1 \approx 2(M_1 - 1), \quad \ell = \frac{8\epsilon a_0^3 \Gamma}{(\gamma + 1)(v_1 - v_2)}.$$

It follows from Eq. (A9.41) that

$$v_1 - v_2 = \frac{2\epsilon a_0}{\gamma + 1}.$$

Substituting this result into the expression for ℓ we obtain

$$\ell = 4a_0^2 \Gamma \approx 2 \frac{\gamma C_v \mu + (\gamma - 1)\kappa}{\gamma C_v \rho_0 v_1}. \tag{A9.43}$$

We obtain Eq. (9.3.26) and the expression for ℓ given by Eq. (9.3.24) from Eq. (A9.42) if we substitute ρ_1 for ρ_0 and x for y in Eqs. (A9.42) and (A9.43).

9.4: Using Eq. (9.4.29) we obtain that

$$\phi = \phi_0(t) \equiv \begin{cases} 1, & t < 0, \\ e^{-\alpha t}, & t > 0, \end{cases} \tag{A9.44}$$

where

$$\alpha = \frac{v_0(\gamma + 1)}{4a_0^2 \Gamma}. \tag{A9.45}$$

To find the solution to Eq. (9.4.29) satisfying the boundary Eq. (A9.44) we use the Fourier transform with respect to t,

$$\hat{\phi}(\omega) = \int_{-\infty}^{\infty} \phi(t)\, e^{-i\omega t}\, dt, \quad \phi(t) = \frac{1}{2\pi} \int_{-\infty}^{\infty} \hat{\phi}(\omega)\, e^{i\omega t}\, d\omega.$$

Applying this transform to Eq. (9.4.29) we obtain

$$\frac{\partial \hat{\phi}}{\partial x} = -\left(\Gamma\omega^2 + \frac{i\omega}{a_0}\right) \hat{\phi}. \tag{A9.46}$$

It follows from this equation and the boundary condition (A9.44) that

$$\hat{\phi} = \hat{\phi}_0(\omega) \exp\left[-\left(\Gamma\omega^2 + \frac{i\omega}{a_0}\right) x\right]. \tag{A9.47}$$

Since $\phi_0 \not\to 0$ as $t \to -\infty$ we need to consider $\hat{\phi}_0$ as a generalised function. Now, using the formula

$$\frac{1}{2\pi} \int_{-\infty}^{\infty} \exp\left[-\left(\Gamma\omega^2 + \frac{i\omega}{a_0}\right) x\right] e^{i\omega t}\, d\omega = \frac{1}{2\sqrt{\pi x \Gamma}} \exp\left[-\frac{1}{4x\Gamma}\left(t - \frac{x}{a_0}\right)^2\right]$$

and the convolution theorem we obtain

$$\phi = \frac{1}{2\sqrt{\pi x \Gamma}} \int_{-\infty}^{\infty} \phi_0(\tau - u) \exp\left(-\frac{u^2}{4x\Gamma}\right) du, \tag{A9.48}$$

where $\tau = t - x/a_0$. Using Eq. (A9.44) we transform this expression into

$$\begin{aligned} \phi &= \frac{1}{2\sqrt{\pi x \Gamma}} \left\{ \int_{\tau}^{\infty} \exp\left(-\frac{u^2}{4x\Gamma}\right) du + \int_{-\infty}^{\tau} \exp\left(-\alpha(\tau - u) - \frac{u^2}{4x\Gamma}\right) du \right\} \\ &= \frac{1}{2}\left\{ 1 - \operatorname{erf}\left(\frac{\tau}{2\sqrt{x\Gamma}}\right) + \exp[\alpha(\alpha x\Gamma - \tau)]\left[1 + \operatorname{erf}\left(\frac{\tau - 2\alpha x\Gamma}{2\sqrt{x\Gamma}}\right)\right]\right\}, \end{aligned} \tag{A9.49}$$

where

$$\operatorname{erf}(y) = \frac{2}{\sqrt{\pi}} \int_0^y e^{-z^2}\, dz$$

is the error function. The solution to the boundary value problem that we are studying is given by Eqs. (9.4.28) and (A9.49). Differentiating the expression for ϕ yields

$$\frac{\partial \phi}{\partial t} = -\frac{\alpha}{2\sqrt{\pi x \Gamma}} \int_{-\infty}^{\tau} \exp\left(-\alpha(\tau - u) - \frac{u^2}{4x\Gamma}\right) du. \tag{A9.50}$$

It follows from Eqs. (9.4.28), (A9.48), and (A9.50) that $v < 0$ everywhere.

Below we use the relation

$$\operatorname{erf}(y) = \operatorname{sgn}(y) - \frac{e^{-y^2}}{y\sqrt{\pi}}\left[1 + \mathcal{O}\left(|y|^{-1}\right)\right], \quad |y| \gg 1, \tag{A9.51}$$

which can be easily obtained using integration by parts. First we consider $|\tau| \gg \sqrt{x\Gamma}$ and $\tau < 0$. Then, using Eq. (A9.51) we obtain from Eqs. (9.4.28) and (A9.49) that $\phi \approx 1$ and $v \approx 0$. Next, we consider $\tau - 2\alpha x\Gamma \gg \sqrt{x\Gamma}$. Then, we obtain $\phi \approx \exp[\alpha(\alpha x\Gamma - \tau)]$ and $v \approx -v_0$.

Finally, we assume that x is so large that $\alpha x\Gamma \gg \sqrt{x\Gamma}$ and consider τ satisfying

$$\sqrt{x\Gamma} \ll \tau \ll 2\alpha x\Gamma - \sqrt{x\Gamma}. \tag{A9.52}$$

Then, using Eqs. (A9.49) and (A9.51), and taking into account the inequality (A9.52) we obtain

$$\phi \approx \frac{2\alpha x\Gamma\sqrt{x\Gamma}e^{-\tau^2/4x\Gamma}}{\tau(2\alpha x\Gamma - \tau)} \approx \frac{\sqrt{x\Gamma}}{\tau}e^{-\tau^2/4x\Gamma}, \tag{A9.53}$$

$$\frac{\partial\phi}{\partial t} \approx -\frac{1}{2\sqrt{x\Gamma}}\left(1 + \frac{x\Gamma}{\tau^2}\right)e^{-\tau^2/4x\Gamma} \approx -\frac{e^{-\tau^2/4x\Gamma}}{2\sqrt{x\Gamma}}. \tag{A9.54}$$

Now, using Eq. (9.4.28) yields

$$v \approx \frac{2a_0}{\gamma + 1}\left(1 - \frac{a_0 t}{x}\right). \tag{A9.55}$$

It is straightforward to verify that this is the expression describing the velocity in a self-similar rarefaction wave where the velocity varies from 0 to $-v_0$. Summarising, we conclude that for large x the solution is close to a self-similar rarefaction wave with two transitional regions of thickness of the order of $\sqrt{x\Gamma}$ connecting the rarefaction wave with two regions with constant velocity.

9.5: We use the expansions

$$\begin{aligned} &I_0(s) = 1 + \frac{s^2}{4} + \mathcal{O}(s^2),\ I_1(s) = \frac{s}{2} + \mathcal{O}(s^3), \\ &I_2(s) = \frac{s^2}{8} + \mathcal{O}(s^4), \qquad I_n(s) = \mathcal{O}(s^n), \end{aligned} \tag{A9.56}$$

where $n = 3, 4, \ldots$ Using these expansions we obtain from Eq. (9.4.38)

$$\begin{aligned} \phi = {} & 1 + se^{-\Gamma\omega^2 x}\cos[\omega(t - x/a_0)] \\ & + \frac{s^2}{4}\left\{1 + e^{-4\Gamma\omega^2 x}\cos[2\omega(t - x/a_0)]\right\} + \mathcal{O}\left(s^3\right). \end{aligned} \tag{A9.57}$$

Substituting this expression into Eq. (9.4.28) we obtain

$$v = -U \left\{ e^{-\Gamma\omega^2 x} \sin[\omega(t - x/a_0)] - \frac{s}{2} e^{-2\Gamma\omega^2 x} \left(1 - e^{-2\Gamma\omega^2 x}\right) \sin[2\omega(t - x/a_0)] \right\} + \mathcal{O}(s^3). \quad \text{(A9.58)}$$

Further Reading

Mathematical Preliminaries

1. H.J. Keisler. *Elementary Calculus: An Approach Using Infinitesimals,* Dover Publ. Inc. (2000)
2. W. Rudin. *Principles of Mathematical Analysis,* McGraw-Hill, New York (1976)
3. M. Itskov. *Tensor Algebra and Tensor Analysis for Engineers: With Applications to Continuum Mechanics,* Springer (2015)
4. K.F. Riley, M.P. Hobson, and S.J. Bence. *Mathematical Methods for Physics and Engineering,* Cambridge Univ. Press, Cambridge (2002)
5. E.A. Coddington and N. Levinson. *Theory of Ordinary Differential Equations,* McGraw-Hill, New York (1955)

General Continuum Mechanics

1. P. Chadwick. *Continuum Mechanics, Concise Theory and Problems,* George Allen and Unwin, London (1976)
2. S.C. Hunter. *Mechanics of Continuum Media,* 2nd Edition, John Wiley & Sons, New York (1983)
3. L.I. Sedov. *A Course in Continuum Mechanics,* Wolters–Noordhoff Publishing, Gröningen (1972)
4. A.J.M. Spencer. *Continuum Mechanics,* Longman, London and New York (1980)

Elasticity

1. L.D. Landau and E.M. Lifshitz. *Theory of Elasticity. Course of Theoretical Physics, Volume 7,* Pergamon Press, Oxford (1970)
2. W.S. Slughter. *The Linearised Theory of Elasticity,* Birkhäuser, Basel (2002)
3. S. Timoshenko and J.N. Goodier. *Theory of Elasticity,* 2nd Edition, McGraw–Hill, New York (1951)

Viscoelasticity and Plasticity

1. J. Lemaitre and J.-L. Chaboche. *Mechanics of solid materials,* Cambridge Univ. Press, Cambridge (1990)

© Springer Nature Switzerland AG 2019

M. S. Ruderman, *Fluid Dynamics and Linear Elasticity*, Springer Undergraduate Mathematics Series, https://doi.org/10.1007/978-3-030-19297-6

2. A.S. Khan and S. Huang. *Continuum theory of plasticity,* John Wiley & Sons, New York (1995)

Hydrodynamics

1. G.K. Batchelor. *An Introduction to Fluid Mechanics,* Cambridge University Press (1967)
2. H. Lamb. *Hydrodynamics,* 6th Edition, Cambridge University Press (1932)
3. L.D. Landau and E.M. Lifshitz. *Fluid Mechanics. Course of Theoretical Physics, Volume 6,* Pergamon Press, Oxford (1987)

Waves

1. J. Lighthill. *Waves in Fluids,* Cambridge University Press (2001)
2. O.V. Rudenko and S.I. Soluyan. *Theoretical Foundations of Nonlinear Acoustics,* Plenum Publ. Corporation, Consultants Bureau, New York and London (1977)
3. G.B. Whitham. *Linear and Non-Linear Waves,* John Wiley and Sons, New York (1974)

Index

© Springer Nature Switzerland AG 2019
M. S. Ruderman, *Fluid Dynamics and Linear Elasticity*, Springer Undergraduate Mathematics Series, https://doi.org/10.1007/978-3-030-19297-6

Printed in the United States
By Bookmasters